Building and Delivering Microservices on AWS

Master software architecture patterns to develop and deliver
microservices to AWS Cloud

Amar Deep Singh

Building and Delivering Microservices on AWS

Group Product Manager: Preet Ahuja

Publishing Product Manager: Niranjan Naikwadi and Suwarna Rajput

Senior Editor: Arun Nadar

Content Development Editor: Sujata Tripathi

Technical Editor: Nithik Cheruvakodan

Copy Editor: Safis Editing

Project Coordinator: Aryaa Joshi

Proofreader: Safis Editing

Indexer: Rekha Nair

Production Designer: Vijay Kamble

Marketing Coordinator: Agnes D'souza

First published: May 2023
Production reference: 2080324

Published by Packt Publishing Ltd.

Grosvenor House

11 St Paul's Square

Birmingham

B3 1RB, UK..

ISBN 978-1-80323-820-3

www.packtpub.com

Writing this book has been challenging to manage between work and my personal life. I would like to thank my loving wife, Nidhi, and my kids, Aarav, Anaya, and Navya, for their dedicated support and encouragement throughout this two-year journey.

I would also like to thank my parents, Mr. Balveer Singh (late), Mrs. Rajbala Devi, and my brother, Dr. Ghanendra Singh, for their blessings and inspiration to do more and keep moving.

I also like to thank my colleagues, friends, walking buddy Kartikey Sharma, and cricket team for their unconditional love and support in making this happen.

And lastly, thanks to Jeff Carpenter for being an inspiration and providing necessary support.

– Amar Deep Singh

Foreword

Microservices have become so ubiquitous in software systems, especially in cloud computing, that it's hard to believe the trend has only been around for a decade. The microservice boom was in itself a response to the challenges of a previous paradigm: **service-oriented architecture (SOA)**.

The initial promise of SOA for producing flexible, composable systems had been gradually drowned out as vendors lured developers toward complex products such as application servers and enterprise service buses, configured by a never-ending flow of XML that proved difficult to understand and maintain. The microservice boom of the 2010s was a welcome reaction against the complexity and bloat of the SOA world, focusing on a simpler set of conventions centered around **representational state transfer (REST)** APIs and more human-readable formats such as JSON.

It was in this season of rapid innovation that Amar Deep and I first encountered the powerful combination of microservices and public cloud infrastructure, with AWS as the early leader. We eagerly followed the maturation of DevOps principles espoused by Netflix and other early adopters as they made various frameworks open source and popularized methodologies such as chaos engineering. These visionaries showed us how to produce microservice-based systems that demonstrated high availability and performance at scale.

However, as might have been expected, microservices have reached that inevitable stage of the hype curve where cautionary tales and articles suggesting "you must be this tall to use microservices" are rampant. These practitioners advise new projects, to begin with a monolithic architecture to minimize development risk and time-to-market and only migrate to microservices when the monolith begins to strain under operational needs. So, how can you know whether microservices are right for your project and whether you're executing them effectively?

This book emerges at an opportune moment to help answer that question, combining the essential patterns and principles to approach microservice development properly from day one with the practical guidance needed to maintain your microservices for the long haul.

Specifically, you'll learn multiple architecture patterns for microservices, including how to identify them, how they interact with each other, and how they make use of supporting infrastructure for data storage and movement. Then you'll learn how to use tools such as Git, AWS CodeCommit, AWS CodeGuru, AWS CodeArtifact, and AWS CodeBuild to write and test your microservices and orchestrate all these tools with AWS CodePipeline to deploy to environments, including AWS **Elastic Compute Cloud (EC2)**, AWS **Elastic Container Service (ECS)**, AWS Elastic Kubernetes Service **(EKS)**, and AWS Lambda. You'll also learn how to extend your pipelines beyond AWS using Jenkins and deploy to on-premises servers.

With tons of practical, relatable domain examples throughout, Amar Deep has provided an outstanding end-to-end resource to help you deliver microservices successfully with AWS. With this guide, you'll design, develop, test, deploy, monitor, and secure your microservices with confidence now and in the future.

Jeff Carpenter

Software engineer at DataStax and the author of Cassandra: The Definitive Guide (O'Reilly) and Managing Cloud Native Data on Kubernetes (O'Reilly).

Contributors

About the author

Amar Deep Singh is an author, architect, and technology leader with 16 years of experience in developing and designing enterprise solutions. He currently works for US Bank as an engineering director and helps digital teams in cloud adoption and migration. He has worked in the banking, hospitality, and healthcare domain and transformed dozens of legacy enterprise applications into cloud-native applications. He specializes in modernizing legacy applications and has expertise in developing highly available, scalable, and reliable distributed systems. He holds several professional certifications, including AWS Certified Solutions Architect at the Professional level, AWS Certified Security and Machine Learning Specialist, TOGAF Certified Enterprise Architect, Certified Jenkins Engineer, Microsoft Certified Professional, and Terraform Associate certification.

I want to thank my wife, Nidhi, my parents, Mr. Balveer Singh, and Mrs. Rajbala Devi, and my kids, Aarav, Anaya, and Navya, for their unconditional love and support during this journey. Nidhi has been a great support and always encouraged me whenever I feel low or discouraged.

About the reviewers

Sourabh Narendra Bhavsar is a senior full stack developer and an agile and cloud practitioner with over eight years of experience in the software industry. He has done a post-graduate program in artificial intelligence and machine learning at the University of Texas at Austin, a master's in business administration (marketing), and a bachelor's of engineering (IT) at the University of Pune, India. He currently works at Rabobank in the Netherlands as a lead technical member, where he is responsible for designing and developing microservice-based solutions and implementing various types of workflow and orchestration engines. Sourabh believes in continuous learning and enjoys exploring emerging technologies. When not coding, he likes to play the tabla and read about astrology.

Kathirvel Muniraj has worked in analytical banking applications for more than eight years in the cloud solutions industry and specializes in DevOps and DevSecOps. His experience includes Kubernetes, MLOps, and DevSecOps.

He has completed many global certifications such as Certified DevSecOps Professional, Certified Kubernetes Security Specialist, Microsoft Certified: DevOps Engineer Expert, Certified Kubernetes Administrator, and AWS Certified Sysops Administrator.

I'd like to thank my family and friends who understand the time and commitment. I'd like to thank my mentor, Mohammad Samiullah Mulla for his guidance, encouragement, support, and motivation over the past 4 years. I am also thankful to my whole family for supporting me and tolerating my busy schedule while still standing by my side. I owe my accomplishments and triumphs in life to the unwavering guidance and assistance of a person who prefers to remain anonymous. I have endearingly nicknamed this individual 'Bujji'.

Table of Contents

6

Automating Code Reviews Using CodeGuru   199

7

Managing Artifacts Using CodeArtifact   233

8

Building and Testing Using AWS CodeBuild   251

Part 3: Deploying the Pipeline

9

Deploying to an EC2 Instance Using CodeDeploy   295

10

Deploying to ECS Clusters Using CodeDeploy   337

14

Extending CodePipeline Beyond AWS   499

Appendix   543

Index   567

Other Books You May Enjoy   578

Preface

This book provides a step-by-step guide to developing a Java Spring Boot microservice and guides you through the process of automated deployment using AWS CodePipeline. It starts with an introduction to software architecture and different architecture patterns, then dives into microservices architecture and related patterns. This book will also help you to write the source code and commit it to CodeCommit repositories, review the code using CodeGuru, build artifacts, provision infrastructure using Terraform and CloudFormation, and deploy using AWS CodeDeploy to **Elastic Compute Cloud** *(EC2)* instances, on-prem instances, ECS services, and Kubernetes clusters.

Who this book is for

This book is for software architects, DevOps engineers, **site reliability engineers** *(SREs)*, and cloud engineers who want to learn more about automating their release pipelines to modify features and release updates. Some knowledge of AWS cloud, Java, Maven, and Git will help you to get the most out of this book.

What this book covers

Chapter 1, Software Architecture Patterns, teaches you about software architecture and about different software architecture patterns.

Chapter 2, Microservices Fundamentals and Design Patterns, describes microservices and different patterns related to microservices. In addition, this chapter explains different strategies and design patterns to break a monolithic application into a microservice.

Chapter 3, CI/CD Principles and Microservice Development, covers different CI/CD principles and explains how to create a sample Java Spring Boot application to be deployed as a microservice and expose a **REpresentational State Transfer** *(REST)* endpoint to ensure that our users can access this endpoint.

Chapter 4, Infrastructure as Code, explains what **Infrastructure as Code** *(IaC)* means and what tools and technologies you can use to provision different resources. We will explain how you can run a CloudFormation template and how you can create infrastructure using Terraform.

Chapter 5, Creating Repositories with AWS CodeCommit, explains what a version control system is and covers the basics of Git-based version control systems. This chapter explains the AWS CodeCommit service and its benefits and then guides users on committing application source code to the CodeCommit repository.

Chapter 6, *Automating Code Reviews Using CodeGuru*, walks through what the AWS CodeGuru **artificial intelligence** (**AI**) service is and how it can be used to review code automatically and scan for vulnerabilities.

Chapter 7, *Managing Artifacts Using CodeArtifact*, explains the AWS CodeArtifact service, its usage, and its benefits. This chapter walks through the different generated artifacts and how they can be securely stored with CodeArtifact.

Chapter 8, *Building and Testing Using AWS CodeBuild*, focuses on the AWS CodeBuild service and explains how you can use this service to customize the build and code testing process.

Chapter 9, *Deploying to an EC2 Instance Using CodeDeploy*, explains the AWS CodeDeploy service and how it can be used to deploy applications to EC2 instances and on-premises servers. This chapter takes a deep dive into different deployment strategies and configurations available to deploy applications.

Chapter 10, *Deploying to ECS Clusters Using Code Deploy*, focuses on explaining what a container is and how you can deploy Docker containers to an AWS ECS service. In this chapter, we configure CodeDeploy to automatically deploy sample applications to ECS containers.

Chapter 11, *Setting Up CodePipeline*, explains what CodePipeline is and how it can help us to orchestrate other AWS services to set up continuous development and delivery of the software.

Chapter 12, *Setting Up an Automated Serverless Deployment*, introduces you to serverless ecosystems and how AWS provides scalable solutions through Lambda, and how you can set up automated serverless Lambda deployment.

Chapter 13, *Automated Deployment to an EKS Cluster*, focuses on understanding Kubernetes and learning about the **Elastic Kubernetes Service** (**EKS**) provided by AWS and automated application deployment to an EKS cluster using CodePipeline.

Chapter 14, *Extending CodePipeline Beyond AWS*, focuses on extending AWS CodePipeline beyond AWS-related infrastructure and services. In this chapter, you will learn to integrate CodePipeline with Bitbucket and Jenkins and deploy to instances hosted outside AWS.

Appendix, focuses on creating **Identity and Access Management** (**IAM**) users and tools needed for the application development such as Docker Desktop, Git, and Maven, which are important but not part of the core chapters.

To get the most out of this book

You need to have some basic understanding of AWS cloud, Java, Maven, and Git to get started. Having some knowledge about Docker, Kubernetes, and Terraform will help you, although we will be covering the basics.

Software/hardware covered in the book	Operating system requirements
Java	Windows, macOS, or Linux
Terraform	
AWS account	

If you are using the digital version of this book, we advise you to type the code yourself or access the code from the book's GitHub repository (a link is available in the next section). Doing so will help you avoid any potential errors related to the copying and pasting of code.

> Note
>
> *This book contains many screenshots with text formatted to fit the page width. Therefore, the text in these screenshots may appear small and difficult to read. However, you can get a better view using the free PDF, which is provided with every purchase. Please refer to the section at the end of the Preface for instructions on obtaining your PDF copy.*

Download the example code files

You can download the example code files for this book from GitHub at `https://github.com/PacktPublishing/Building-and-Delivering-Microservices-on-AWS`. If there's an update to the code, it will be updated in the GitHub repository.

We also have other code bundles from our rich catalog of books and videos available at `https://github.com/PacktPublishing/`. Check them out!

Download the color images

We also provide a PDF file that has color images of the screenshots and diagrams used in this book. You can download it here: `https://packt.link/B4nWn`.

Conventions used

There are a number of text conventions used throughout this book.

`Code in text`: Indicates code words in text, database table names, folder names, filenames, file extensions, pathnames, dummy URLs, user input, and Twitter handles. Here is an example: "The following `buildspec.yml` file describes how CodeBuild will run the `maven package` command to get the artifacts and include the Java JAR file, Dockerfile, `appspec.yml`, and other files in the output."

A block of code is set as follows:

```
version: 0.0
os: os-name
files:
    source-destination-files-mappings
permissions:
    permissions-specifications
hooks:
    deployment-lifecycle-event-mappings
```

Any command-line input or output is written as follows:

```
terraform destroy -auto-approve
```

Bold: Indicates a new term, an important word, or words that you see onscreen. For instance, words in menus or dialog boxes appear in **bold**. Here is an example: "Now, click on the **Create Deployment group** button."

> **Tips or important notes**
> Appear like this.

Get in touch

Feedback from our readers is always welcome.

General feedback: If you have questions about any aspect of this book, email us at customercare@packtpub.com and mention the book title in the subject of your message.

Errata: Although we have taken every care to ensure the accuracy of our content, mistakes do happen. If you have found a mistake in this book, we would be grateful if you would report this to us. Please visit www.packtpub.com/support/errata and fill in the form.

Piracy: If you come across any illegal copies of our works in any form on the internet, we would be grateful if you would provide us with the location address or website name. Please contact us at copyright@packt.com with a link to the material.

If you are interested in becoming an author: If there is a topic that you have expertise in and you are interested in either writing or contributing to a book, please visit authors.packtpub.com.

Share Your Thoughts

Once you've read *Building and Delivering Microservices on AWS*, we'd love to hear your thoughts! Scan the QR code below to go straight to the Amazon review page for this book and share your feedback.

https://packt.link/r/1803238208

Your review is important to us and the tech community and will help us make sure we're delivering excellent quality content.

Download a free PDF copy of this book

Thanks for purchasing this book!

Do you like to read on the go but are unable to carry your print books everywhere?

Is your eBook purchase not compatible with the device of your choice?

Don't worry, now with every Packt book you get a DRM-free PDF version of that book at no cost.

Read anywhere, any place, on any device. Search, copy, and paste code from your favorite technical books directly into your application.

The perks don't stop there, you can get exclusive access to discounts, newsletters, and great free content in your inbox daily

Follow these simple steps to get the benefits:

1. Scan the QR code or visit the link below

https://packt.link/free-ebook/9781803238203

2. Submit your proof of purchase
3. That's it! We'll send your free PDF and other benefits to your email directly

Part 1:
Pre-Plan the Pipeline

You will learn about software architecture and microservices development and the challenges that microservices bring, then about **continuous integration/continuous delivery (CI/CD)** and how it can help to deliver microservices. We will create a sample Spring Boot Java microservice to deploy in the AWS environment.

This part contains the following chapters:

- *Chapter 1, Software Architecture Patterns*
- *Chapter 2, Microservices Fundamentals and Design Patterns*
- *Chapter 3, CI/CD Principles and Microservice Development*
- *Chapter 4, Infrastructure as Code*

1

Software Architecture Patterns

In this chapter, you will learn about software architecture and what it consists of. You will learn about software architecture patterns and how to use these different patterns to develop software. After reading this chapter, you will have a fair understanding of layered architecture, microkernel architecture, pipeline architecture, service-oriented architecture, event-driven architecture, microservices architecture, and a few other major architectural patterns. This chapter will discuss real-world examples of each of these patterns, as follows:

- What is software architecture?
- Architecture patterns overview
- Layered architecture pattern
- Major architecture patterns

What is software architecture?

Before we start learning about microservices architecture and the different patterns related to it, we need to learn what software architecture is.

If we have to construct a building, a structural architect needs to lay out the design of the building and think about building capacity, the weight the foundation needs to hold, the number of floors the building will have, staircases, elevators for easy access, and the number of entry and exit gates.

Similar to a construction architect, a software architect is responsible for designing software and defining how software components will interact with each other.

Software architecture defines how your source code should be organized and how different elements in your code interact with each other in a software application using proven design patterns to achieve a business outcome.

The following diagram shows the interaction between source code components and their organization with the help of design patterns, which is what software architecture is built on:

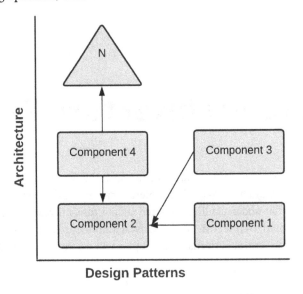

Figure 1.1 – Software architecture structure

Now that we have defined what software architecture is, let's compare it with the building example we discussed earlier. If you look closely, you will find that the construction project is very similar to an application and that each area in a building resembles a different aspect of software architecture. Each floor in a building can be thought of as a software layer in an application.

Building capacity is very similar to the load or number of requests your application can handle. The building's foundation can be compared to the software infrastructure or hardware on which the application is deployed, while load capacity is directly related to the memory and space needed by the application. Staircases and elevators can be thought of as items used for your users to access your application and entry and exit gates can be treated as endpoints exposed by your application to outside systems. Design patterns can be thought of as a method you use to mix the concrete or how many iron rods you need to lay out a solid foundation.

Software architecture is built by organizing the source code components and their interaction with each other, constructed with the help of design patterns.

> **Important note**
>
> A design pattern defines a way to solve a common problem faced by the software industry. Design patterns don't provide any implementation code for the problem but provide general guidelines on how a particular problem in a given context should be solved. These solutions are the best practices to use in similar situations. An architect has the flexibility to fine-tune solutions or mix different patterns or design new patterns to solve their specific problems or adjust solutions to achieve certain results.
>
> Fine-tuning a solution or design pattern is known as an architectural trade-off, where you balance out your parameters to achieve a certain result. For example, let's say you need to have your building foundation a few meters under the Earth to make a skyscraper, which increases your construction cost; however, if you want to make a building with only a few floors, then you don't need to make your foundation so solid. Similarly, you can make adjustments to your software architecture/design pattern to achieve certain results. To mitigate risk to a project, the **Architecture Tradeoff Analysis Method** (**ATAM**) is used in the early phase of an architectural design.

How you build your design pattern or interaction is based on your use case and the architectural trade-offs you have made to achieve certain software goals; there is no right or wrong architecture – it is something that keeps evolving.

> **The ATAM process**
>
> The ATAM process collects quality attributes such as business goals, functional requirements, and non-functional requirements by bringing the stakeholders together. These quality attributes are used to create the different scenarios; then, architectural approaches and decisions run through these scenarios to create an analysis of risks, sensitivity points, and trade-offs. There can be multiple iterations of this analysis and each iteration fine-tunes the architecture. The solution proceeds from being generic to more specific to the problem and risk is mitigated.

Architecture patterns overview

Now that you have a fair understanding of software architecture and design patterns, let's talk about architecture patterns and their types.

An architectural pattern defines a repeatable architectural solution to a common problem in a specific use case. In other words, an architecture pattern describes how you arrange your functional blocks and their interaction to get a certain outcome.

An architectural pattern is similar to a software design pattern, but the scope of an architectural pattern is broader, while a design pattern is focused on a very specific part of the source code and solves a smaller portion of the problem.

There are different types of architectural patterns to address different types of problems, so let's learn about some of those that are widely used. We will provide a high-level overview of these architectural patterns as diving too deep into them is outside the scope of this book.

A layered architecture pattern

A layered architecture is made up of different logical layers, where each layer is abstracted from another. In a layered architecture, you break down a solution into several layers and each layer is responsible for solving a specific piece of that problem. This architecture pattern is also known as an N-tier architecture pattern.

In this architectural pattern, your components are divided into layers and each layer is independent of the other but connects with its immediate layer through an interface to exchange information. Layered architecture focuses on the clean separation of concerns between layers and these layers are usually closed to other layers other than their immediate top layer. In a layered architecture, a closed layer always connects to its immediate layer, while an open layer architecture allows you to skip a layer in the layered chain. The following diagram shows an example of a closed-layered architecture:

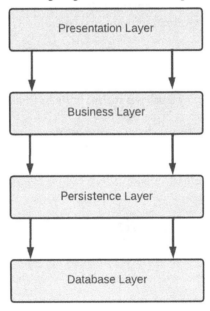

Figure 1.2 – Layered architecture pattern

In the preceding figure, we have divided an application component into four layers. The presentation layer contains any application code needed for the user interface. The business layer is responsible for implementing any business logic required by the application. The persistence layer is used for manipulating data and interacting with vendor-specific database systems. Finally, the database layer stores the application data in a database system.

All of these layers in a layered architecture only interact with the layer immediately below; this semantic is known as a closed layer. This approach gives you an advantage if you have to replace or change a layer; it doesn't affect other layers in an application as there is no direct interaction between one layer and another non-immediate layer.

This style brings some complexity as well because some of the layers for certain scenarios can be just pass-through layers. In closed layers, you don't do anything for a particular model or method but still have to go through all the layers, which can be a performance issue in certain scenarios. For example, while looking up a customer record on the user interface screen, no business logic is needed, but you still have to go through that layer to get a customer record.

To handle this kind of scenario, you can keep your layers open by allowing a direct call from the business layer to the persistence layer or a direct call from the presentation layer to the service layer, but that will bring more complexity to your architecture. Looking at the example shown in the following diagram, we have a five-layer architecture with an open style. Here, calls from one layer go directly to another layer by skipping some layers:

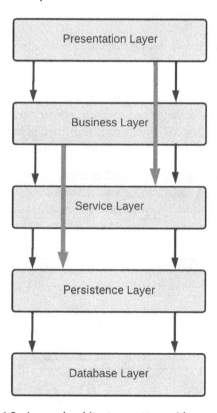

Figure 1.3 – Layered architecture pattern with open layers

In the preceding layered architecture pattern, if you need to change your persistence layer, then it is going to affect your service layer and, as a ripple effect, this will also affect your business layer. So, whether you want to keep your layers closed or open is an architectural trade-off you need to make.

A layered architecture pattern is very simple and easy to understand, and it can be easily tested as all the layers are part of the same monolithic application. Testing this architecture is easy as you don't have to have all of your layers available at the same time – you can stub or mock the layer that isn't available and simply test the other layers. For example, in the preceding diagram, if the presentation layer is not ready, you can test the other layers' flow independently or if any other layer isn't ready, you can simply stub that out and mock the behavior that is expected from that layer and later on, that stub can be replaced by actual logic.

This architecture does scale well but it is not very agile as changing one layer will have a direct effect on the immediate neighboring layer and sometimes have a ripple effect on other layers as well; however, this is a great starting point for any application.

A microkernel/plugin architecture pattern

Microkernel architecture is also known as plugin architecture because of its plugin-based nature. Microkernel architecture is suitable for product-based development in which you have a core system or **Minimum Viable Product** (**MVP**) and you keep adding more functionality as needed so that it works to customize the core system and can be added or removed based on your needs. In this architecture, you need a registry in the core system that knows about the plugin you added and handles the request by utilizing the newly added module. Whenever you remove a plugin from the architecture, it will remove the reference from the registry.

Eclipse or Visual Studio Code IDEs are very good examples of microkernel architectures. As a core system, both of these are text editors and allow you to write code. Both of these IDEs can be extended by adding plugins to the core system; take a look at the following diagram, which explains the architecture of the Eclipse IDE:

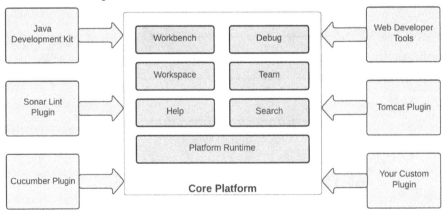

Figure 1.4 – Eclipse IDE plugin-based architecture

In the preceding diagram, you can see that Eclipse is made up of hundreds of plugins; its core system is built with some basic functionalities that are extendable and allows any new plugin to be registered with the core system whenever you install a new plugin. Once a plugin has been installed, its functionality is available to use as a feature of the core platform, but if you remove that plugin, it only removes the functionality provided by the plugin.

Let's take another example of a microkernel architecture pattern to understand it better. In a fictitious financial institution, a core system is designed to open an account by creating an account, opening the application, creating a user profile, generating an account statement, and using a system to communicate with the customers. A customer can see their application/account details on the account dashboard:

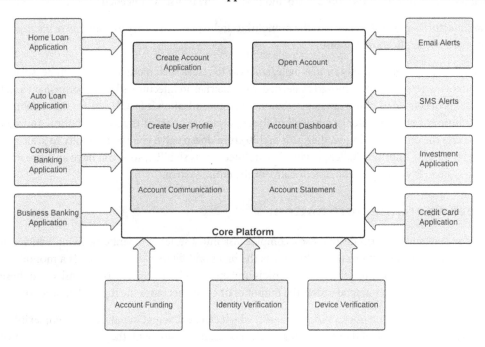

Figure 1.5 – Microkernel architecture for a financial institute

In this financial system example, the architecture core module can do normal account-opening activities, but this architecture is expandable, and we have added other functionalities in the form of plugins. This core module is being extended via plugin architecture to support credit card, home loan, auto loan, consumer banking, and business banking account types, and these plugins handle their specific type of functionality. By using a plugin architecture, each product type can apply its own rules.

The account communication module utilizes a microkernel architecture pattern and is being extended by adding support for email and SMS alerts. If an institution doesn't want these features, they can simply not register/install these modules.

Similarly, validating a user device for login and verifying their identity and eligibility for a particular banking product has also been added in a microkernel fashion to the core system. Similarly, some banking products might have funding requirements, so you can add a funding plugin to supply that functionality to the core system.

To minimize the impact on the core system, plugins utilize a standard interface provided by the core system. This core system exposes an interface; all plugins must follow the same integration interface to be part of the core system.

If you want to utilize a third-party plugin, which doesn't have a similar interface, you have to develop an adapter that connects to the core system and perform any translation needed.

This architecture pattern also falls into the monolithic architecture category as your entire package is part of the same application runtime.

This architecture pattern is pretty simple, agile, and easy to test as each module change can be isolated and modules can be added or removed as needed. Microkernel architecture is highly performant as all components become part of the same installation, so communication is fast. However, the scalability of the solution is still a challenge since all of these plugins are part of the same solution and you have to scale the entire system together; this architecture doesn't give you the flexibility to scale any individual plugin. Another challenge with this architecture is stability, so if you have an issue with the core system, it will impact the functionality of the entire system.

A pipeline architecture pattern

Pipeline architecture patterns decompose a complex task into a series of separate elements, which can be reused. The pipeline pattern is also known as the pipes and filters pattern, which is a monolithic pattern. In this architecture pattern, a bigger task is broken down into smaller series of tasks, and those tasks are performed in sequential order; the output of one pipe becomes the input of another one.

For example, an application does a variety of tasks, which can be different in nature and complexity, so rather than performing all of those tasks in a single monolithic component, they can be broken down into a modular design and each task can be done by a separate component (filter). These components connect by sharing the same data format, with the output of one step becoming the input of another one. This pattern resembles a pipeline where data is flowing from one component to another like water flowing through a pipe, and due to this nature, it is called pipeline architecture. Pipes in this architectural style are unidirectional and pass data in only one direction:

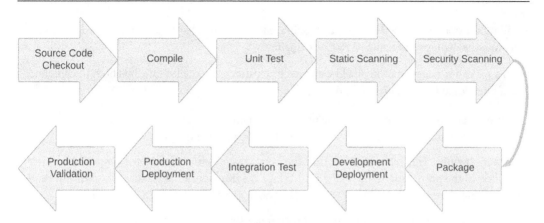

Figure 1.6 – A deployment pipeline example

The preceding diagram shows a typical deployment pipeline for a Java application, which is a great example of pipeline architecture.

A development pipeline starts with a developer committing code into a source code repository and performs the following tasks:

1.   The pipeline checks out the source code to the build server from the source code repository and then starts building it in the next stage.

2.   The source code is built and, if it is a compiler-based language, then code is compiled in this step.

3.   Once the source code has been built, unit test cases are executed and, on successful execution, the build progresses to the next step.

4.   The static scanning step performs the static (non-running) source code analysis to find any vulnerabilities or code issues. SonarCube and PMD are two famous static code analyzer tools.

5.   In the next step, the pipeline performs security scanning using tools such as Black Duck, Fortify, and Twistlock to find any runtime vulnerabilities.

6.   Once these quality gates have passed, the pipeline builds the final package for deployment and passes it to the deployment phase.

7.   In the deployment phase, the pipeline deploys the packaged application into the dev environment and passes control to the integration test phase.

8.   The integration phase runs the test suite, which is designed for validating the application, and makes sure that the end-to-end application is running with no issues.

9.   Once the integration phase has passed and the application has been validated, it is promoted to the higher environment for deployment. In this case, it is the production environment.

10. Once the production deployment has been successful, the application is validated and verified for a successful production rollout and the pipeline is marked as completed.

We will talk more about the deployment pipeline in upcoming chapters as this book focuses on automating the deployment using AWS CodePipeline.

Now that we've had a refresher on pipeline architecture patterns, let's talk a little bit about the different types of filters used in a pipeline architecture:

- **Producer**: Producers are the starting point of this architecture and have an outbound pipe only. In the preceding example, the **Source Code Checkout** stage is a producer.

- **Transformer**: Transformer filters take the input, process it, and send the output to another filter. In the preceding example, the **Build** stage is a transformer as it takes the source file and generates a compiled version of the source code.

- **Tester**: Tester filters are usually pass-through filters that take the input and call the other filters in the pipeline; sometimes, they can also discard the output. In our example, **Static Scanning** is a tester as it will perform scanning, and if anything fails, it will fail the phase without going through the different stages.

- **Consumer**: Consumer filters are the ending point of a pipeline architecture and can be recognized by the fact that they do not feed the output into another stage of the pipeline. In our example, the **Production Validation** stage is a consumer.

The pipeline or filter and pipe architecture is very simple and flexible and is easy to test, but scalability and performance are a problem since each stage of the pipeline must be completed sequentially. Small blockers in one of the pipeline steps can cause a ripple effect in subsequent steps and cause a complete bottleneck in the system.

A space-based architecture pattern

Scaling an application is a challenging task, so to scale this, you might need to increase the number of web servers, application servers, and database servers. However, this will make your architecture complex because you need high performance and scalability to serve thousands of concurrent users.

For horizontally scaling a database layer, you need to use some sort of sharding, which makes it more complex and difficult to manage. In a **Space-Based Architecture** (**SBA**), you scale your application by removing the database and instead have memory grids to manage the data. In an SBA, instead of scaling a particular tier in your application, you scale the entire layers together, known as a processing unit.

> **Important note**
> In vertical scaling, you add more resources such as memory and compute power to a single machine to meet the load demand, while in horizontal scaling, you join two or more machines together to handle the load.

SBAs are widely used in distributed computing to increase the scalability and performance of a solution. This architecture is based on the concept of tuple space.

> **Tuple space**
>
> Tuple space is an implementation of the associative memory paradigm for parallel/distributed computing. There's a processing unit, which generates the data and posts it to distributed memory as tuples; then, the other processing units read it based on the pattern match.

The following diagram shows the different components of an SBA:

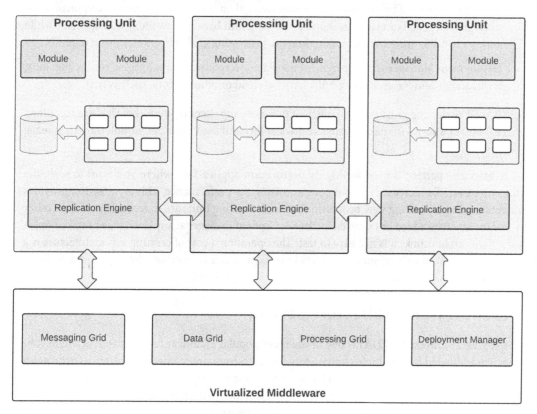

Figure 1.7 – SBA pattern

SBA comprises several components:

- **Processing Unit**: A processing unit is your application deployed on a machine and backed by an in-memory data grid to support any database transactions. This in-memory data grid is replicated to other processing units by the replication engine.

- **Messaging Grid**: This is a component of the virtualized middleware and is responsible for handling client requests and session management. Any request coming to the virtualized middleware is handled by the messaging grid, and it redirects that request to one of the available processing units.

- **Data Grid**: The data grid is responsible for data replication between different processing units. In a SBA pattern, the data grid is a distributed cache. The cache will typically use a database for the initial seeding of data into the grid, and to maintain persistence in case a processing unit fails.

- **Processing Grid**: This is an optional component of a space-based architecture. The processing grid is used for coordinating and combining requests to multiple processing units; if a request needs to be handled by multiple processing units, then the processing grid is responsible for managing all that. For example, if one processing unit handles inventory management while another handles order management, then the processing grid orchestrates those requests.

- **Deployment Manager**: The deployment manager is responsible for managing processing units; it can add or remove processing units based on load or other factors, such as cost.

In an SBA, all processing units are self-sufficient in processing client requests, but they are combined to make it more powerful for performance and scalability and use virtualized middleware to manage all that.

This architecture pattern is used for highly performant applications where you want to scale the architecture as traffic increases without compromising the performance. This architecture provides horizontal scaling by adding new processing units. Processing units are independent of each other and new units can be added or removed by the deployment manager at any time. Due to the complex nature of this architecture, it is not easy to test. The operational cost of creating this architecture is a little high as you need to have some products in place to create in-memory data grids and replicate those to other processing units.

An event-driven architecture pattern

Event-driven architecture (**EDA**) is one of the most popular architecture patterns; it is where your application is divided into multiple components and each component integrates using asynchronous event messages. EDA is a type of distributed system, where the application is divided into individual processes that communicate using events. This architecture pattern is made up of loosely coupled components that integrate using these messages. These messages are temporarily stored in messaging queues/topics. Each message is divided into two parts – a header and the actual message payload.

Let's look at a simple example of an EDA for a hotel chain called "Cool Hotel," which has chosen to deploy its reservation system using the EDA. A hotel **Property Management System** updates its inventory for available rooms, which is updated in **Inventory System** using an event message:

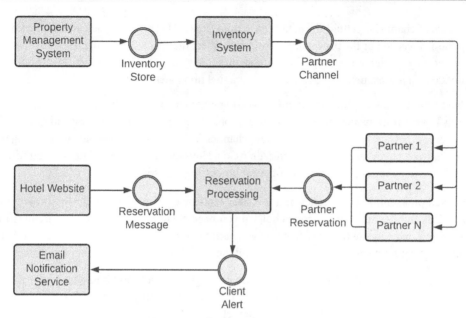

Figure 1.8 – Cool Hotel EDA example

Once **Inventory System** has been updated, it communicates that change to its partners through an event called **Partner Channel**, where different partners consume the message and update their listing.

Whenever a reservation is made either directly through the **Cool Hotel** website or through a partner listing, a reservation message is pushed to the **Reservation Processing** system, which then has the responsibility for multiple tasks. It updates the availability of the room by sending an event to **Inventory Store** and also generates an event, which is sent to the **Client Alert** system. As we have shown, all of our system components are very loosely coupled and are not aware of each other; they only communicate with each other through events.

There are several components in an EDA:

- **Event generator**: This is the first logical layer in the architecture. A component in an architecture that produces an event for other components to consume is known as an event generator or producer. An event producer can be an email client, a sensor, or an e-commerce system.

- **Event channel**: The event channel is the second logical layer in the EDA. An event channel propagates the information from the event generator to the event processing engine. This component temporarily stores the information in a queue and hands it over for processing asynchronously whenever the processing engine is available.

An event channel can hold the data for a certain time and remove it when either a retention period is reached or the processing engine can collect the event. This depends entirely on the underlying implementation of the event channel; Active MQ, AWS Kinesis, AWS SQS, and Kafka are a few examples of popular event channel implementations.

- **Event processing engine**: The event processing engine is the third logical layer in the EDA. This layer is responsible for selecting the appropriate event and then filtering and processing the event message received from the event channel. This layer is responsible for executing the action triggered by an event. In the preceding Cool Hotel example, the email notification service and inventory service are examples of event processing systems.

EDA is based on eventual consistency and it is near real-time due to the asynchronous nature of its events. This architecture is not a good choice if you need changes in real time. This architecture increases your availability but there is a trade-off you need to make in terms of consistency as your data can be inconsistent for a small amount of time until the asynchronous message has been processed.

This architecture brings agility to your product and it is highly scalable and improves performance, but it is hard to test these systems end to end. EDA is not simple to implement and you have to rely on a messaging system to find a reliable event channel.

EDA brings more complexities into the architecture as you have to deal with message acknowledgment, message re-delivery in case of failure, and dead letter queue analysis and processing for any corrupted events.

> **Dead letter queue**
>
> In a messaging system, a dead letter queue is a specialized queue that handles the messages that have been rejected or have overflowed from the main queue due to a message format issue, non-existing queue, message size limits, message rate limit, or software issues because of which a message is not processed successfully. Usually, dead letter queues work as a backup for main queues/topics and are used for offline processing and analyzing messages to identify any potential issues in the consumers or producers due to which messages are rejected.

A serverless architecture pattern

This is one of the more modern architectural patterns and has gained more and more traction recently. The name would make you believe that there is no server involved in this architecture, but that is not true. In this architecture pattern, rather than being responsible for all aspects of server infrastructure and management, you are utilizing the infrastructure owned by a third party and paying for the service you utilized. Serverless architecture is different from **Infrastructure-as-a-Service** (**IaaS**) offerings; in this architecture pattern, you are not responsible for managing servers, operating system updates, security patches, scaling up or down to meet demand, and so on.

Serverless architecture falls into two categories. Let's look at what they are.

Backend as a Service (BaaS)

In this architecture style, instead of creating a backend service, you are utilizing a third-party provider. For example, your application requires you to validate the identity of a user. You could develop an authentication scheme and spend time and money on hosting the service, or you could leverage a third-party identity system such as Amazon Cognito to validate users in your system.

Taking this approach, you save on development efforts and costs. You reduce your overhead by not owning and managing the servers used for authentication; you only need to pay for the usage of the service.

This architectural model is not very customizable, and you have to rely a lot on the features provided by the third party. This approach may not be suitable for all use cases since the degree of customization required by some workloads may not be available from third-party providers. Another disadvantage of using this approach is that if your provider makes an incompatible change to their API/software, then you also need to make that change.

Functions as a Service (FaaS)

In this architectural style, you don't have to use a third-party service directly, but you must write code to execute on third-party infrastructure. This architecture style provides you with great flexibility to write features with no direct hardware ownership overhead. You only pay for the time when your code is being executed and serving your client. The FaaS pattern allows you to quickly create business services without focusing on server maintenance, security, and scalability. This service scales based on the demand and creates the required hardware as needed. As a consumer, you just have to upload your code and provide a runtime configuration and your code will be ready to use.

With the increasing popularity of cloud computing, serverless architectures are a hot topic; AWS Lambda, Azure Functions, and Google Cloud Functions are examples of the FaaS architecture style.

This serverless pattern provides a lot of scalability and flexibility for customizing the solution as needed, but it has some downsides, and it is not suitable for all use cases. For example, as of December 2022, AWS Lambda runtime execution is limited to 15 minutes. If you have a task that takes longer than 15 minutes, then a Lambda function isn't a good choice.

Let's look at an example where AWS Lambda is a good solution. In the hotel industry, inventory rollover and rate rollover are scenarios where property owners want to set their inventory the same as last year and also want rates to be pretty much the same. Suppose you have a big event – for example, on July 4, which is Independence Day in the USA, your room rate will be high due to occupancy, so you want to maximize your profit and don't want to sell your rooms for a lower rate than you might have a week before. So, property owners want to automatically set the same rate and room inventory as last year, but still want the flexibility to make a manual change or override, if necessary, through their property management system. For this kind of problem, you don't need to have a service that is running 24/7 because you could instead have a scheduled job, which can be run on a daily or a weekly basis.

AWS Lambda can be a very good solution to this problem because it doesn't require a server to do this job. An AWS Lambda function can be triggered by a CloudWatch schedule or EventBridge schedule event and call rate/inventory service to perform the rollover after providing the authentication token retrieved from the Amazon identity service (Cognito). Once this has been completed, the server that ran your function can be terminated by AWS. By using the serverless approach, you didn't need to run your server all the time, and when your function was done executing, AWS was able to return the underlying hardware to the available pool. As of December 2021, sustainability is an AWS Well-Architected Framework pillar, so by using the Lambda approach here, you are not just able to control your costs but you are also helping save the environment by not running your server and burning more fuel:

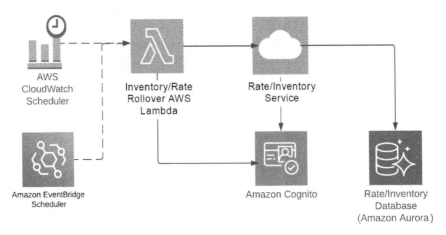

Figure 1.9 – Serverless architecture example for hotel inventory/rate rollover

In the preceding example, we use three AWS serverless resources. The first is an AWS Aurora database as a service, which helps write and read the data so that you don't have to worry about how database servers are being managed or backed up.

The next service is AWS Lambda, which executes your code that performs the actual update to the database. You are not maintaining any server to run this code; you simply write a function in Python, Go, .NET, Java, or any of the other available runtimes, upload the code to AWS Lambda, and configure an AWS CloudWatch/EventBridge trigger (which can use either cron-style static repetition or can be set up to look for events in a log) to invoke the code. The final service is Amazon Cognito, which is used by AWS Lambda to retrieve an authorization token for the rate/inventory service; this is an example of the BaaS serverless architecture we discussed earlier.

Serverless architecture is a good design choice when you need scalability and your tasks can be executed within 15 minutes, and you don't want to run your servers. It is also a very cost-effective architecture and brings agility to your solution because you only pay for the memory you provision for the Lambda, as well as the execution time. The initial request for these functions can be challenging as underlying infrastructure provisioning can take some time and you might need to prewarm your functions before they can take requests.

A service-oriented architecture pattern

Service-oriented architecture (**SOA**) is one of the widely used architecture patterns in enterprises and promotes the usage of services in an enterprise context. SOA is an enterprise integration pattern that connects different heterogeneous systems to carry out business functions.

In SOA, services communicate with each other in a protocol-agnostic way, but they may use different protocols such as REST, SOAP, AMQP, and so on. They share the information using a common platform known as the **Enterprise Service Bus** (**ESB**), which provides support for protocol-agnostic communication.

> **An ESB**
>
> An ESB is a centralized software component that is responsible for integration between different applications. Its primary responsibility is to perform data transformations, handle connectivity, messaging, and routing of requests, provide protocol-neutral communication, and potentially combine multiple requests if needed. An ESB helps in implementing complex business processes and implementing the data transformation, data validation, and security layers when integrating different services.

SOA is different from microservices architecture, which I will be talking about later in this chapter.

SOA is enterprise-focused since services target the entire enterprise. There are five types of services we focus on in an SOA; let's take a look.

Business services

Business services are designed around the functionalities an organization has to perform to run its business operations. These services are coarse-grained and not very fine-grained:

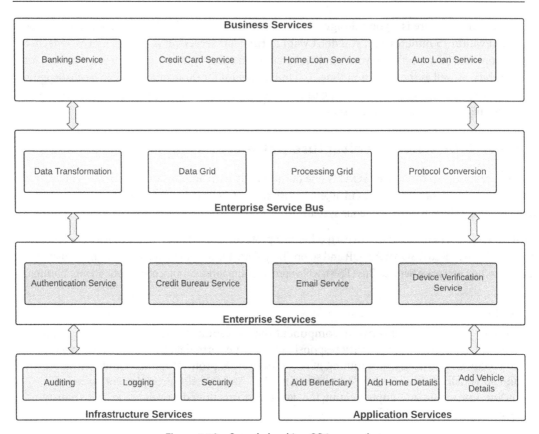

Figure 1.10 – Sample banking SOA example

You can identify a business service by filling in the blank in the following sentence: "*we are in the business of _____.*" For example, concerning a financial company, you can say the following:

- We are in the business of *home loans*
- We are in the business of *credit cards*

So, to support these two businesses, you need to create customers, so you might have a service in your architecture to create the customer but you can't say that *we are in the business of creating customers*. So, any service that identifies your business is considered a business service in an SOA. If you take an example of a banking SOA architecture, as shown in the preceding figure, you can place **Auto Loan Service**, **Credit Card Service**, **Banking Service**, and **Home Loan Service** in the business services category.

Enterprise services

In an SOA, services that are used across an organization or enterprise are known as enterprise services. For example, authenticating a user is an organization-level issue, and you don't want to have several applications implementing their logic to authenticate users if this can be done by a single service throughout the enterprise.

Another example of an enterprise service is managing customer information that can be shared by all other business services. For example, in a banking institution, you don't want to maintain the customer information in home loan, auto loan, and core banking systems differently; the customer information might be different in each business unit. So, you should have an enterprise or shared service that is being utilized to maintain the customer information at a central place and being used by all of the business services across the enterprise. Enterprise services are usually owned by a shared service group or by an enterprise architect group.

In a banking SOA architecture, you would create capabilities that would be shared by multiple business applications. For example, many applications need to send emails to customers, so it's sub-optimal for each one of them to "reinvent the wheel" of sending out emails; it's better to have a dedicated email-sending task responsible for sending emails, handling undeliverable emails, and so on.

Application services

Application services are scoped to each application level and are fine-grained to a specific application. The scope of these services is specific to the individual application – for example, opening a business account, checking a customer's credit card balance, or generating a credit card statement.

Application services have a very specific purpose for doing something related to the application context. Application services are owned by application teams within a line of business. In the banking SOA architecture example, you can see that adding a vehicle is a very specific service and limited to just doing one thing, so this is a very specific and application-level service.

Another example is adding home details for a home loan application. This is also very specific and categorized into an application-level service.

Infrastructure services

Infrastructure services are common concerns for the applications that don't provide any features to the business application or application services but play a supporting role by providing platform-level features. Those services implement platform-level functions that are very generic and can be used by enterprise services, as well as by application services.

For example, logging is an integral part of any application, so that service will fall into the platform- or infrastructure-level services; any enterprise service or application service can use this for its logging requirements. Similarly, auditing data is not the core function of an application service, but governmental oversight or internal compliance organizations may require it, so an auditing service becomes a part of infrastructure services and can be invoked by application services as needed.

The ESB

In the middle of our sample banking architecture, we have an ESB, which is responsible for business process choreography, translating business data, changing message formats, or providing protocol-agnostic transmission. It is possible to have an SOA architecture without the need for an ESB, but then your services are dependent on each other directly and become tightly coupled with no abstraction.

SOA is not very simple to implement or easy to test, but this architecture is very scalable. The cost of implementing an SOA is a bit high as you have to be dependent on third-party software to have an ESB in the middle, although you might not need all the features provided by the ESB. Therefore, this architecture has a downside, and ESB can be a single point of failure.

A microservices architecture pattern

As the name suggests, the microservices architecture pattern promotes the usage of smaller services in a bounded context. Microservices architecture is a distributed architecture and is very similar to SOA, with some exceptions. In a microservices architecture, services are fine-grained and serve a specific purpose. In other words, microservices are lightweight and serve a very specific purpose while in SOA, services are more in the enterprise scope and cover a segment of functionality.

In the microservices architecture pattern, an application is divided into loosely coupled smaller self-contained components known as services. Each service runs in a process and connects to other services in a protocol-aware synchronous or asynchronous fashion if needed. Each microservice is responsible for carrying out a certain business function within a bounded context and the entire application is a collection of these loosely coupled services.

Unlike SOA, no message bus is involved in this architectural pattern. Microservices provide protocol-aware interoperability, so the caller of the service needs to know the contract and protocol of the service it is calling.

The following diagram explains the characteristics of a microservices architecture pattern. It explains how a microservices architecture is distributed and that each service is a separately deployed unit and designed around data domains within a bounded context exposed through APIs, where communication can happen with other services through API endpoints in synchronous mode or through events in asynchronous mode:

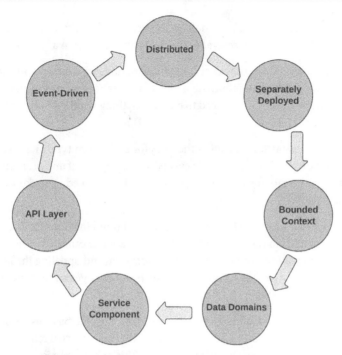

Figure 1.11 – Microservices architecture characteristics

I will provide more details about microservices in the upcoming chapters, but for now, let's look at an example from the hotel industry:

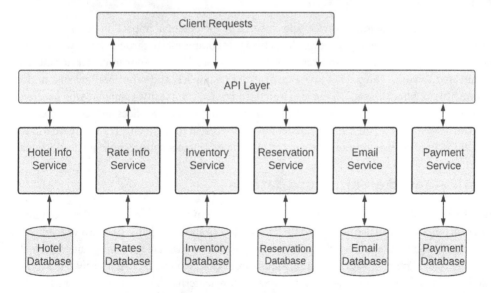

Figure 1.12 – Microservices architecture example

In this diagram, instead of having a single monolithic application taking care of the entire reservation process, the application is divided into several domain-specific services in a bounded context.

Hotel Info Service has a bounded context of hotel information, which keeps all the information related to a hotel, such as demographic information, the number of rooms, information about facilities, contact information, and so on. So, if you need to change anything within the context of a hotel, you need to make changes to this service.

Rate Info Service is responsible for managing the rates for each room type in a specific hotel for a specific night, based on demand or any specific events coming up. So, if any property management system needs to change the rate for any rooms in any hotel, they need to make a call to this rate information service.

Similarly, **Inventory Service** keeps track of the available hotel room inventory and is responsible for adding or removing a room from the inventory when a reservation is canceled or made. **Reservation Service** is responsible for creating a reservation within the system and updating the hotel's inventory. Although the reservation system has its own database, those changes need to be reflected in the inventory database as well.

Each of these services has a bounded context and provides a specific business function. Services provide an API layer, which is how they interact with each other and external requestors. The division of functionality is done so that each service is autonomous. This architecture is very agile and scalable because it allows you to develop each service individually without affecting others. It can make testing simpler as well because a service will have a defined API, which your test harness can utilize to confirm that your service is not only service accepting valid requests and rejecting invalid ones but the response from the service is also matching the API.

One negative aspect of the microservice pattern could be the potential impact on performance; each service must invoke another service over the network or communicate through a messaging system asynchronously, which is subject to latency, and this can cause issues. Your microservice must also be architected to withstand packet loss, out-of-order responses, and other realities of a network backbone. This architecture pattern can also lead to additional complexity, as anyone who has tried to *break apart the monolith* can tell you. We will learn more about microservices and their benefits and drawbacks, as well as how you can write better microservices, in the next chapter, where we will learn about a monolithic application and how you can overcome the issues a monolithic application creates.

A service-based architecture pattern

Service-based architecture is a distributed architecture pattern and is based on service taxonomy. In the previous sections, you learned about the microservices and SOA patterns. Both of them sound very similar, with one exception – microservices architecture services are very fine-grained while SOA services are coarse-grained, with messaging middleware. This architecture pattern solves the complex problems faced in the SOA and microservices architecture patterns, so it is a kind of balance between both of those architecture patterns.

When you first try to "break apart the monolith" to convert it into a series of microservices, you will soon discover that isolating functionality and determining the area of responsibility for each function can be quite difficult. Remember, in a true microservices architecture, there is no direct access from one function/process to the data store of another. Deciding where data needs to reside, and the degree of data duplication you're willing to accept, can be daunting. Since each microservice has a database, aggregating that data and accessing it from different services is a challenge.

Additionally, when designing a microservice, you should incorporate DevOps principles from the very beginning; using an automated pipeline may be more difficult when starting, but will pay off handsomely later when you have 20 or more microservices that need to be managed. We will learn more about this in the next chapter, where I will explain some design patterns related to microservices.

So, implementing the microservices architecture from ground zero is not an easy task and requires rewriting the entire application and defining the communication and orchestration patterns; on the other hand, SOA is focused more on the enterprise level and requires the messaging bus, which adds additional cost and complexity to the architecture. So, we need a middle ground as a basis to move away from a monolithic architecture to a microservices architecture, and SOA provides that to us.

In service-based architecture patterns, services are coarse-grained and based on the application domain, instead of a specific purpose-bounded context. In this architecture style, services share the same database and interact with each other in a protocol-aware fashion, so we don't need an ESB to work as middleware. In this architecture, we divide the application into parts of related functionality to create services. The service granularity is considered macro and contains a bunch of modules that make sense based on the transaction scope of the application and avoid the service orchestration overhead we have in a microservices architecture.

Service-based architecture focuses on extracting the related functionality from an application into the services so that the transaction boundary of the functionality can be done within the same service and no distributed transaction, service orchestration, or choreography is needed between these services. You can create these services around the complex functionalities within your application. These services are called macro-services or application services as they take a portion of the full application, while microservices target a single purpose. This architecture style creates tight coupling between complements within a service, but a lot of organizations use this architecture style as a stepping stone toward a full microservices-based architecture.

In this architecture style, all the services access the shared database, so any issues or downtime to the database can impact the entire application. In a service-based architecture, communication between services is protocol-aware and services share contracts; any change to these contracts can impact the other services. Since no middleware is involved between services, it makes it simple and easy to implement. Some variations of this architecture can have a lightweight integration hub in between services, which can help in translation and service orchestration.

For example, an e-commerce site has several modules to carry out its business functions, so breaking it down into a microservices architecture will take significant effort and create additional overhead in setting up automated pipelines and dividing each module into a single feature-focused, bounded context service. So, the functionality is divided into portions – order-related functions are handled by **Order Management Service**, vendor-related functions are carried out by **Vendor Management Service**, and all customer-related functions are separated by **Customer Management Service**. However, all of these services access the same database, and communication between these services is done through REST-based APIs:

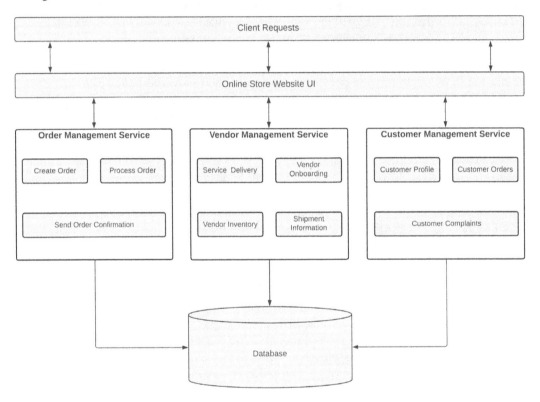

Figure 1.13 – Service-based architecture of an online store

The application user interface is deployed as a separate service. Communicating with the required services through their APIs is similar to a microservices architecture; the only difference is that these services handle a bigger responsibility and are not focused on a single purpose.

In a service-based architecture, there are fewer services compared to microservice-based architecture, so it makes it simple and you don't have to have automated deployments from day one. This makes deploying these services a bit easier in the sense that not thousands of them are involved and you can eventually build your release pipelines. The performance of this architecture is higher compared to a microservices-based architecture since you are not communicating with several services to perform a single business function. This architecture is also cheaper compared to SOA as no heavy messaging bus is needed to implement this architecture. These services are a little hard to develop as they contain the entire functionality of a complex business feature, which needs to be understood well by the development team and poses a challenge in making frequent releases.

Summary

In this chapter, you gained some background knowledge about software architecture and some of the widely used architectural patterns, including layered, pipeline, event-driven, service-based, and others. This chapter also provided you with a good understanding of the two most popular architectural patterns: SOA and microservices architecture and the differences between the two. You can apply this knowledge to design a good solution based on your use cases, confident that the architecture pattern you selected will be a good choice for a given problem.

In the next chapter, we will dive deep into microservices architectures and investigate different strategies to decompose a monolithic application. You will also learn about the different design patterns for developing robust microservices and how you can automate their deployment.

2

Microservices Fundamentals and Design Patterns

This chapter will help you to understand microservices fundamentals and different design patterns related to microservices. You will learn about the software development problems that microservices solves and the challenges it brings to an application architecture. Understating microservices design patterns will help you in real life to develop good microservices.

In this chapter, we will be covering the following topics:

- Monolithic versus microservices
- Microservices design patterns overview
- Microservices decomposition patterns
- Microservices database patterns
- Microservices integration patterns
- Microservices observability patterns
- Cross-cutting concern patterns

Before you start learning about microservices, you need to understand what a monolithic architecture is and why you need a microservices architecture. Let's first focus on a monolithic architecture, and then we will work to understand different ways to decompose a monolithic application into a microservices application.

A monolithic application architecture

To understand a monolithic architecture, let's take an example of a new banking application. Suppose you are starting a new banking business and you want to develop an application to support the checking and saving account business functions at this new bank. You developed a single application that carries out all banking transaction functions. Now, your business is increasing and you want to offer your customers investment opportunities, so you make necessary enhancements to your application and add features to support investment functions. You made adjustments to the application architecture so that both investment and banking transactions can be supported, which increases some load on the application as well, so you change the size of your application servers and database servers so that the additional load can be handled.

With this successful launch of an investment portfolio, banks now need to offer home loans and auto loans to customers, so you have to make necessary adjustments to the application and add these new features to it, as well as enhance the current architecture. Everything is part of the same application, and it is taking the load of all four business segment users.

These kinds of applications, which are performing more than one function, are considered monolithic applications. Here is a more formal definition of a monolithic application.

An application is considered monolithic if it is responsible to carry out not just a particular task but also completing all the required steps needed for a business function. Usually, in these applications, the user interface and code required to access the data are packaged together and these applications are self-contained.

In our banking application, you can see that the same application is responsible for performing banking, investment, and loan functions, so this is classified as a monolithic application. In fact, a single business function can also be classified as a monolithic application if it is carrying out multiple tasks, such as accepting banking applications, opening a bank account, creating customers, generating bank statements, funding bank accounts, and managing credits and debits for accounts.

Now that you have an understanding of what a monolithic application is, let's try to understand what the problems are with being monolithic. Let's take an example, In 2020–2021, due to the COVID-19 pandemic, the US government cut down the interest rates to almost zero, which had an impact on the mortgage rates, and homeowners started refinancing their homes due to these lower mortgage rates. Subsequently, banks started getting a lot more home loan applications than at any time in history. To handle such traffic demand, we would need to scale up our application, but scaling just the home loan part of our application is not possible because our entire business functions are deployed as a single monolithic application. Therefore, if you are scaling the home loan segment of an application, you are scaling the entire application, and that's not a very good usage of our resources.

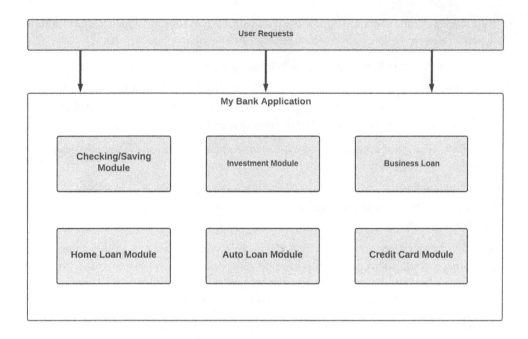

Figure 2.1 – An example of my bank's monolithic application architecture

Another impact of the pandemic was that a lot of small businesses faced challenges and needed money to keep running their payroll. Therefore, the US government launched a specific **Paycheck Protection Program (PPP)** through which a business could apply for a loan to keep their employees on the payroll. This was a new program for small business banking customers, so banks immediately needed to make a change to their application to support these loan applications. However, making sudden changes isn't easy for our monolithic application, due to the large code base supporting multiple business functions, and making these kinds of rapid changes would risk the stability of the application. Also, the deployment of this change would cause a system outage for investment, auto, and home loan customers who have nothing to do with PPP. So, here, we see multiple issues due to the monolithic nature of our banking application, one being application stability and another one being an outage in the non-relevant business sections.

The following are some of the disadvantages of having a monolithic architecture:

- The size of the code base is big, and it takes time for developers and test engineers to understand the code and business functionality. Code complexity also increases over time.

- Due to the large code base size, code compile time is higher and developers spend a lot of time building source code even for small changes they are making, which indirectly reduces productivity.

- Application components are tightly coupled with each other and, usually, splitting those dependencies can be tough, and sometimes, those dependencies are circular.

- Application code is not very reusable, as everything is part of the same application.

- Monolithic applications can also be difficult to scale when different modules have conflicting resource requirements – for example, one module needs high CPU demand while another module needs to do a lot of disk operations, so you have to scale both, as you have a single application.

- Monolithic applications cause reliability issues in an application because a memory leak in one part of it can bring it down entirely.

- Monolithic applications cause downtime to an entire application, so if you are making a change to one part of it as a single deployment unit, it will be deployed altogether, causing downtime to other modules.

- Monolithic applications are difficult to test, as making a change to one module can cause issues in another module, as everything is part of the same code base, and at times, you don't know the ripple effects of a change. So, testing these applications is difficult and requires quality engineers to test the entire application, instead of just that particular change.

- Monolithic applications are usually big and you are tied to a single technology stack, as rewriting these applications isn't easy, due to the large code base and complexity involved.

- Due to the large size of an application, deployment time and application startup time is large, and automating testing is not easy and feasible.

- Managing changes with a monolithic application is difficult, as a lot of coordination is needed between teams so that the changes do not conflict, and one team/developer doesn't override the changes of another.

Now that we understand why it is challenging to have a monolithic application, let's go ahead and see how we can solve these challenges using a microservices architecture.

Understanding a microservices architecture

Now that we have an idea of what a monolithic application is and what problems it causes, let's focus on microservices and how these services solve the problem posed by a monolithic architecture.

A component of your entire application stack that focuses on a single capability and is deployed and managed separately is known as a microservice. Your entire application is a series of these microservices. A microservice is autonomous in nature and can be developed and deployed independently without causing service disruptions to other parts of the system.

Let's take an example to better understand microservices. Suppose you have an e-commerce website where you sell online goods to consumers. In your application, you will have separate features or functions to support your business, such as vendor management, product catalog, order placement, billing, payment order confirmation, shipment, and delivery tracking. Each of these features in your application can be a separate microservice so that changes to shipment tracking or disruption to a payment service is not affecting your entire application. That is one of the greatest advantages that microservices offer.

Now, let's look at this example in a diagram, showing how a microservice is different from a monolithic version of the same application.

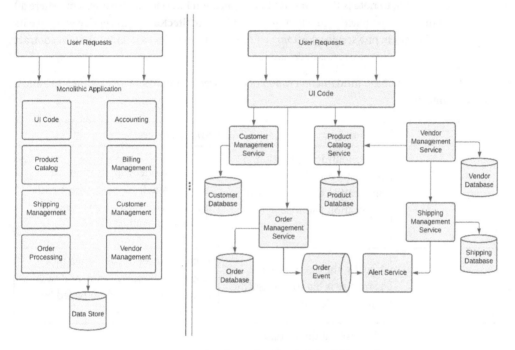

Figure 2.2 – An example of an online store monolithic application architecture versus a microservices application architecture

In this example, on the left side, we have the monolithic style architecture, and on the right side, we have the microservices style architecture. In a monolithic style application, the entire code base and all features are part of the same deployment unit, while on the microservices side, all the application's components are divided into a series of services that are autonomous in nature and can be deployed separately from each other. These services interact with each other in a protocol-aware fashion, either synchronously using an API endpoint or asynchronously through a messaging queue/topic.

If an application team did a great job designing a monolithic architecture, then the entire code might be organized in a module fashion and components might not have a dependency on each other, but still, deploying and testing the entire application brings many risks. In a microservices architecture, you divide your application in such a way that your services have a bounded context and related functionality is in a single service. There are different strategies to effectively break down your application, which we will discuss later in this chapter.

Another important thing to note is that a monolithic application has one database system, where all data is being captured in one place, while in a microservices architecture, each service will have its own database, so changes in one service don't impact other services unless a communication contract is changing.

Let's take a look at some of the fundamental differences between a microservices architecture and a monolithic architecture.

	Monolithic	Microservices
Focus area	A monolithic application is a full-fledged big application that contains all application features.	A microservices application is a series of small services, where each microservice focuses on a single business capability.
Team size	Usually, the development team size is big to manage the entire application.	Each microservice is owned by a small development team, and each team is small enough to be fed by two medium pizzas.
Application complexity	The application becomes complex over time, due to all business capabilities in a single unit, and developers keep moving in and out of teams, creating knowledge gaps.	Since each microservice focuses on a single business capability or business sub-domain, it is easy to understand.
Deployment	Since it is a single binary, the deployment of these applications is easier.	An application is built from several microservices, and deploying so many services makes deployment time-consuming, so microservices demand automated deployment from scratch.

	Monolithic	Microservices
Change complexity	Making a change in this architecture style is complicated because a change in one module can impact other parts of the application, and the entire application will then need to be tested.	Making a change in this architecture style is easy because each microservice focuses on a bounded context, so changes impact that area only, and understanding the change impact is easy for the development team.
Fault isolation	Failure in one part of the application can cause downtime or crash the entire application, as it is a single binary.	Microservices are designed for fault tolerance and resiliency, and usually, any issue with a service impacts only that service.
Scalability	Scalability is an issue with this architecture style, as some code might be CPU-bound while other parts of the application require more disk operations, so scaling these applications is not easy and can utilize system resources a lot quicker.	Each microservice is deployed individually and may have different resource requirements, so each service can be scaled individually as needed, and you don't have to scale each microservice if it is not required.
Technology	These applications are big, so once you are committed to one technology stack, such as .NET or Java, you have to stick with that technology, as re-writing the entire application is not easy.	In this architecture style, there is no dependency on technology; each service can have a different technology stack, and they just need to connect through an agreed protocol and schema. Therefore, teams with different skill sets and technology stacks can collaborate on the same application.

Table 2.1 – The differences between a monolithic architecture and a microservices architecture

Now that you have a fair idea of a microservices architecture and the difference between a monolithic application and a microservice-based application, let's take a look at some of the best design considerations for a microservices architecture in the next section.

Microservices design patterns

Microservices come with some challenges as well – for example, in a monolithic application, you have to take care of only one application deployment, but in a microservices architecture, you might have hundreds of services to do the same job, depending on the size of the application. Let's discuss some of the design patterns related to microservices to effectively utilize this architecture style. Microservices design patterns are classified into different categories, based on the nature of the work they handle, so let's focus on each category one by one.

Decomposition patterns

When you are working to transform an application from a monolithic to a microservices architecture, or writing a new application using a microservices architecture, the biggest challenge you are going to face is how you are going to divide your application into microservices. Scratching the surface of the application and dividing that into microservices requires following certain strategies to logically isolate application functionality into autonomous services, as well as defining a data access and communication strategy. So, let's focus on the different microservices decomposition patterns and learn how to divide monolithic applications into smaller autonomous microservices.

Decomposing by business capability

In this decomposition strategy, a monolithic application is divided into autonomous microservices using business capabilities identified for the application. This decomposition is based on the business architecture modeling concepts that a business capability is what a business does to generate value for its business. For a credit card application, the marketing of credit card products, opening a credit card account, activating a credit card, and generating credit card statements are a few examples of its business capabilities.

Over time, the way you do business might change but what your core business doesn't change much, so decomposing your application using business capabilities doesn't change the business capability, although it might change the manner in which business is conducted. For example, today you might be accepting credit card applications through paper copies, and tomorrow you might start accepting them online or through a mobile application, or via phone, but the way a credit card account needs to be opened and how the credit card will be activated in the backend won't change. So, when you decompose your application using a business capability, it is pretty much aligned with your business the way it currently works, so your business users won't feel any change.

To decompose an application around business capabilities, you need a team that has insight into an organization's different business units and subject matter experts who understand all the business capabilities and can help to decompose the application.

In this strategy, you identify a business capability by identifying the purpose from a business aspect of doing something. For example, if I see creating a customer as a business capability and I want to create a service for that, that's not correct, as we are not in the business of creating customers, but we are in the business of providing a credit card to customers, and we might have some services that support our business capabilities. Therefore, you have to be very careful while identifying microservices based on this decomposition pattern.

In the following diagram, we have broken down a monolithic credit card application around different business capabilities that you would find in any credit card application. For example, when opening a credit card account for a customer, it doesn't matter how you open this account, whether via a banker working in a bank branch, a customer through your website, a mobile app, or a customer calling a phone number to apply for a credit card. The way the credit card account was opened doesn't matter; this service fulfills the business capability.

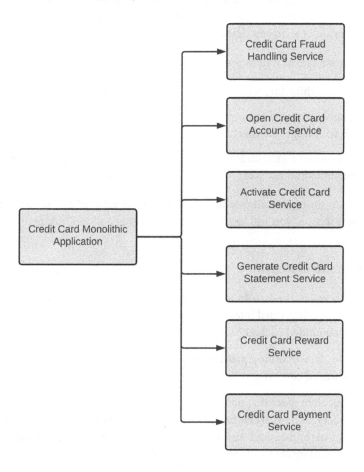

Figure 2.3 – Credit card monolithic application decomposition using a business capability

Another capability of this monolithic application is activating a credit card to start allowing purchases on it, so one microservice is decomposed to allow a card activation business capability. How a credit card gets activated doesn't matter, either by phone or by vising the website; this service handles the activation.

There are a few advantages of using this strategy to decompose your monolithic application to a microservices architecture:

- This approach creates a very stable microservices architecture if business capabilities are relatively stable

- The individual components in your architecture may evolve over time as aspects of business change, but your architecture remains unchanged, so this approach gives you a very stable architecture

- Another benefit of this approach is that your development teams are trained to be cross-functional and organized around delivering business value, instead of delivering technical features, so the people in the product team can relate to them easily

There are a few disadvantages as well to using this approach:

- Your application architecture will be tightly coupled with the business model

- Over time, people get used to doing business in a certain way, and if that process has any anomaly, then your target architecture will reflect it

- This approach requires a solid understanding of an overall business, and it is difficult to identify business capabilities and services

Now that you have an understanding of how to decompose an application using the business capability, in the next section, you will learn about decomposing a monolithic application using the sub-domain model.

Decomposing by sub-domain

This decomposition strategy is based on **domain-driven design** (DDD) principles. In this decomposition strategy, a business domain is divided into subdomains, and microservices are created based on these sub-domains within a bounded context.

To understand how the decomposition pattern works, we need to understand what exactly a business sub-domain and a bounded context are.

Your business is the problem space that needs to be solved, and this is known as a domain. A domain consists of multiple subdomains, where each sub-domain corresponds to a different part of your business entity. For example, in an online retail store business, vendor management, customer management, order management, and so on are the business sub-domains.

There are different classifications for sub-domains, based on their relationship with your business domain:

- **Core**: This is the most important part of the application and a key attribute of your business
- **Supporting**: These sub-domains are related to your business but are not the core of your business
- **Generic**: These sub-domains are not specific to the business domain and can be used across domains

Before we define this strategy further, we need to understand what a bounded context is.

In DDD, a bounded context is a boundary in which a particular domain model applies. For example, a policy domain model can have different meanings in different domains; a policy can be an auto policy in an auto insurance problem space, while in another context, it can be a life insurance policy. So, a bounded context defines the boundary of a domain model with a domain and explains exactly what a domain model means within that context, and what rules apply to it.

Understanding sub-domain and bounded concepts can be confusing, so to illustrate the difference between these two, let's look at the following diagram:

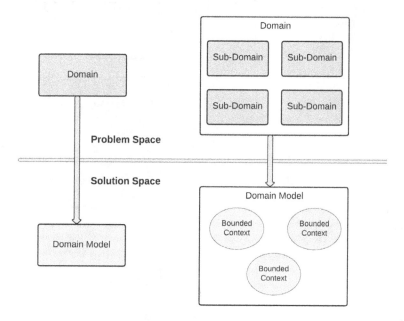

Figure 2.4 – Business sub-domains versus bounded context

In this diagram, the top area shows the problem space, the left side presents a domain or a business, and the right side shows how that business domain is divided into smaller problems, known as sub-domains. A domain represents the problem to be solved, while a domain model represents the model that solves the problem, similar to how a sub-domain represents a part of the problem, while a bounded context is a segment of the solution.

Now that you understand what this decomposition pattern is, let's see an example of it.

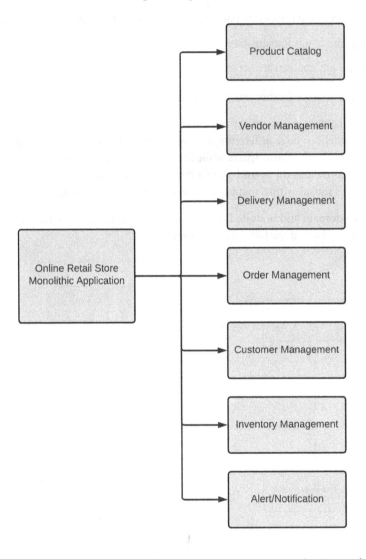

Figure 2.5 – Online store monolithic application decomposition using a business sub-domain

In the preceding example, we decomposed an online retail store application using a business sub-domain technique and identified sub-domains such as customer, vendor, order, product, inventory, and so on in our application. Out of these domain models, some are key to our business domain, some are related, and some are generic sub-domains.

In this example, any functionality related to customers – creating a customer, activating a customer, updating a customer, deleting a customer, changing a customer address, and so on – belongs to the customer sub-domain. Similarly, anything related to the vendor will go under the vendor management sub-domain. So, in this strategy, you clearly identify the key aspects of a business domain and then you divide that into sub-domains, and anything related to a particular sub-domain within a bounded context belongs to that microservice.

Decompose by transactions

When you switch from a monolithic application to a microservices architecture, one of the challenges you will face is how to manage the transactions that span multiple microservices. In microservices, it often happens that your business functionality is divided in such a way that one part is done by one microservice and another part is done by another service, and each service also has its own database, so managing the database-level transactions is not possible.

Let's understand this problem with an example. Suppose we have an e-commerce application that we decomposed using a business sub-domain strategy. Now, our customer has placed an order for one of the products, so we need to process the payment for the customer, adjust the product inventory, send confirmation to the customer, and at the same time, send an order for fulfillment. So, in our architecture, this business transaction is spreading across microservices – inventory management, order management, alert notification services, and so on. So, as part of the necessity to ensure a smooth business function, we need to make sure that either all of these microservices successfully process these requests or all of them are able to roll back the transaction.

To overcome these cases, you have two options; either you implement a database-level two-phase commit pattern or implement the **Segregated Access of Global Atomicity (SAGA)** pattern to overcome these microservice limitations. A two-phase commit pattern is not advisable for microservices architecture due to the performance issues it poses to the architecture, and we will talk more about the SAGA pattern later in this chapter.

> **Note**
> The **two-phase commit protocol (2PC)** is a technique for implementing a transaction across multiple databases, where changes are either committed or rolled back from all involved components.

Decomposing services using previous decomposition strategies has the challenge of transaction boundaries, as you have to communicate with other services to complete a business transaction.

Therefore, using business transactions as a basis to divide a monolithic application into microservices is another strategy. In this strategy, you divide your business functions in such a way that each service within your architecture is responsible for carrying out a complete business workflow, so you don't have to worry about the 2PC or SAGA patterns at all. An entire business transaction is carried out by the same microservice, and either it is completed or rolled back.

This pattern is very useful when response time is an important aspect of your application. Since back-and-forth communication between services and then coordinating that transaction takes time, it adds network latency. Therefore, this pattern helps to reduce the latency, as your entire business transaction is within the same service, and you don't have to make a call to other services.

Let's take an example of a trading platform to understand how this monolithic application can be decomposed, based on a transaction. In a real-world scenario, trading platforms are more complex and have a lot more features and compliance requirements, but our feature set is small, and we are only supporting the buying/selling of securities and account functions, such as the deposit and withdrawal of funds.

In the following diagram, you can see that we have broken down this trading application into six services, based on the business transaction – for example, customer registration is quite complex, and you have to collect their personal details, run through some credit checks, validate their identity, send agreements, create a customer profile, send a welcome email/mail, and so on. Therefore, all of those functions are carried out by the customer registration service.

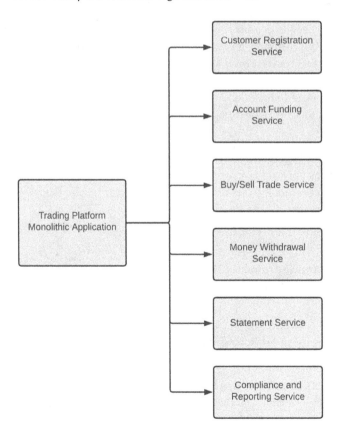

Figure 2.6 – Trading platform monolithic application decomposition using a transaction boundary

Another service in the same application is a buy/sell trade service, so when you are buying/selling a security, you have to place your order, fulfill that order, update an account balance with a debit/credit, update a security holding in the customer account, send a trade confirmation, and so on. All of this happens within a transaction, and all those functions are carried out by the same service.

Similarly, adding funds to a customer account, withdrawing money from the account, generating reports, and sending trade transactions to stock exchanges are all different services, which carry out those business transactions on their own without the need for distributed transactions.

There are some great advantages to using this strategy to decompose your monolithic application:

- This strategy has less response time compared to other decomposition strategies, as there are no remote calls to multiple services to carry out a business function, so this pattern saves time on network trips and makes overall service response time fast

- This improves the availability of business functions, since all dependencies of that business transaction are part of the same service, so overall service availability increases

- In this approach, you don't have to worry about data consistency, since all changes related to that business feature happen on the same service database, and implementing rollback will be easy

There are some disadvantages as well to using this strategy:

- Since we design our services around business transactions, we had to package multiple modules together that participate in the transaction, which can create a monolithic application in itself.

- When deploying a microservice, there can be two versions of the same service around deployment, which can cause issues if dealing with transaction-based services.

- The complexity of the service increases, and you might need a bigger team to manage the services, which loses a few of the biggest benefits of having a microservices architecture.

- The size of a service can be really big, as some business transactions have multiple domains and dependencies with each other.

- In this pattern, it is possible that your microservice will be doing more than one thing, which might be unrelated from a domain perspective, but that is part of the same business transaction. For that reason, this pattern violates the Single Responsibility Principle.

This decomposition strategy helps us to understand how we can get a similar response time as a monolithic application, while taking full advantage of the microservices architecture.

Strangler pattern

Due to the consistent innovation in hardware/software technology, changing business requirements, and the addition of new features to original applications, any system, development tools, hosting technology, and even application architecture gets obsolete over time. So, basically, every application has a shelf life, and once that shelf life is reached, you have to rewrite the application, but rewriting an application isn't that easy and involves a complex migration process.

Strangler is a decomposition pattern and provides a way to slowly migrate away from a monolithic or legacy application to a newly designed application. In this pattern, what you have to do is to create a strangler façade around your legacy application and route all your requests through that, migrating your functionality to your newer architecture over time.

Since the API contract is not changing and functionality is slowly moving from a legacy to a new service stack, there wouldn't be any impact on consumers. Only your façade knows where a particular functionality is located, so that's the only change you will have.

Let's take an example to understand this pattern. In the following example, we have a monolithic credit card application, which is wrapped around a strangler service façade. In the initial stage (**A**), when we just start the migration, there is no functionality on the microservice side.

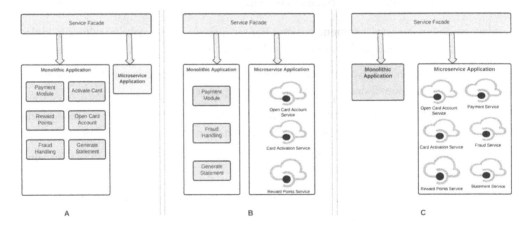

Figure 2.7 – Credit card monolithic application decomposition using the strangler pattern

In phase **B**, we move 50% of the functionality to the microservices-based architecture, and the remaining 50% is still running on the monolithic side, but it doesn't impact our consumers, since we have a strangler façade, so changes are abstracted from our clients. In phase **C** of our migration, we move the entire functionality to our new microservices-based architecture and shut down our legacy monolithic application.

The strangler pattern is widely used for rewriting legacy applications or mainframe systems, but you can't use this pattern for a greenfield application, as there is no existing application to migrate from. This pattern provides a quick way to start migration without impacting consumers.

Bulkhead pattern

This pattern is inspired by ship design. The base of the ship is divided into partitions known as bulkheads, so if there is somehow a leak in the hull, it doesn't affect the entire ship, and only that portion of the ship is filled with water, and the sinking of the entire ship is prevented. This design pattern is focused on fault isolation while decomposing a monolithic architecture to a microservices architecture.

In this pattern, you divide your service in such a way that if there is an issue with a portion of the service, it doesn't affect its consumers. The service should still be available to other consumers by containing the issue to the affected consumer or a portion of the functionality.

Let's use an example to understand this pattern better. Suppose we have a microservice in an e-commerce application that is used to send email notifications to our clients. This microservice is used by different consumer applications within an organization.

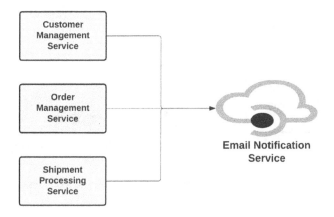

Figure 2.8 – An e-commerce application email notification service

This service is used by customer management, order management, and shipment processing services to trigger an appropriate notification. Now, the problem is that during a sales event, we process a lot of orders, so the email service is heavily used by the order management service.

Due to the high traffic sent by the order management service, the email service is not able to take the load and starts failing, or has degraded performance. Due to the degraded performance of the email service, the shipment processing service and customer management service now start seeing failure. So, failure from one service now starts affecting another service, meaning that we are not able to send shipment notifications and also not able to send account alerts to a user about any changes that occur in the customer data.

So, how do we solve this issue? To increase the availability of the email service, we need to make sure that one consumer of the service should not be able to take down the entire service. The following diagram shows how we can isolate failure on an affected service only and ensure that it doesn't affect other services. In this diagram, we have deployed the email service in such a way that each consumer connects to a different instance of it. Although we are running the same version of code on each email service instance, we have divided it into consumer groups so that if the order management service takes down the email service instance, it doesn't affect the customer management service and shipment processing service.

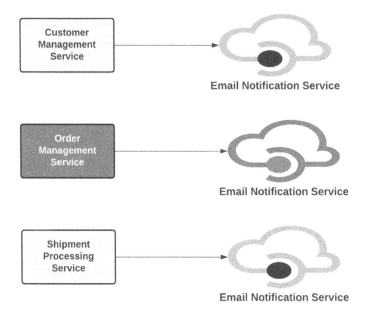

Figure 2.9 – Email service fault isolation implementation using a bulkhead pattern

Let's use another example to understand this pattern better. In the same e-commerce application, the order management service calls the payment service to process payments, the inventory service to update the product inventory, and the email service to send the order confirmation. The following example shows how a problem in one service can cause cascading effects in other services:

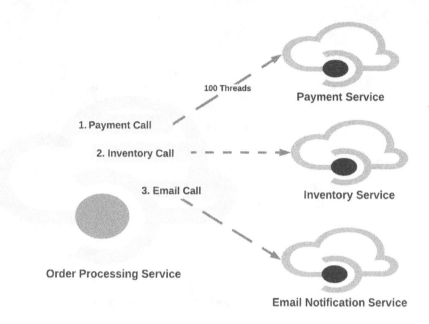

100 Threads

Payment Service

1. Payment Call

2. Inventory Call

3. Email Call

Inventory Service

Order Processing Service

Email Notification Service

Figure 2.10 – An example of an order processing service failure cascading effect

In this example, let's suppose there are 100 application threads to process a client request. When a client request comes in, the order processing service assigns a thread to process the payment and calls the payment service. The payment service is slow in responding, so the thread wait time is higher. Once the payment is processed, it calls the inventory service and assigns a thread to make the remote call, and subsequently calls the email service. Over time, the number of threads waiting for payment service grows, and now the order service runs out of threads to process client requests and is unable to update inventory and send email notifications because all threads are waiting for the payment service to complete the request. So, you see that the slowness of the payment service now has a cascading effect on the order management service and does not let it process the inventory updates, even though the payment was processed successfully for that particular order.

The bulkhead pattern suggests that we should isolate behavior where the effect of one service failure gets propagated to other services. The following diagram shows a solution for the problem we just highlighted. We have assigned the maximum number of threads that can be assigned to each service call. In our example, requests to the payment service can take a maximum of 50 threads, and if the payment service is not able to respond, the order service will start blocking the client requests or respond with an alternative mechanism, but it will not completely go down, as 50 other threads are available to process inventory and email service calls.

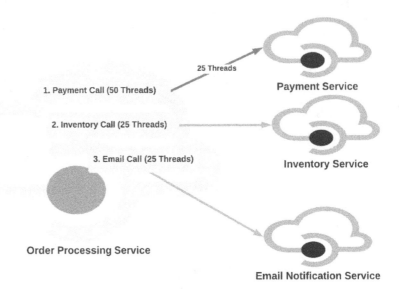

Figure 2.11 – Order processing service fault isolation using the bulkhead pattern

So, by using the bulkhead pattern, we have isolated the payment service to consume a maximum of 50 threads from the order processing service. Due to this configuration, the impact of a failure in the payment service will not have cascading effects on the email notification and inventory services.

The following are some of the benefits we get by using this pattern:

- We should consider using this pattern to protect an application from cascading failures
- It isolates critical consumers from standard consumers to increase service availability
- We should isolate the resources used to call other services so that if the backend service is not available, at least some functionality of our service is working

This pattern has the following downsides as well and may not be suitable for all situations:

- This pattern adds some complexity to your architecture, and you shouldn't use it if you don't need it
- This pattern is not very efficient, as it may underutilize some of the resources that are reserved for certain services that might not get enough traffic for the reserved capacity, but other services might be in need of more resources

Sidecar pattern

This pattern is called "sidecar" because it resembles a sidecar attached to a motorcycle. In this pattern, an application component is attached to the parent application, like a sidecar, and provides supporting features to the parent application. This component follows the same life cycle as the parent application. This is a decomposition pattern and is sometimes referred to as a sidekick as well.

In a microservices architecture, services often require related functionality, such as logging, monitoring, configuration, and proxy services. These tasks can be implemented as separate components or services.

These tasks are tightly integrated into an application and can be deployed along with application code, but if there is an outage in any of these components, then it can impact the entire service. That means these components are not well isolated.

At the same time, we can't deploy these tasks as an independent service, as each microservice can have its own technology stack. Accessing these services often requires their own dependencies and language-specific libraries to access the underlying platform, and having a separate service will add latency.

So, we are in a catch-22 position, where putting these tasks together with our service has isolation problems, and separating these as different services has language and platform-specific accessibility problems with increased latency, so these tasks need to be isolated and closely related to the services that use them.

The sidecar pattern co-locates a cohesive set of tasks with the primary application but places them inside their own process or container. The sidecar provides a homogeneous interface for platform services across languages.

So, a sidecar service is not necessarily a part of an application but is connected to the parent application and follows its life cycle. It runs as a separate child process for the parent application and provides a language-independent interface.

The following diagram shows how an application is hosted on a single host, where the sidecar is attached to the main application and provides necessary features that are not part of the core application:

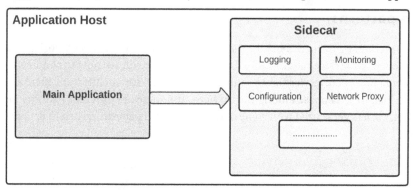

Figure 2.12 – A sidecar pattern example

In this diagram, you can see that the sidecar is not part of the core application but instead follows the same life cycle as the core application.

The following are the benefits of using this pattern:

- The sidecar pattern can help you to extend the functionality of the main application by attaching the sidecar when needed

- The sidecar is deployed along with your application, so there is no network latency and the performance of the application is better

- The sidecar has the same resources as the primary application, so it can monitor the resources used by the sidecar and core application

- The same sidecar can be used by different applications without the need of redeveloping it, as it provides a language-neutral interface

- The sidecar provides an abstraction on the library or tool provided by a third party by not directly integrating it with your application

This pattern has the following downside as well, and the following are some considerations of when this pattern shouldn't be used:

- This pattern shouldn't be used when inter-process communication between a parent and child is overly complex or can cause significant latency

- This pattern shouldn't be used when a parent and child process has a different scaling requirement

- The sidecar pattern shouldn't be used when isolation of the child process isn't needed, or if the child process doesn't need to be a reusable component and can't be used in more than one place

The sidecar pattern is widely used in container technology, where you associate additional features with the main container. Now that we understand the microservices decomposition patterns, in the next section, we will learn about the different database strategies for microservices.

Database patterns

In the previous section, we learned about the different patterns involved in decomposing a monolithic application into a microservices architecture. In this section, we will see how to handle the data and database structure in a microservices architecture. In a microservices architecture, there are different strategies to distribute a database among the services, and there are different patterns involved to access that data. So, let's get started with some of the widely used patterns related to data distribution and collection.

A shared database per service

This is an anti-pattern for microservices architecture; in this approach, all the services share the same database. This approach sounds a little weird in regards to a microservices architecture, but this is a great starting point to decompose a monolithic application. In this approach, instead of trying to do everything in one go, the first application layer is broken down into a microservices architecture and then, over time, the database layer is divided into service-specific schemas.

The following diagram shows an application where multiple services access the same monolithic database:

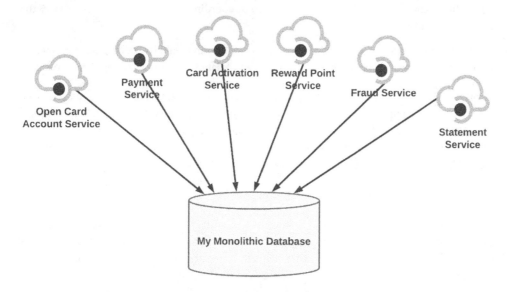

Figure 2.13 – An example of a credit card shared single database schema

You have the following advantages of using this approach:

- The application layer is divided into separate services, so your components are loosely coupled and developers can quickly work independently and deliver functionality faster.

- Since there is no database-level separation, your local transactions will still work, and you can easily satisfy the ACID requirements of your applications. You don't need to worry about 2PC.

- This is a work-in-progress architecture, from monolithic to microservices, so developers don't feel many structural changes and quickly adapt to the paradigm shift.

- A single database is simple to operate, so you don't have to worry about data backup, archival, and restoration of multiple databases.

- In a shared database approach, there is only one database platform, so you don't need people to maintain different database technology platforms.

The shared database approach has some disadvantages as well:

- Sharing the same database layer is a risk to all services if there is an outage to a platform
- Each service might have different scaling requirements for the database layer, so in this case you can't scale database layer independentely
- Any schema change to a database table needs to be coordinated with all the microservices that might have an impact
- Using the same database and accessing the same tables can create issues between two services – for example, one service might be waiting to write data while others are holding the lock

As we mentioned earlier, this is an interim approach to decompose an application into a microservices architecture. Depending upon the nature of the application, this can be a suitable interim approach for a giant legacy application, but for a greenfield application, this is not a recommended approach, as it creates the monolithic architecture at the database level. With this approach, you will not be able to receive all the benefits provided by the microservices architecture.

Database per service

One of the core advantages a microservices architecture provides is bringing agility to your business. It allows us to make rapid changes to our application as business demands and deliver a business feature at the speed required. This architecture helps us to reduce the time to market for a product feature. To achieve this, we need to have a separate database schema for each service, and only the relevant service should access the relevant database directly, and if any other services need that data, they should connect using the API interface that the microservice provides.

Let's try to understand the problem of multiple services accessing the same database schema. In the following example, we have a credit card application, which is designed in a microservices fashion, but the database layer is using the same database schema for all the services. This application will scale well on the application tier, but the database tier would encounter a problem when you need to change a database table. Suppose you want to add a new column to the product table. Now, since your database schema is changing, you will create an outage to all the services when you apply this change, and you also need to test all of the microservices, as all are accessing the same database tables and schema. So, with this design, you give away one of the core benefits of a microservices architecture:

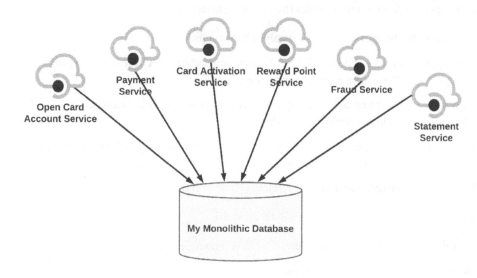

Figure 2.14 – A credit card application with a shared database

Database-per-service patterns emphasize the usage of a private data store for each service, and no two or more services should access the same data store to bring agility to the architecture. Each data store is private to each service, and other services access this data through the APIs provided by the related service. For example, for relational databases, we can use private-tables-per-service, schema-per-service, or database-per-service. So, by using the database-per-service strategy, your database will look like the following example for the same credit card application.

In the following example, each microservice has its own database schema, but if one service needs any data from another one – for example, if a statement service needs card transaction details – it can call the card account service through their API but can't access the database directly.

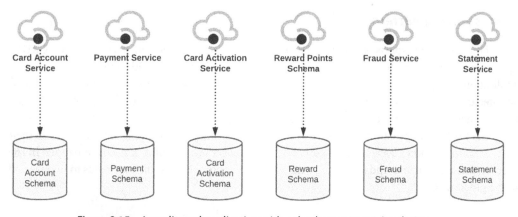

Figure 2.15 – A credit card application with a database-per-service design

The database-per-service pattern provides the following benefits:

- You can make changes within each database schema independently
- Your application can choose a different database solution that best suits the problem – for example, NoSQL versus a relational database
- It promotes decoupling between applications, and you can deploy changes independent of other services without impacting functionality
- You don't need to test each and every microservice when making a change in one particular application schema, since there is no change in the contract of the applications
- You can scale the database tier independently, based on the need of each service

There are a few disadvantages as well to using this pattern:

- Your data is scattered between different services, and access through services adds latency.
- Data consistency is a challenge, with each service having its own copy of the same data, and you have to rely on the orchestrator process to eventually sync the data.
- You can't query data easily, which is hosted by different services; you have to rely on the APIs provided by related services.
- You have to maintain more database systems, as types of databases and the number of schemas can increase.
- You may have to pay to run a database server that isn't heavily used – for example, on Amazon **Relational Database Service (RDS)**. If you launch an RDS instance for a service, that's a fixed cost even if the service is mostly idle.

In most modernization projects and any greenfield projects, database-per-service is widely used and is an ideal approach over shared database-per-service.

Event sourcing pattern

One of the important aspects of handling data is providing **Create, Read, Update, and Delete (CRUD)** operations on it. In traditional applications, it is easy to handle the data within the same monolithic application, and you can either commit the transaction or roll back if needed. In microservices-based architecture, your application is divided into different services, and usually, data state changes are required across different services and data stores. Any update involving two or more services needs to update the local transaction and send a message/event to other participating services to finish their part. Sometimes, other services can't complete the transaction, so rollback needs to happen in the first service. Concurrent updates of the data also make it complex, as the data needs to be reverted to the previous state.

To understand this problem better, let's dig a little deeper. Service A needs to update a database and send a message to Service B in order to ensure data consistency. However, it is not viable to use a distributed transaction across different services and data stores. In absence of a distributed transaction, it is possible that once you send a message to the service B, your database transaction in Service A might fail, and if you try committing the database transaction first, it is quite possible that sending a message to a Server B might fail, which will leave your data inconsistent.

To solve this problem, an event sourcing pattern can be used. In the event sourcing pattern, you sequence your changes as events and you don't update the data. Whenever you need to read the data, you basically play back the history of events and receive the current state of the domain object. In this approach, operations on data are driven by a sequence of events, each of which is recorded in an append-only store. A new event is added to the list of events whenever there is a change to the business entity, since saving an event is a single database operation, so it is atomic. Whenever an application needs to read data, it play back those events and constructs the business entity for the current state.

Application code sends a series of events that explain every action that has happened on the data to the event store, where they are persisted. Each event represents a set of changes to the data. These events are immutable, so once an event is published to the event store, you can't change it, and if you need to change it, you need to record another event representing the compensating change.

In the event sourcing pattern, the event store is used as a source of truth and can be used to audit the changes that happened in the business entity over time.

Let's take an example to understand this better. Suppose you have an online e-commerce system, and recorded orders go through several status updates – for example, created, approved, shipped, delivered – so instead of trying to update the same record each time, these actions are recorded as events and all status updates are taken as a sequence of events, and whenever we need the current status, we can create a materialized view of the data to find the current state.

In the following diagram, an order service stores the events in an appended fashion as those actions occur. A payment service and shipment service are subscribers of the appropriate event.

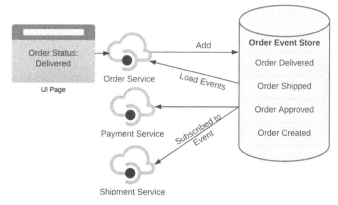

Figure 2.16 – An order processing application event sourcing design pattern

Let's suppose a user requested a UI page, which needs the current status of the order. The order service loads the order events from the event store and merges the records to get the current status of the order, sending it back to the user.

Over time, some of the domain objects might have a big list of events, so loading a big list of these events can be a time-consuming process and cause performance issues. In those scenarios, a snapshot of data is created, stored as an event, and referenced for faster loading, and events are appended on top of that snapshot. Multiple snapshots of data can be created over time.

In this pattern, you learned how you can use the event sourcing pattern to receive the current state of an object and how you can get the full history of changes. In the next database-related pattern, let's see how to scale an application that has different requirements for scaling on the read and write sides of the same domain data.

Command Query Responsibility Segregation (CQRS)

In the microservices world, data is mostly separated into multiple services, and often, we need to join data from different services to show it on a UI screen or for any auditing requirements. Implementing queries that join data from multiple services and their own databases is a real challenge.

Getting data from multiple services can also have performance issues, as it requires aggregating data from different services. In certain applications, there can be different scaling requirements as well for reading and writing data. In traditional applications, to address scaling issues, the solution is to split an application into two parts – the command side and the query side.

CQRS is a pattern that separates read and update operations for a data store so that different parts can be scaled separately.

In this approach, the command side handles create, update, and delete operations and emits events when data is changing, whereas the query side of CQRS handles queries. Events published by the command side are consumed by the query side, which creates a materialized view of the different events and returns the data needed for the read side. CQRS in your application can maximize its performance, scalability, and security.

Let's use an example of an order processing system to understand the CQRS pattern better. In the following diagram, you can see that we have an order history service, which basically listens to the events generated by the order service, payment service, and shipment service.

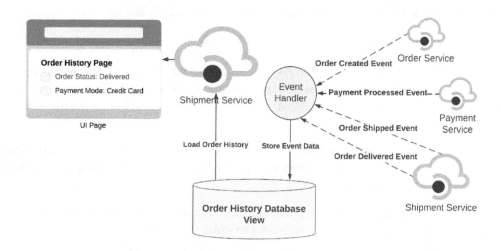

Figure 2.17 – A CQRS design pattern implementation for an online order processing application

Any incoming update request related to order, payment, or shipment is handled by these respective services, but if there are any incoming requests to read the data, that read is handled by the order history service. In the order processing service, updates are received through different events and stored in a local database, and a materialized view is created to easily access this data, so you don't have to load data from three different services to display it on the UI page. In our example, when we need the order status and payment information, a call is made to the order history service, and a single call serves all the required data on the UI page.

As you can see here, since we have different access patterns to update and read data, the order history service and individual services can be scaled differently as there are required.

SAGA pattern

The database-per-service model provides a lot of benefits in a microservices architecture by speeding up development and keeping all model dependencies within a service, but it also brings the challenge of keeping data consistent when your transaction goes beyond one service.

You can use 2PC distributed transactions, but that makes services slow and requires all participants to commit or roll back a transaction. 2PC also has limitations for NoSQL database and messaging systems.

SAGA is a design pattern to ensure data consistency when your business transaction spans multiple services. The SAGA pattern uses a sequence of local transactions to ensure data consistency, and once a service is finished committing a local transaction, it publishes a message or event to trigger the next local transaction in the SAGA pattern. If a local transaction fails, then SAGA executes a series of compensating transactions to undo the changes made in a previous local transaction.

Based on the approach taken to complete the local transaction, SAGA is divided into two types.

Choreography-based SAGA

In choreography-based SAGA, there is no central coordinator for transaction management; each service participating in SAGA commits its local transaction and produces an event to the next service in line, and then listens to the events produced by other services in sequence to see whether there is any action required.

The following diagram shows an e-commerce order processing transaction that is spread across four microservices – an order service, an inventory service, a payment service, and an email service. Here are the steps followed in this process:

1. The order service receives the order processing request, creates the order in its local database, completes the local transaction, and creates an order in a pending state. Once the order is created, the service sends a message as an event to the inventory service.

2. The inventory service is subscribed to the order-created event. It processes this event and updates the inventory for the sold product.

3. The inventory service generates an inventory-updated event, indicating that the inventory is processed for the given order.

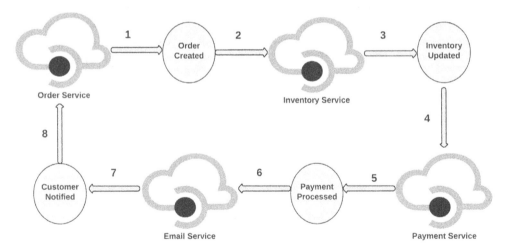

Figure 2.18 – The process-based SAGA pattern implementation for an online order system

4. The payment service processes the payment and commits the local transaction in its database.

5. The payment service generates a payment-processed event to notify the email service of the processing payment.

6. The email service listens to the payment-processed event and sends a confirmation to the customer.

7. The email service sends a confirmation message to the order service.

8. The order service listens to the customer-notified event and updates the local database order status to confirmed.

Note here that there is no central coordinator, and each service is communicating with others through an event to start processing their part in the transaction.

Orchestration-based SAGA

In this approach, there is a central SAGA orchestrator that is responsible for managing all the transactions and directing the participant services to execute local transactions based on events. This orchestrator calls the different services involved in each transaction, and if there is any failure, this orchestrator is responsible for the rollback of the local transactions. This orchestrator makes sure that either all local transactions are committed by each participant service or they all are reverted if there's a failure. This orchestrator can also be thought of as a SAGA manager.

The following diagram shows orchestration-based SAGA for the same e-commerce online order processing system:

1. In this example, the order service plays the role of the orchestrator. It receives an order and creates it in the local database, with a pending status, and it then sends a message to the inventory service to update the inventory.

2. The inventory service processes the inventory in its local database and commits the local transaction on the successful process of the inventory update. It then sends a reply back to the order service.

3. The order service, on the successful process of the inventory, sends a message to the payment service to process the payment.

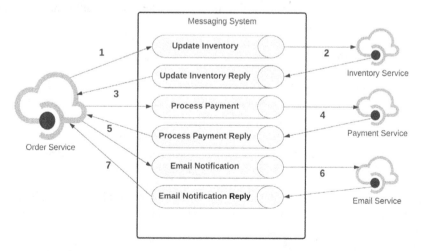

Figure 2.19 – An orchestrator-based SAGA pattern implementation for an online order system

4. The payment service processes the payment and then responds with a reply to the order service for further processing.

5. The order service receives the reply from the payment service and then sends a message to the email service.

6. The email service sends a notification to the customer and replies to the order service with a message.

7. The order service changes the order status to confirmed and completes the transaction.

Since you can't ignore the communication between different services within a microservices architecture, understanding communication patterns is really crucial to deliver a production-grade application. SAGA is one of the most popular design patterns, and understanding this in depth is important.

The usage of orchestration SAGA or choreography SAGA depends on the use cases, but in general, orchestration-based SAGA creates more tight coupling with participating services. Since orchestration-based SAGA is heavily reliant on API-based invocation and an orchestrator directly calls each participating service, one service failure or slowdown creates a cascading effect on other services. On the other hand, in choreography-based SAGA, communication is asynchronous with the use of a messaging broker, which makes it more agile and loosely coupled.

In this section, we learned about different design patterns related to data storage. In the next section, we will learn how to aggregate data that is distributed across different services in a microservices-based architecture.

Integration patterns

When you decompose your monolithic application into a microservices architecture, your application is divided into multiple services and your data is distributed within each service, so you need a mechanism to integrate this functionality or data to make meaningful decisions, or provide a way for your clients to use this functionality. In this section, we will learn about different patterns related to the integration of data in a microservices-based application.

The API gateway pattern

In a microservices architecture, you often interact with your consumer or clients through APIs. Since an entire business functionality is divided into a set of services, these services interact with each other through APIs. Sometimes, the granularity of an API is a problem; microservices are very fine-grained APIs in nature, and at times, a client or consumer may have a different requirement. Sometimes, clients use a different data formatas compared to what our consumers; some are required to authenticate to access an API, while internal consumers can access the API with just an API key.

In widely used APIs, some consumers need few fields, while others need more fields. If you implement this logic in your microservice, then it becomes complex, and customer-specific behavior propagates to your API, which is not a good design practice.

The API gateway is an integration pattern that offers a single point of entry to our services and provides the functionality of authentication, authorization, aggregation, and a fine-grained API for each specific type of client, helping to exchange data with different protocol types. The API gateway works as an additional layer on top of your service and provides these features as a Plug and Play device.

Your APIs don't even need to know about the API gateway; you just configure the API gateway by defining the different rules, the aggregation mechanism, and the API's endpoints to reroute requests to get data.

The API gateway also works as a proxy to your service and routes your request to the actual API endpoint. Conversely, it can work by fanning out requests to multiple services to aggregate data.

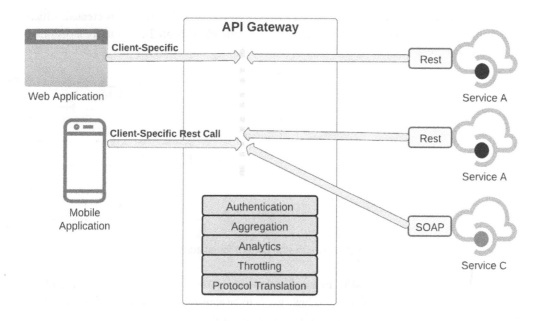

Figure 2.20 – An API gateway pattern example

The API gateway pattern is widely used to enforce organization-level security on all incoming API requests at the entry point and also map an external endpoint to an internal API endpoint. Amazon API Gateway and Google's Apigee API Gateway are two very popular, widely used API gateway products.

The aggregator pattern

The aggregator pattern is another integration pattern that helps to aggregate data from different microservices. The aggregator pattern takes the requests from the client and then calls multiple services to collect the data, sending a complete response back to the client. In this pattern, a composite service is created that takes the responsibility of merging data from multiple services and works as a frontend for these services. Making a direct call from the UI or client to multiple services sometimes can make the architecture complicated, and the client has to deal with different endpoints and dependencies. Sometimes, the aggregator can cache the data required by different clients and return from the aggregator service itself, which can speed up things.

Let's use an example to understand the aggregator pattern and why we need it. Suppose in a banking system that customer data is available in the customer service, while all related accounts for that customer need to be loaded from the account service. So, instead of calling multiple services from your mobile/web application, a composite service is created to front-face this request, which saves one network trip from the client browser to your services and additionally reduces the complexity in your UI.

In the following example, a composite service called the customer profile service is created, which basically calls the customer service and accounts service to provide all related accounts for a customer with customer personal details.

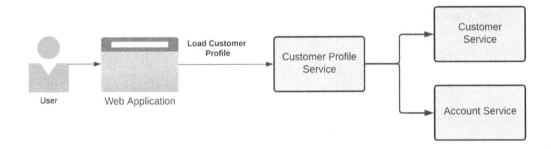

Figure 2.21 – Aggregator design pattern for customer profile

This pattern is very similar to the API gateway, but the composite pattern is preferred instead of the API gateway when some additional business logic needs to be applied to a client request. So, you need to do your analysis and make a decision that best fits your use case.

The proxy pattern

The proxy pattern is an integration pattern. In this pattern, client requests are passed through a proxy service. The proxy may be a dumb proxy that works like a pass-through layer to call other services. Alternatively, it may be a smart proxy, where some data transformation is applied before or after the response is served to the client.

This pattern is very similar to the aggregator pattern, but here, you are not calling multiple services to aggregate the data. Here, the proxy service is just sitting in front of your service. The proxy pattern enforces loose coupling of a microservice to its consumers, so if you have a major change in your microservices, you can migrate your consumers from one version to another using a proxy. What version of your API needs to be called can be decided by the proxy.

An example of using a proxy service is loading the different UI pages for the mobile and desktop platforms based on the header value received from the client.

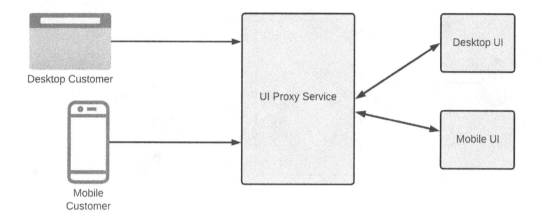

Figure 2.22 – Proxy design pattern example

The proxy pattern is widely used to decouple consumers from a backend service and also perform some cross-cutting tasks, such as adding audit logs, authenticating user requests, and injecting any context data into a backend service.

Gateway routing pattern

The gateway routing pattern is an integration pattern where you have a gateway in front of all your services, and through a single URL, you expose all of your services. This pattern uses the OSI model's seventh layer (the application layer) to route requests to the appropriate backend microservice.

In this pattern, your client doesn't need to know the address where each of your services is deployed. The client always communicates with a single endpoint, and with the help of a URL pattern, your request is routed to the correct service instance.

This pattern is useful for keeping the services abstract from the client, as they always interact with a single endpoint, so you can make changes to backend services or completely remove or migrate them and the client wouldn't be aware, as you just need to change the routing information on the gateway.

In the following diagram, a client makes calls to the same URL, but in the backend, which service handles the request is decided by the gateway, based on the URL Routing rules configured on the gateway have those URL patterns and the addresses of the services defined, and based on those, the request is handled by the appropriate service.

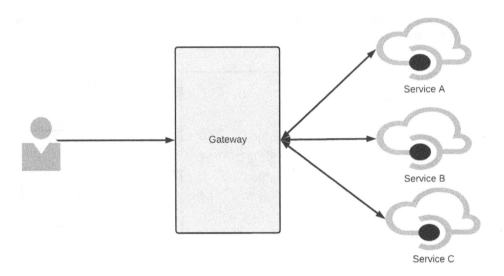

Figure 2.23 – A gateway routing design pattern for a web server

The gateway routing pattern is just used to route the request to the appropriate service. It doesn't implement any service-specific behavior; how the client request needs to be handled is the responsibility of the service. Using a web server such as NGINX as the router is an example of the gateway routing pattern.

The chained microservice design pattern

In this design pattern, as the name suggests, a chain of services is used to fulfill a client's request. In this design pattern, a single response is produced for the consumer by a series of microservices, which are called sequentially.

For example, a client calls service A to get some data, but service A doesn't have the full data, and it needs to call service B to get additional data. Service B is dependent on service C to get the data. So, once service C responds to service B and service B responds to Service A, service A has all the consolidated data and responds to the client. Since a chain of services is used to fulfill the client request here, this pattern is known as the chained microservice design pattern. The following diagram shows this communication between microservices and how a client request is served by three services working in a chained fashion.

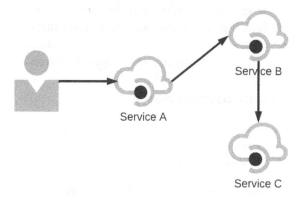

Figure 2.24 – An example of a chained microservice design pattern

In this design pattern, all the services likely use a synchronous HTTP request/response message. Since service A can't respond to the client until service C and service B process the client's request and respond, the client is basically blocked. This is one of the main concerns when using this pattern – that each dependency you add in a chain will add latency to the client request, and the client is blocked until all the services in the chain finish processing.

The branch pattern

The branch microservice design pattern is a combination of the aggregator and chained microservices pattern. The branch pattern extends the aggregator design pattern and allows you to call multiple chains of services simultaneously, returning an aggregated response. Branch patterns can also be used to call different chains, or a single chain, based on business needs.

The following example shows service A using the branch pattern. It uses multiple chains of services, from service A to service B to service C, and another branch from service A to service D to service E.

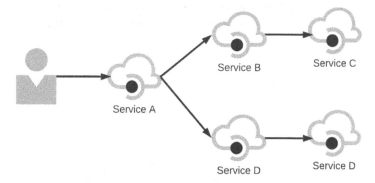

Figure 2.25 – An example of the branch design pattern

Service A can call both branches simultaneously, and once the response from both branches is received, it can aggregate the response and return it to the client. Since this pattern uses both chaining and aggregation and the request path is branched into multiple paths from Service A, this pattern is known as the branch pattern. This pattern may or may not have a chain of services.

The client-side UI composition pattern

For enterprise applications, developing a frontend is a complex task, as these applications are not specific to a single product or service. UI landing pages usually represent an entire organization, and each section on the page might be owned by a specific product or team.

In a microservices environment, no single team owns the entire page, as data comes from hundreds of microservices. This makes the architecture very complex and the page hard to maintain. So, as per the client-side UI decomposition pattern, the UI has to be designed as a skeleton, with multiple sections/regions of the screen/page, where each UI section can be loaded separately and maintained by different teams. Each UI section can load its data from an individual backend microservice.

This pattern hides the complexity and promotes the development of each section on its own and at runtime, all the sections of the pages coming together to generate the entire user experience. A UI team is responsible for implementing the page skeletons that build pages/screens by composing multiple, service-specific UI components.

With the advancement of UI technologies such as AngularJS and ReactJS, this development approach becomes really easy. These applications are known as **single-page applications** (**SPAs**).

Let's use an example of a financial institution. When a customer is logged in, first the user dashboard is shown, which pulls data and contents from several different components.

The user dashboard has different sections showing different information, such as a checking account balance, a debit account balance, credit card transactions, auto loan information, and an inbox for messages unread by the customer.

The following diagram shows a UI skeleton page created with separate sections, and each section is backed by a microservice to render that section. These services can call other services to get the required data.

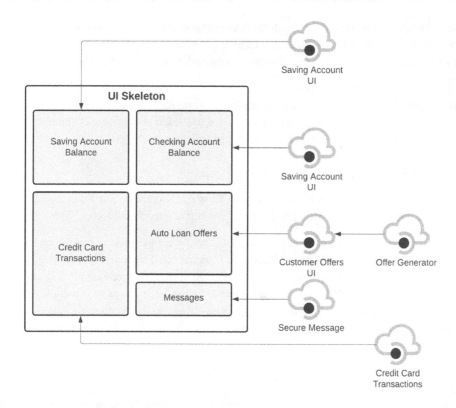

Figure 2.26 – Client-side UI decomposition for a banking application

This pattern makes it very simple for individual teams to just focus on their part or product, and they don't have to deal with the complexity of the entire UI page. This design pattern speeds up the development of UI components, which can be plugged into any other place as needed.

Observability patterns

In microservices, observability makes sure that a developer has access to the data they need to identify problems and detect failures.

Decomposition of a monolithic application into microservices is one challenge, but once you have achieved this and deployed your application in production, running it successfully is a bigger challenge.

In a microservices world, one monolithic application is divided into hundreds of microservices, and these services interact with each other to carry out business transactions. Data is flowing in and out of these different services, so it becomes complex to troubleshoot issues in this architecture, as your transaction is spreading across different microservices, and each service collects log and metrics information independently and in isolation, which doesn't make sense as a whole.

Data collected by individual services without the knowledge of full application context and the scale at which it grows makes it more complex. To address these issues, you need to follow certain design practices so that you have the appropriate logging data and metrics required to troubleshoot and resolve an issue.

If a microservice is not observable for effective debugging and diagnosis, then it is hard to run that system in a production environment. In this section, we will learn about some of the observability design patterns, which are a must for any microservices architecture.

Log aggregation

The log aggregation pattern specifies that in a microservices architecture, we should aggregate the logs generated by microservices in a central place for easy lookup and troubleshooting.

In a microservices architecture, your application runs in a multiple instance mode so that your application can scale horizontally if a load increases. Logs collected by individual service instances don't provide much context, as one request for a user can go to one instance and another one can go to some other service instance. When you are troubleshooting an issue where your request is handled by multiple service instances, searching through logs is not an easy task, so you need some aggregator tool to collect the logs in one place so that you can generate meaningful information from them.

Each service instance generates a log file in a standardized format. We need a centralized logging service that aggregates logs from each service instance. Users can search and analyze the logs. They can configure alerts that are triggered when certain messages appear in the logs. A log aggregator stores the information in a time series fashion and provides the capability to quickly search through these logs and generate meaningful observations from them for troubleshooting, and it also helps to monitor a service if any failure message is detected in the logs.

In the following diagram, service A is running in a two-node cluster, front-faced by a load balancer. When a user sends a request, the load balancer intercepts that request and transfers it to one of the service A nodes to handle it. Service A nodes log the information to their local log file. Having that log file locally doesn't help much, as in a microservices architecture, these instances can go on and off, based on scaling needs. Another problem with local log data is that to search for specific information, we have to log in to each node and then aggregate the information, which is not an easy task. So, the need for a log aggregator arises. On the right side of the diagram, there is a log aggregator, which basically collects these logs from each instance and aggregates them, based on the timestamp in a single place.

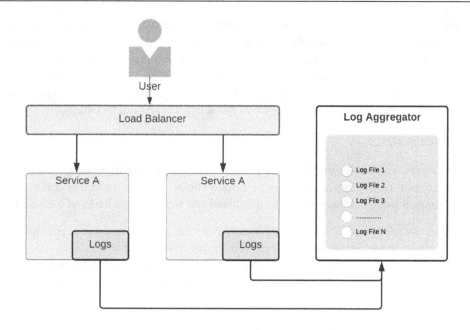

Figure 2.27 – The log aggregator design pattern

Once the log data is aggregated in one place, you can use it for analysis, troubleshooting, or alerting across all the services and you can retain this log data for auditing requirements as well. There are several commercial log aggregator tools available that are scalable and provide a lot of features. Amazon CloudWatch, Splunk, and the ELK Stack are some of the most popular log aggregator solutions.

Distributed tracing

In a microservices architecture, requests often go through multiple services before a response is returned to a client. Each service handles a portion of the request by performing one or more operations across multiple services. Tracing these requests across services is difficult, and while troubleshooting an issue, correlating these requests is a big challenge. In order to solve this challenge, we need to trace a request end to end. While dealing with a user issue, it is important for the operations team to trace each user request across services. Distributed tracing allows us to trace each request end to end for easy troubleshooting and visibility. In distributed tracing, you have a transaction ID that is included in a request, which travels through each service, so you can easily and uniquely identify the request path. Distributed tracing is implemented in the following ways:

- By assigning each external request a unique external request ID
- By passing the external request ID to all services, preferably as a header value
- By including the external request ID in all log messages

Let's consider a use case of an online e-commerce retail site. When an order is placed through an order processing service, it makes a call to a payment service to process the payment, an inventory service to change product inventory, and then sends a confirmation message to the customer using an email service. So, when an order request is received by the order processing service, a unique request tracking ID should be injected in the header and logged in each originating request as a response to this user request, ensuring that the end-to-end flow of the user request can be traced.

Performance metrics

Within a microservices architecture, the number of services increases, and monitoring these services for performance and errors is important. When a service portfolio increases, it becomes critical to keep a watch on transactions so that patterns can be monitored and alerts are sent when an issue happens.

In this pattern, you need to collect the metrics from each service instance running in your workload and gather statistics about individual operations. Usually, you have a separate service to collect these metrics and analyze your memory and CPU usage, along with application-level monitoring. This service collects application usage data, memory usage, error rate, and other custom metrics published by an application, and it provides reports and alerts on aggregation.

The following diagram shows a performance metrics aggregator service that collects the metrics published by each service through agents, it centrally collects these metrics to generate application dashboards and graphs, and sets up alerts whenever any metrics go beyond the set limit value.

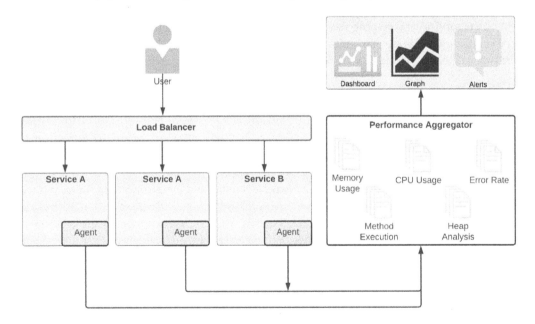

Figure 2.28 – A centralized performance metrics design pattern

An application can be scaled up and down based on these metrics. The availability and reliability of a system can be increased with these metrics, and the cost of operating a system can be lowered. There are two types of performance metrics aggregators, one being push-based and another pull-based:

- **Push**: In a push-based monitoring system, metrics are periodically published by the agent running on the host of a central aggregator system, which generates meaningful results from the aggregated data. Examples of push architectures include Graphite and CloudWatch.

- **Pull**: In a pull-based monitoring system, a central aggregator periodically requests metrics from each service host. Examples of pull architectures include SNMP and JMX.

There are so many services available commercially to monitor your application and infrastructure. AWS provides CloudWatch metrics, and you can monitor any application using it. CloudWatch integrates well with other services in the AWS cloud to provide an auto-scaling capability.

Health check

In a microservices architecture, there are thousands of services running to support business needs. These services are designed to be resilient, and service instances can go on and off anytime. Therefore, you need to have your system designed so that, if an instance is down, you don't forward live traffic to that particular instance, and if the instance returns, it automatically registers to the load balancer and can start accepting traffic to serve the client requests.

As per this design pattern, each microservice needs to implement a health endpoint to indicate that the service is ready and can handle client requests. The load balancer uses this health endpoint to determine whether a client request can be forwarded to that particular service instance or not. This health endpoint is usually exposed with the `/health` URI and returns an HTTP 200 status code to indicate success. This exposed API endpoint should check the status of the host, the connection to other services/infrastructure, and any specific logic.

In the following diagram, a service is front-faced by the load balancer, which checks the heartbeat of each service instance on a periodic interval by calling the health endpoint, `/health`. If a service instance returns a 200 status code, that means the service is healthy, and the load balancer can forward client requests to that particular instance.

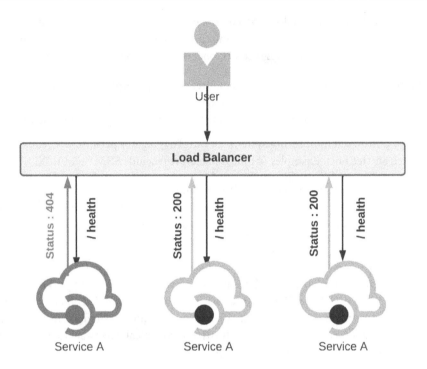

Figure 2.29 – A health check design pattern

In the given example, the first instance of service A returns a 404 status code back to the load balancer, so the load balancer doesn't forward any subsequent client requests to this instance, marks this node dead, and forwards requests to the other two remaining instances, since their health endpoint is returning a 200 status code. Whenever instance one is healthy and able to return a status code of 200, the load balancer will return it to the service and start forwarding the client requests. These health checks can be as simple as just checking the status of the process, or they can be more complex, where you run a dummy end-to-end liveness to ensure that all your service dependencies are running as well.

Cross-cutting concern patterns

There are some microservices design patterns that are very generic in nature and don't fall under the decomposition, integration, or observability category. These design patterns are very important and generic so that they can be applied to any application beyond a microservices architecture. These patterns are not specific to any parts of microservices and are instead generally applied to all parts of the architecture.

External configuration

Configuration is an integral part of any system or service. A microservice typically calls other services and databases as well. Each microservice runs in different environments, such as development, **Quality Assurance (QA)**, **User Acceptance Testing (UAT)**, and production. For each of these environments, the configuration parameters and URLs are different. You can't keep the same value for your development and production application.

A change in any of these properties might require rebuilding and redeploying services, which causes downtime or service interruption. To avoid code modification or redeployment to change a configuration parameter, this pattern suggests externalizing the configuration properties.

As per this pattern, you should define all your changing values as configuration parameters and load them, including service endpoint URLs, credentials, or feature flags at application startup from a configuration server. The application should be able to refresh these values whenever you publish a change without a service restart.

Usually in this implementation, your configurations are stored in a central configuration server or repository, and your application thread loads these properties on startup. After startup, your application periodically makes a call to the configuration server or repository to load any updates to these parameters, and if you find any change, it immediately gets applied to your environment without a service restart.

The following diagram shows an implementation of a centralized externalized configuration system using the Spring Boot Config Server.

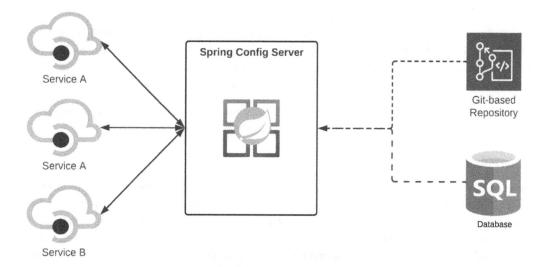

Figure 2.30 – An example of an externalized configuration pattern

In this diagram, the Spring Config Server is deployed, which manages the configuration for us, and it is backed by a Git repository. The Spring Config Server loads the configuration in its memory from the Git repository and keeps the updated configuration.

Conversely, when a microservice is started, it requests the configuration from the config server and sets the initial values. Microservices run a config thread that periodically makes a request to the config server and loads updated configuration. Whenever there is any change in the configuration, the operations team commits the change to the Git repository, and that change gets picked up by the config server and is subsequently updated on each microservice's instance.

Service discovery pattern

In a microservices architecture, you have so many services running, and they interact with each other. The way microservices are designed is that they can go down at any time and new service instances can appear to help serve requests. You need to know your service dependencies and how to connect to them. Whenever any service instance restarts or a new instance gets deployed, its IP address changes, and you should know the address of your dependencies in order to call them.

The evolution of container technology increased the utilization of resources, and you can run multiple containers on the same machine, which makes the architecture more complicated, as you get multiple dynamically allocated IP addresses to the service instances. Every time an address changes, a consumer service can break and require manual changes. In addition, all the consumers of your service need to know the URL or IP address of your service in order to connect, and if your service address changes, all your consumers will disappear until they fix the new service address in their configuration. With this setup, your application consumers become tightly coupled with your application. A service discovery pattern is used to address this kind of problem.

In the service discovery pattern, you need to create a service registry, which basically keeps the metadata of each producer service available on the network and serves the request. On application startup, each microservice registers itself with the service registry. A service instance should register to the registry server when starting and should de-register when shutting down.

Amazon ECS Service Discovery, Consul, etcd, Eureka, and Apache ZooKeeper are some examples of service discovery products. There are two types of service registries, which we will see next.

Server-side service discovery

In server-side discovery, the client usually makes a request via a load balancer that runs on a pre-defined address. The router then makes a call to the service registry, which receives the current address of the service and serves the requests. These service registries can be part of the router (the load balancer) itself. The router can transfer the request to any of the service instances to serve the request. More sophisticated routers provide more configuration parameters to distribute traffic on matching service instances. Round robin is one of the popular strategies for a load balancer to transfer traffic.

In the following diagram, a client API or application makes a request to the load balancer on a pre-configured, known URL, but what instance is going to serve that request or where it is located is not known to the client.

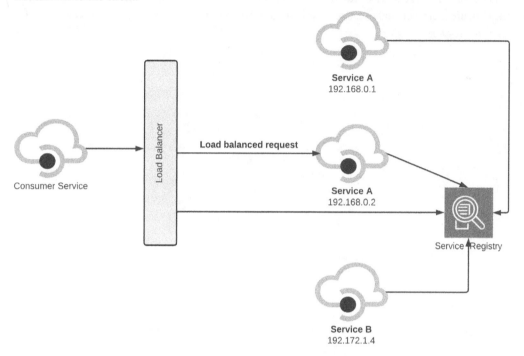

Figure 2.31 – An example of a server-side service discovery design pattern

When the load balancer receives the request for this service, it looks through the service registry to find where the service instances are located and then redirects the client request to a particular instance. These service instances are pre-registered with this registry; in most cases, this registry is part of the load balancer. Since, with this approach, there is no change or specific implementation for the client or consumer, and all discovery work is done by the load balancer and service register, it is known as server-side discovery.

Client-side service discovery

In the client-side discovery pattern, you need to create a service registry, which basically keeps the metadata of each producer service available on a network, with their name and an address where the service is located. Once a consumer application needs to access another service, the consumer application makes a call to the registry and passes the friendly name of the service (for example, `aws-code-pipeline-service`) that it needs to connect. In return, the registry provides them with the physical IP address of the service where it is available. Then, the consumer application makes a request to that producer application on the provided address; these calls can be load-balanced on the client side to send requests to different available application instances.

The following diagram shows an example of client-side service discovery. When a consumer service needs to request some data from another service, it makes a call to the service registry with the provider service's friendly name. A service registry provides the current address of the available service, and then the consumer service makes a client-side, load-balanced call to the dynamic address provided by the service registry.

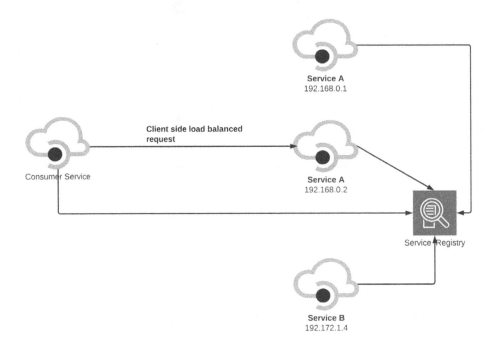

Figure 2.32 – An example of a client-side service discovery design pattern

The provider service processes the request and returns an appropriate response. Whenever a new service instance starts or goes down, it updates the service registry, so the service registry is kept aware of the current address of the service.

Circuit breaker pattern

Microservices are designed so that each service can be deployed at any time or independently of each other. The circuit breaker design pattern name comes from an electric switch, and it provides similar functionality in service interactions. This pattern helps to prevent the cascading effects of a service failure. Suppose you have a performance degradation on service B, which is used by service A, so service A should be careful in making a request to service B and shouldn't be making things worse by calling it again and again when service B is not able to handle the load, or returning error responses, which would also make things worse.

The circuit breaker pattern allows us to build a fault-tolerant and resilient system that can survive when key service dependencies are not available. This pattern allows the calling service to receive the required data through an alternative way or fail fast, without making the key service completely out of service by making repetitive calls.

When a circuit breaker pattern detects a certain number of failures from a service call, it will automatically trip and stop sending new requests to the service, and in the fallback method, it allows you to either fail fast and let your users know about the issue or load the data from an alternative system.

After the timeout, the circuit breaker allows a limited number of test requests to pass through, and if there are no more errors from the service, it will resume the operations as normal; otherwise, if there is a failure, the timeout period begins again.

The circuit breaker can also help to protect downstream services from a DDOS attack. Whenever it will see more failures from the downstream service due to a DDOS attack, it will stop sending more traffic, so your downstream service can still work for other consumers, as your compromised service wouldn't be sending more traffic for the time being, helping you to reduce your impact area.

Let's use an example to understand circuit breaker pattern, suppose you are implementing a hotel room reservation feature in your application that calls a hotel information service to get hotel information , payment service to receive payment for booking, inventory service to update the room avaiability, and email notification service to complete the reservation process. In your room reservation service you have implemented circuit breaker pattern to communicate to the hotel service. If your hotel service is going through maintenance or running on a lower capacity, so your reservation calls might not be able to get the hotel information when requested and starts failing to make reservations. So, in these scenarios, instead of going completely out of service in the hotel reservation application, you can look to get the hotel data from an alternative source, such as a cache.

By not hitting the hotel service for a certain period of time during performance degration you will give hotel service an opportunity to become avaiable again to serve the requests. Circuit breaker patterns helps to facilitate all of that by loading the data from alternate sources during service unavaiability and slowness and avoid putting more load to a slow services, until regain its health. If you keep hitting the service, you will put more load on it, and it will become overwhelmed with the number of requests. Follwing diagram shows that how circut breaker open the circuit and loads data from alternate source in case of a service downtime.

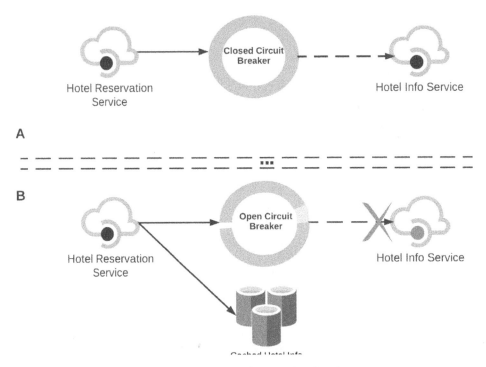

Figure 2.33 – An example of a circuit breaker design pattern

This pattern is suited to prevent an application from the cascading effects of failure in a remote service, and where chances of failure in accessing a shared resource are high.

Blue-green deployment pattern

In a microservices-based architecture, you have lots of services, and each one of them runs several instances for high availability. Automation is required from the ground up in a microservices architecture; otherwise, dealing with so many services and their instances can be a nightmare. Even if you have all the capability to automatically deploy your services and set up CI/CD pipelines, you still need a deployment strategy to minimize the impact on your production workload during deployment.

If you are trying to deploy a newer version of your service and stop all the current service instances, the downtime will be huge, and your customers will see outages during that time, which will definitely impact the business revenue. Also, if you find an issue with a new version of the application, then rolling back that version to the previous one will be also a huge effort and cause an outage to customers. The blue-green deployment pattern solves this problem by minimizing the impact on your services during deployment.

The blue-green deployment strategy can be implemented to reduce or completely remove downtime during the new version deployment of your services. In the blue-green deployment pattern, there are two identical production environments created for the services called blue and green. In this strategy, one of the environments is live and takes the production traffic, and when you are rolling out a new version of the service, you deploy it to the blue environment. Since the blue environment is not taking any live traffic, there is no outage caused in the environment. After deployment, you validate your blue environment and ensure that there are no issues with the service. Once you are satisfied with the blue environment deployment, you make that environment live in production and switch your traffic from green to blue. Usually, traffic is switched from a DNS or load balancer, and you start scaling down your non-live environment and scale up your live environment.

If there is a rollback, you take the same strategy; you scale up the non-live green environment and start bringing down the blue environment, and then switch traffic from a load balancer or a DNS. This way, since you are not doing deployment directly to any live production environment, there is no impact to live traffic; your consumers wouldn't even notice any change.

The following diagram shows a blue-green environment deployment model, where the on the left side, the green environment is live in production and takes traffic from users through load balancing. Our blue-green deployment pipeline performs offline deployment in the blue environment.

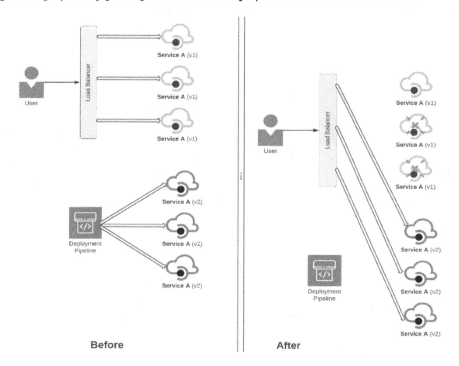

Figure 2.34 – An example of a blue-green deployment design pattern

Once we validate the blue environment deployment, the load balancer will start sending traffic to the blue environment, and green will be scaled down, as shown on the right side of the diagram. All cloud platforms provide options for implementing a blue-green deployment.

Summary

In this chapter, you learned about microservices architecture-based applications and how they are different from monolithic applications. You learned different strategies to break down a monolithic application into a microservices-based application.

You learned about different approaches for microservices decomposition, aggregation, and observability, as well as the different design patterns related to them.

You also got a fair understanding of what challenges microservices bring to the table and, with the help of different design patterns, saw how we can address those challenges.

In the next chapter, we will explain what CI/CD is and what benefits it provides for automating the software delivery of microservices. You will learn about CI/CD pipelines and how they are used in software delivery. You will learn how to create a Spring Boot microservice. We will also download the necessary software and then develop a sample microservices application.

3

CI/CD Principles and Microservice Development

Developing a microservice is one part of solving the puzzle, but how we automate the delivery of these services is another challenge. To automate software delivery, we need continuous integration and continuous delivery, so this chapter will explain to you what they are and how they can be used to automate our release process.

This chapter will give an overview of the tools and technologies used for microservice development, and then it will walk you through the development environment setup. Further, we will be developing a sample microservice application, which we will be deploying to the AWS cloud in later chapters. This chapter will take you, step by step, through how to write a microservice application, compile it using Java, and build an artifact/binary using the Maven build tool. Finally, you will learn how to run this application using an IDE or from the command line.

In this chapter, we will be covering the following topics

- Continuous integration

- Continuous delivery

- Continuous deployment

- A sample microservice overview

- Tools and technologies used

- Setting up a development environment

- Implementing the application

- Running the application

Understanding continuous integration and continuous delivery

Now that you have a good understanding of the microservices architecture and related patterns, let's focus on how we can deliver good microservices to production. In a microservices architecture, your application might have hundreds of services, so delivering those to production requires automation from inception. The concept of integrating your software and releasing it to production through automation is known as **Continuous Integration** (**CI**) and **Continuous Delivery** (**CD**). We will be taking this one step further and understanding continuous deployment and release pipelines. Let's understand these concepts in a bit more detail.

CI

In software development, multiple developers or teams work on a single deliverable, so they are often required to integrate changes and share updates with each other. This process is known as integration. When multiple developers are updating the same portions of code, you can experience issues if there are conflicting changes, leading to what is known as a merge conflict Due to these merge conflicts, your code compilation can begin to fail, and it can take time and effort to resolve the issue, reducing the developers' productivity.

CI is a DevOps practice to increase developer productivity and reduce merge conflicts. It requires each developer to make commits to a shared central repository and takes updates from the same repository multiple times a day, and each developer commit needs to be validated by an automated build process.

The following diagram shows how CI focuses on the automation of the following aspects of software development.

Figure 3.1 – CI

The following are the characteristics of a good CI practice:

- The development team must use a central source code repository
- Each developer must take updates from the central repository multiple times a day

- Each developer should commit their changes multiple times a day without any compilation errors

- Each developer's commits must be validated with an automated build process

- When a build fails, the affected developers should be informed, and the person responsible for the build failure should immediately work to fix the issue to reduce effects on other developers

- CI automatically compiles, executes test cases, and integrates your code

The following are the benefits of practicing CI:

- CI brings agility to your software development, working toward making smaller code changes, so finding issues in smaller changes is easy and you can deliver faster.

- Since you make smaller changes and integrate multiple times a day, integration becomes less complex.

- CI provides a faster feedback loop to the development team, and automated builds and tests increase confidence in the release process.

- CI focuses on smaller code changes, speeding up the code review process and making it easy to roll back any change or feature that causes any issues. Code changes are traceable using modern source code repositories.

- CI is the first step in release automation, and it is more of a practice that developers within the team need to follow, making a habit of frequent check-ins to get the true value of the process.

CI is a fundamental building block to automated deployments; the sooner a development team embraces CI, the better your deployment process will become.

CD

CD is the next step in your journey to full automation. While CI focuses on the automated building and testing of software, CD is the process by which software is delivered to a production environment with a minimum number of manual steps.

CD is the ability of a development team to produce software in short cycles and ensure that it can be reliably released at any time once a business approves it. In CD, the software is automatically built, packaged, and deployed to non-production environments where it can be tested. Once you have the go-ahead from your business users, with the click of a button you should be able to deploy your software to a production environment.

The following diagram shows how CD focuses on the automation of the following aspects of software development and how it is one step further than CI. In CD, you have a ready-to-release software package, which may or may not be released to production.

Figure 3.2 – CD

Here are some of the benefits that CD provides:

- With the help of CD, you can make your releases smaller. Small bug fixes and feature enhancements can go to production quickly with the help of automated testing. With smaller releases, finding faults in code is also easy.

- CD improves the **mean time to resolution** (**MTTR**) metrics for your application. It is very quick to find a broken feature in a smaller release, which helps in fixing it fast due to smaller changes.

- CD reduces long release cycles and speeds up release timings and frequency with the help of automation.

- A team that uses CI/CD for deployment usually has a smaller backlog, since changes can be released faster to production with reliability.

- A team with CI/CD capability has more agility and customer satisfaction due to automated testing, since most bugs or issues get caught before releasing the code to production, which causes more reliability in your product and more customer satisfaction.

At this point, you should have a good understanding of the CD process and how it helps us in automating our release process.

Continuous deployment

Continuous deployment is the desired state of a microservices deployment. You need to improve your automated test cases and inspire confidence in your software quality so that you can directly deliver changes into your production. Some organizations have a complex release process that slows down the delivery of code to production.

You can deal with a complex release process when you are delivering a bunch of changes in quarterly or biannual release cycles. Things get complicated when you have thousands of microservices to deploy and changes are happening several times a day, and then the manual deployment process doesn't scale well. To address that scaling problem and maximize your **return on investment** (**ROI**), you need to adopt continuous deployment for your application.

Continuous deployment is one step further than CD. In CD, you may have one or two manual steps, but with continuous deployment, each successful build goes to production.

The following diagram shows what continuous deployment looks like and how it helps us in delivering changes to production:

Figure 3.3 – Continuous deployment

Continuous deployment is the ability to get your code changes or configuration changes into production reliably and sustainably, with no manual interventions.

The following diagram shows the difference between CI, CD, and continuous deployment. With the help of this diagram, you can easily distinguish what stage of CI/CD your team is in. If each successful build is made to production, then you can say that you are doing continuous deployment.

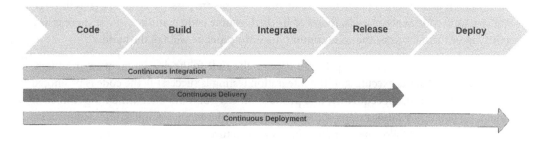

Figure 3.4 – CI versus CD and continuous deployment

If your application still requires a manual approval step before deployment to production but all other aspects of testing are fully automated, you are doing CD.

CI/CD pipeline

In DevOps, a CI/CD pipeline is a series of steps that must be performed in order to deliver a new version of software to production. In a CI/CD pipeline, the output of one step becomes an input to the next step, similar to a water pipeline system.

The following diagram shows a typical pipeline of a software development project across different environments.

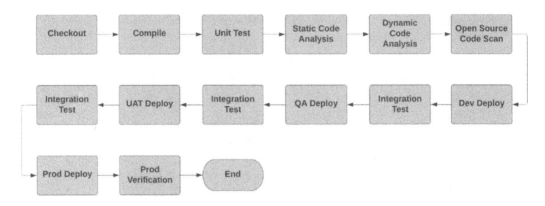

Figure 3.5 – A typical software delivery pipeline for a Java project

In the preceding pipeline, once a developer commits code to a source code repository, it gets automatically compiled, and then unit testing is performed. If all unit test cases are passed, the binary goes through static and dynamic code analysis tools, followed by open source code scanning, and if there is any new security issue introduced by the code commit, then the pipeline will fail the build.

Once quality gates are passed for this pipeline, it deploys the code to the development environment, and an integration test starts executing. If the integration testing is successful, the code is promoted to a QA environment and, subsequently, to a production environment after validating it in each environment. Different organizations have their own tools for building, scanning, and deployment, but more or less, pipelines follow pretty much the same structure.

There are several products that offer pipeline capabilities; Jenkins, Travis CI, TeamCity, and AWS CodePipeline are some of the most popular tools that allow you to create pipelines. Later in this book, you will learn more about AWS CodePipeline and how we can deliver microservices using AWS developer tools.

Now that we understand how we can deliver a microservice using CI/CD pipelines, let's focus on our sample application functionality and its implementation. In later chapters, we will keep improving our application to automate deployment to the AWS cloud.

Microservice development

In this section, we are going to develop a simple web application, listening on TCP/80, with a REST endpoint to return the application name and version information. When a user makes a call to the application /info endpoint, it will read the application name and version number from the properties file and return that information to the user's browser. Before we start setting up the development environment, let's look at the tools and technologies that we are going to use.

Tools and technologies used

We will be using the Java programming language to develop this web application. Spring Framework is a leading web development framework, so we will be using this to develop our REST endpoint. Maven is a dependency management tool that we will use to build our application. We need a source code editor to write the source code; here, we will be using the **Visual Studio Code** (**VS Code**) provided by the Microsoft community. Now that we have a fair idea of what tools we are going to use, let's get started setting up our development environment.

Setting up the development environment

In this section, we will learn how to install the different software and tools required to build our sample application. For this development, we are using a Windows-based system, but the same instructions can be used for macOS or Linux-based systems with a little variation.

Installing Java

At the time of writing, Java 17 is the latest version, but that hasn't yet been adopted by enterprise organizations widely, so we are going to use Java 11 for this sample application. However, the instructions are the same for any version of Java:

1. To install Java, go to the following link and scroll down for the Java SE Development Kit. We are using Java 11.x.x here, but you can select any version of Java you prefer: https://www. oracle.com/java/technologies/downloads/:

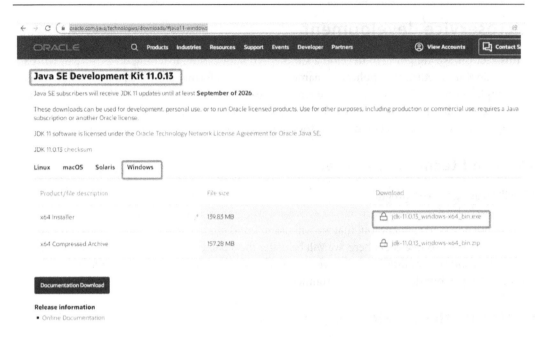

Figure 3.6 – The Java 11 download from the Oracle website

2. On this page, select your OS-related tab and then the appropriate binary. For this example, we are using a Windows system. Click on the download link, which will redirect you to the Oracle login screen. If you don't have an Oracle account, create one and accept the terms and conditions.

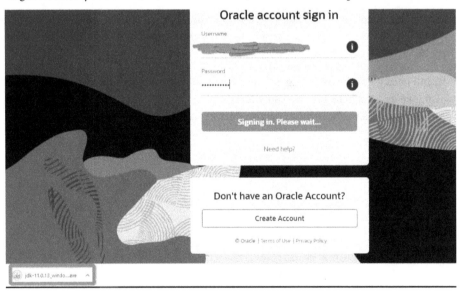

Figure 3.7 – The Oracle login screen to download the Java binary

3. Once you log in to the Oracle account, the Oracle Java binary will be downloaded to your `Downloads` folder. Double-click on the Java executable, and it will ask for your permission to allow the Java binary to make changes to your device. Click **Yes** to proceed:

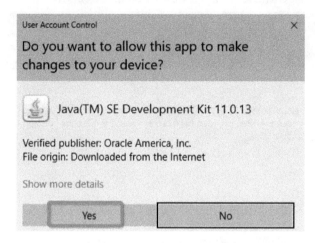

Figure 3.8 – User account control permission – the Java installation warning

Once you allow it to make changes to the system, the Installation Wizard will start.

Figure 3.9 – The Java Installation Wizard

4. Click **Next** and choose the installation directory for Java. We are not changing the directory and keeping the default, as shown in the following screenshot.

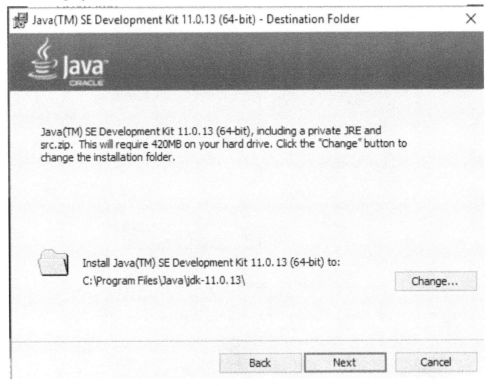

Figure 3.10 – Java installation – the target directory selection

5. Once you click **Next**, the installation will start.

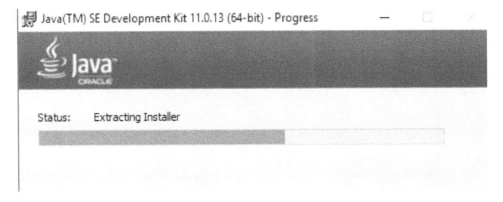

Figure 3.11 – The Java installation progress

The following screen will appear once Java is installed on your Windows system.

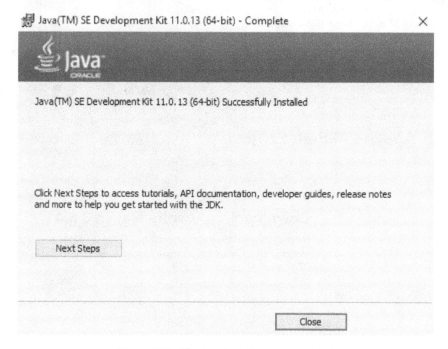

Figure 3.12 – The Java installation completed

Now, the Wizard confirms that the installation is successful, so let's close the window and open Command Prompt to confirm that Java has installed successfully.

6.   Type `Java -version` and press *Enter*; a response similar to the following will confirm that your installation of Java has completed.

```
Command Prompt
Microsoft Windows [Version 10.0.14393]
(c) 2016 Microsoft Corporation. All rights reserved.

D:\Users\amardeep>java -version
java version "11.0.13" 2021-10-19 LTS
Java(TM) SE Runtime Environment 18.9 (build 11.0.13+10-LTS-370)
Java HotSpot(TM) 64-Bit Server VM 18.9 (build 11.0.13+10-LTS-370, mixed mode)

D:\Users\amardeep>
```

Figure 3.13 – Java version information

Now that we have Java 11.x.x set up on our Windows machine, let's go ahead and set up other tools.

Installing Maven

In order to install Maven, follow these steps:

1. First, go to the following URL to download Maven from Apache: `https://maven.apache.org/download.cgi`:

Figure 3.14 – The Apache Maven download site

2. Scroll to the **Files** section and select the appropriate binary for installation. We are selecting the ZIP version of the binary, which will work on Windows as well as any other macOS- or Linux-based systems.

Files

Maven is distributed in several formats for your convenience. Simply pick a ready-made binary di

In order to guard against corrupted downloads/installations, it is highly recommended to verify the

	Link
Binary tar.gz archive	apache-maven-3.8.4-bin.tar.gz
Binary zip archive	apache-maven-3.8.4-bin.zip
Source tar.gz archive	apache-maven-3.8.4-src.tar.gz
Source zip archive	apache-maven-3.8.4-src.zip

Figure 3.15 – The Apache Maven ZIP binary file

3. Once you click on the link, it will be downloaded to your `Downloads` folder. Extract the Maven ZIP file and copy it over to your favorite location; in our case, we have copied it to the `D:\apache-maven-3.8.4` location on our Windows system.

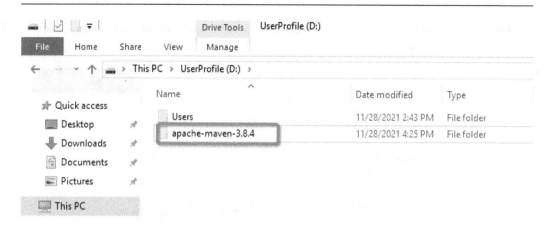

Figure 3.16 – The Apache Maven extracted binary

4. Now, open up the environment variables setup screen by typing `environment` in the Windows search bar, as shown in the following screenshot:

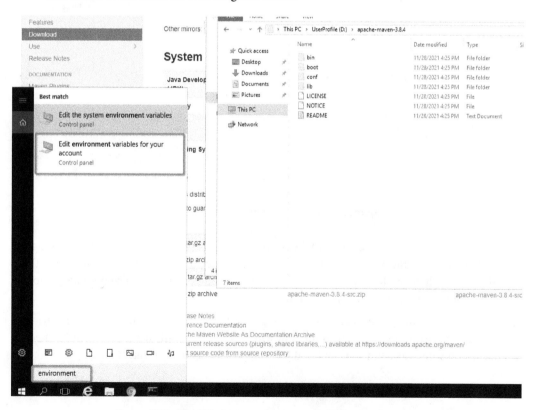

Figure 3.17 – The Windows environment variable search screen

5. Once the edit environment variable screen is open, go to the **User variables for <user>** section, click on the path, and edit it.

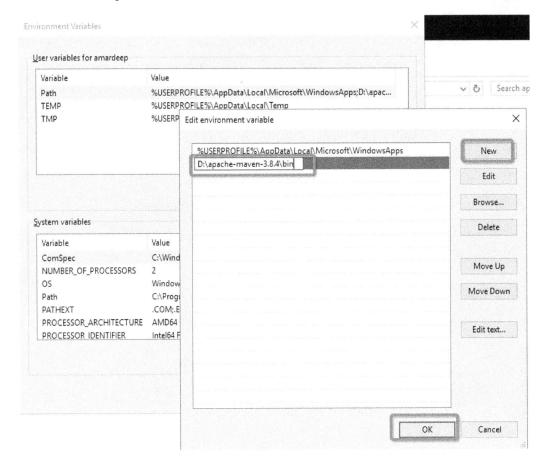

Figure 3.18 – The user environment variable

6. Add a new section to the existing path by clicking **New**, paste D:\apache-maven-3.8.4\bin, as shown in the preceding screenshot, and then click the **OK** button. This will place the Maven bin directory to your system path so that you can execute Maven commands without typing or including the Maven path explicitly.

7. Let's verify the Maven installation by opening Command Prompt again and typing mvn –version. A response similar to the following screenshot confirms that Maven has been successfully installed, and you can build your projects using the command line.

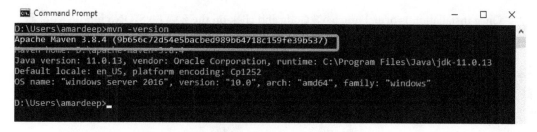

Figure 3.19 – Maven version information

Now that we have our Java and Maven installed successfully, let's go to the next section and install a VS Code IDE for development.

VS Code installation

We are going to use a VS Code IDE to develop the microservice application. The steps to install VS Code are as follows:

1. Go to `https://code.visualstudio.com/download` and download the binary appropriate for your OS. We are using a Windows system here:

Figure 3.20 – VS Code download

Now, double-click the downloaded binary in your `Downloads` folder. The next few screens will ask you to accept agreements, an installation location, a shortcut, and other information. Keep the default options, click **Next**, and install the VS Code IDE.

2. Once installation is over, launch VS Code and click on the **EXTENSIONS** button icon, as shown in the following screen on the left panel:

Figure 3.21 – The VS Code EXTENSIONS screen

3. Now, we need to install a few extensions for Java development. Search for Java in the search box and install the extensions, as shown in the following screenshot. **Extension Pack for Java** and **Spring Initializr Java Support** are two important ones that we need for development.

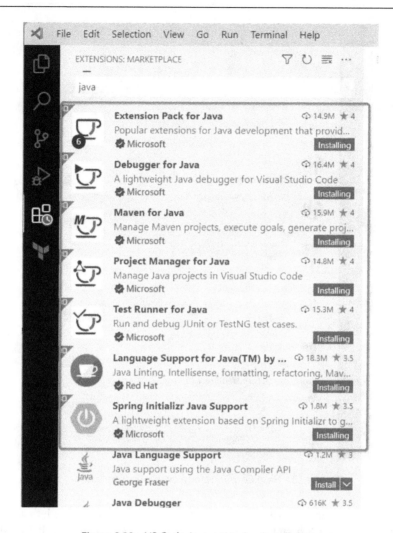

Figure 3.22 – VS Code Java extension installation

4. In the next chapter, we will be learning about Terraform to provision infrastructure. A Terraform extension is available, provided by HashiCorp; to support syntax highlighting for Terraform templates, let's add the Terraform extension to VS Code. Search for `hashi`, select **HashiCorp Terraform**, and then click on the **Install** button next to it.

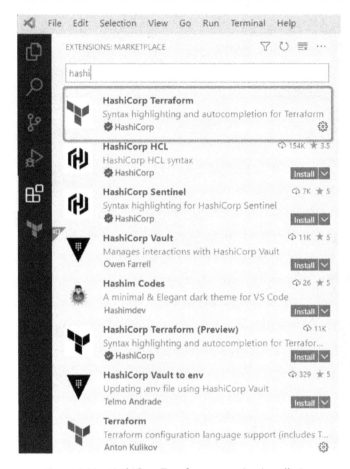

Figure 3.23 – HashiCorp Terraform extension installation

5. In upcoming chapters, we will learn about containers and Docker, so let's add a Docker extension to VS Code. Search for `Docker`, select the extension, and then click on the **Install** button next to it:

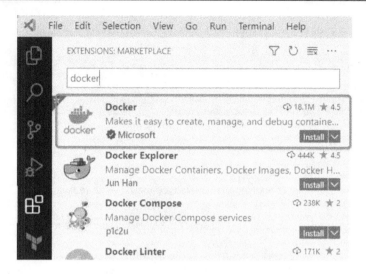

Figure 3.24 – VS Code Docker extension installation

Now, we have all the required extensions we need for development. Once you restart the VS Code editor, it will be ready for development. Let's begin creating our sample `aws-code-pipeline` microservice project.

Implementing the application

In this section, we will create a Spring Boot microservice project, and then we will write some code and try to build and run the project:

1. Go to the VS Code home page, click on the **Create Java Project** button, and then select **Spring Boot** as the project type, as shown in the following screenshot.

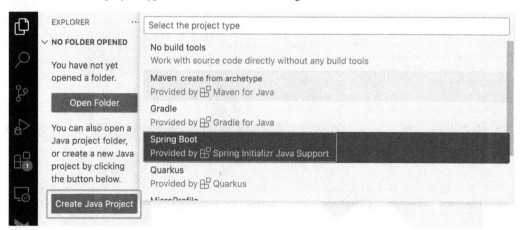

Figure 3.25 – VS Code new Java project creation

2. Select the project build type as **Maven Project**, as shown here.

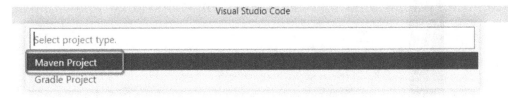

Figure 3.26 – Selecting the project build type for the VS Code Java project

3. Specify the Spring Boot version; we will select the latest release version here.

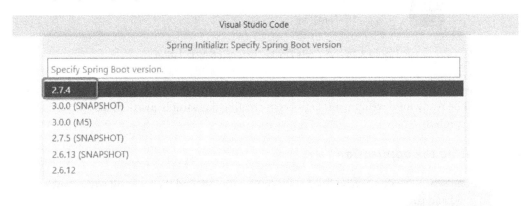

Figure 3.27 – Selecting the Spring Boot version for the Java project

4. Specify the project language as **Java**.

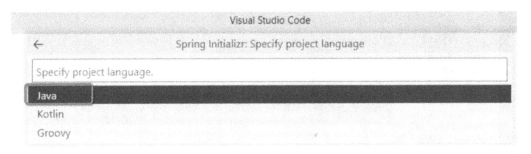

Figure 3.28 – Selecting Java as the VS Code language

5. Now, provide the Maven group ID details as `com.packt.aws.books.pipeline` and press *Enter*.

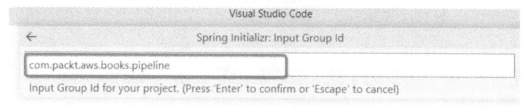

Figure 3.29 – The VS Code Maven project group ID

6. Provide the Maven artifact ID for our application as `aws-code-pipeline` and press *Enter*.

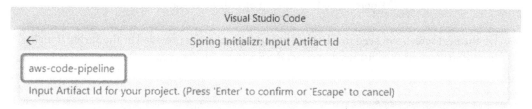

Figure 3.30 – VS Code Maven project artifact ID

7. In the next screen, we need to select the application packaging type. Let's select the **Jar** packaging type:

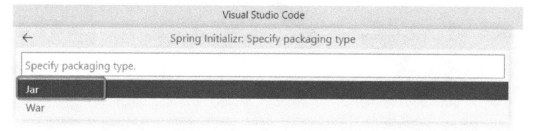

Figure 3.31 – The Maven project packaging type

8. Specify the Java version you will be using for this application:

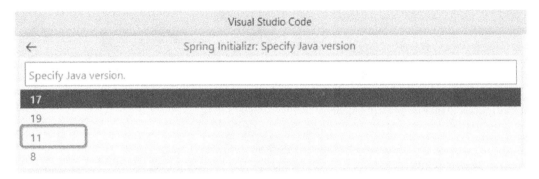

Figure 3.32 – The Maven project Java version

9. On the next screen, we need to select the spring dependencies provided by the Spring Initializr for the application. We just need to select the **Spring Web** dependency here and press *Enter*.

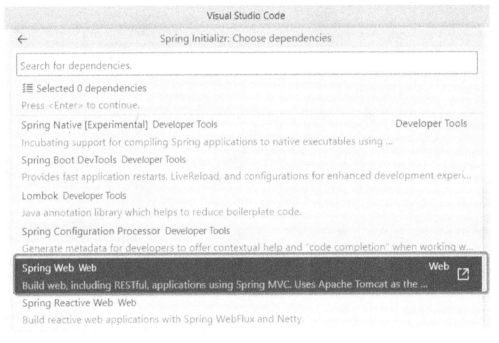

Figure 3.33 – Selecting the Spring Boot dependency

10. On the next screen, we need to select the location on our system where we want our application code to be generated.

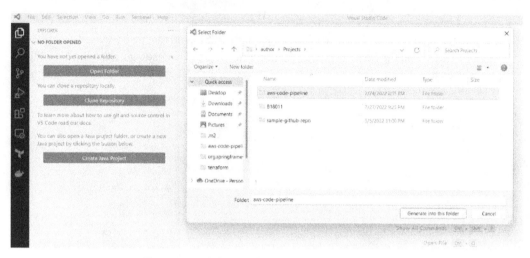

Figure 3.34 – Selecting the VS Code project location

Select any suitable location on your system and open the project.

11. VS Code will prompt you to open the project, or you can open it using the **Folder** option from the **File** menu. The following screen shows the project structure generated by the Spring Initializr.

```
1  package com.packt.aws.books.pipeline;
2
3  import org.springframework.boot.SpringApplication;
4  import org.springframework.boot.autoconfigure.SpringBootApplication;
5
6  @SpringBootApplication
7  public class AwsCodePipelineApplication {
8
   Run | Debug
9      public static void main(String[] args) {
10         SpringApplication.run(AwsCodePipelineApplication.class, args);
11     }
12
13 }
14
```

Figure 3.35 – The aws-code-pipeline project screen

Now that our project is ready in the IDE, let's start editing the project to meet our needs. As we discussed earlier in this chapter, we will be adding a REST endpoint in this project, which can be called from our browser:

1. In Spring, you can create a REST endpoint by defining a REST controller, so let's create a controller by right-clicking on the `com.packt.aws.books.pipeline` package, choosing **New File**, providing a name for this new controller, `WelcomeController.Java`, and pressing *Enter*.

Figure 3.36 – Creating a new controller class

2. This generates a `WelcomeController.Java` file in the `com.packt.aws.books.pipeline` package, as shown in the following screenshot:

Figure 3.37 – The WelcomeController class

3. Let's edit this class and add the code shown here:

```
package com.packt.aws.books.pipeline;
import org.springframework.beans.factory.annotation.Value;
import org.springframework.web.bind.annotation.GetMapping;
import org.springframework.web.bind.annotation.RestController;
@RestController
public class WelcomeController {
@Value("${app.name}")
private String appName;
@Value("${app.version}")
private String appVersion;

@GetMapping({"/","/info"})
public String sayHello() {
return appName+"-"+appVersion;
}
}
```

4. In this code, we have marked the Java class as a controller with the help of @RestController, so Spring can register this class to handle the REST requests.

```
 1  package com.packt.aws.books.pipeline;
 2
 3  import org.springframework.beans.factory.annotation.Value;
 4  import org.springframework.web.bind.annotation.GetMapping;
 5  import org.springframework.web.bind.annotation.RestController;
 6
 7  @RestController
 8  public class WelcomeController {
 9
10      @Value("${app.name}")
11      private String appName;
12
13      @Value("${app.version}")
14      private String appVersion;
15
16      @GetMapping({ "/", "/info" })
17      public String sayHello() {
18          return appName + "-" + appVersion;
19      }
20  }
```

Figure 3.38 – WelcomeController with a REST endpoint

We have created a sayHello() method with a URL pattern of / and /info, which returns the application name and version information. The @GetMapping annotation translates any HTTP GET request with a URL containing / or /info, and it routes user requests to this sayHello() method.

5. Now, we will define the app.version and app.name environment variables in the application.properties file found in the resources section, as shown in the following screenshot, and the @Value annotation loads those values in WelcomeController.

```
1  app.name=aws-code-pipeline
2  app.version=1.0
3  server.port=80
```

Figure 3.39 – The application.properties file

Note in the preceding screenshot that we have added one more property, `server.port`, which is a Spring Boot property that starts the Spring Boot application on the specified port. Here, we have set the server port to `80`.

Now, we have everything that we currently need, so let's try to build the project. Open the terminal. We have two ways to build the project, either through a VS Code-embedded terminal or through the command line Maven we set up earlier in this chapter. Let's first try to build the project using the VS Code terminal:

1. Navigate to the **Terminal** menu and click on **New Terminal**. This opens up a terminal in the lower section of the IDE, where we can specify the Maven goals we want to execute.

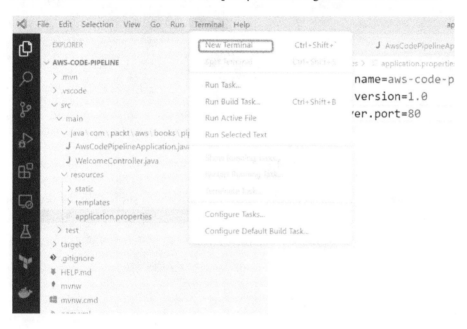

Figure 3.40 – Opening a new terminal in VS Code

2. Here, we want to first clean the project and then install it. Type mvn clean install in the terminal and press *Enter*.

Figure 3.41 – The Maven clean installation

3. This will start building the project, and the following screen shows that the project build was successful.

Figure 3.42 – The Maven build result

4. Alternatively, to build the project using the command line, go to the project directory and type the mvn clean install command:

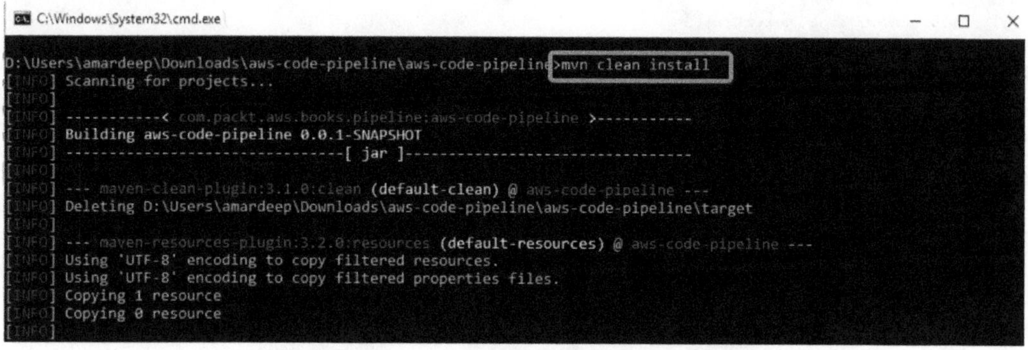

Figure 3.43 – The Maven build using the command line

This command builds the project successfully and generates the application artifact. Now that we have our application code ready and the build is completed, let's learn how to run this application.

Running the application

To run the Spring Boot application, follow these steps:

1. Right-click on the main application file, AWSCodePipelineApplication, and run it; alternatively, you can open the file and click on **Run**, as follows:

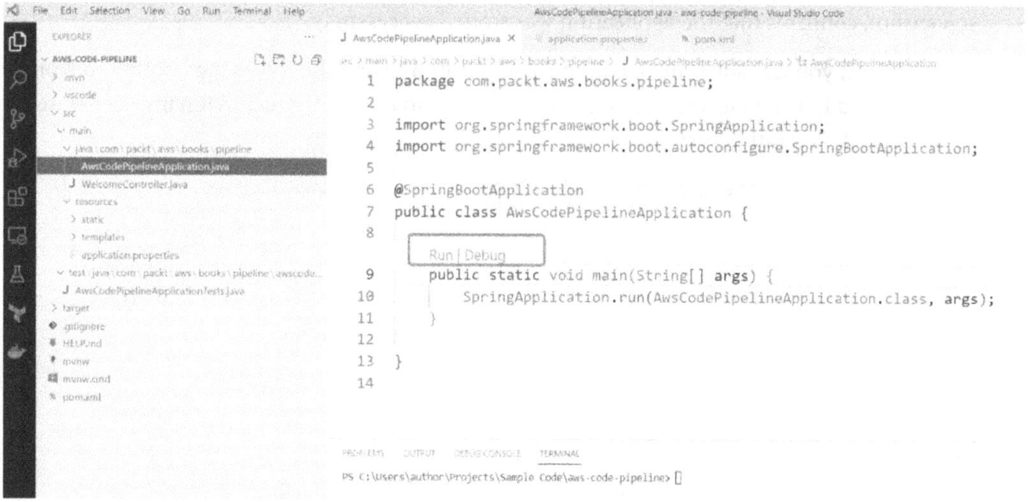

Figure 3.44 – Starting the aws-code-pipeline Spring Boot application

2. The following screen shows a successful startup of the Spring Boot application.

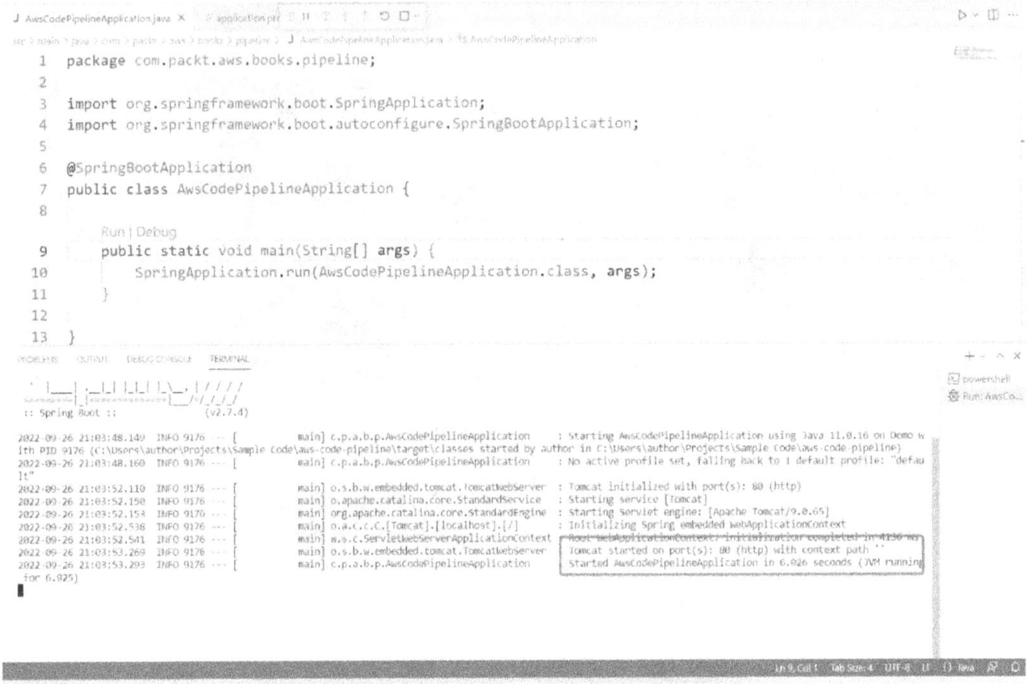

Figure 3.45 – aws-code-pipeline Spring Boot application startup in VS Code

A message, **Started AWSCodePipelineApplication in 6.026 seconds**, as shown in the preceding screenshot, confirms that the application was started successfully.

3. Alternatively, you can start the Spring Boot application using the following command from the command line in your project directory. You need to use the correct JAR name, generated in the project target directory:

```
Java -jar target/aws-code-pipeline-0.0.1-SNAPSHOT.jar
```

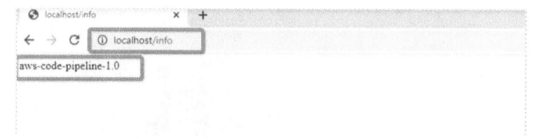

Figure 3.46 – aws-code-pipeline spring boot application startup from the command line

4. Once the application startup is successful, open the browser and type the address `http://localhost:80/info`; you will see a screen similar to the following screenshot:

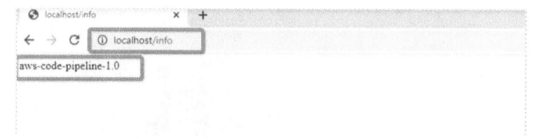

Figure 3.47 – The aws-code-pipeline info endpoint response

As you can see, our sample application, `aws-code-pipeline`, has responded to the HTTP GET request and returned the application name, prepended to the version number specified in the name.

So, now we have our application code ready and the application running on port `80` on `localhost`.

> **Important note**
>
> All the source code and example code we used in this chapter can be downloaded from the GitHub repository: `https://github.com/PacktPublishing/Building-and-Delivering-Microservices-on-AWS/tree/main/chapter_3`.

Summary

In this chapter, we explained CI/CD and all the benefits it provides for automated software delivery of microservices. You learned about CI/CD pipelines and how they are used in software delivery.

You learned about the `aws-code-pipeline` sample microservice application and what functionality we will implement. You also learned about the different tools and technologies used for the development of the sample microservice and installed them for it. We also implemented our microservice and were able to start it successfully.

In the next chapter, we will focus on different strategies to automatically provision AWS infrastructure for the deployment of our `aws-code-pipeline` microservice.

4

Infrastructure as Code

In this chapter, we will learn about the different cloud infrastructure provisioning techniques. We will learn about some good ways to create infrastructure in the AWS cloud and how we can automate the creation of infrastructure using infrastructure provision tools.

This chapter explains what infrastructure as code is and what tools and technologies you can use to provision different resources you require to deploy a sample application in the AWS cloud. For this book, we will use a mix of the AWS console and Terraform to create the required infrastructure. In this chapter, we will explain what tools are available to automate infrastructure provisioning and how to use them.

We will learn about CloudFormation and Terraform and how we can use them to create AWS resources. In this chapter, we will cover the following topics:

- What infrastructure as code is and its benefits
- CloudFormation and its execution
- Terraform and its execution

In the previous chapter, we created a sample microservice application and learned how to run it on our local system. As a next step, we need to take this microservice and deploy this to the AWS cloud in an automated way. In order to do that, we need to create a infrastructure on which we can deploy the service. The easiest way to create the infrastructure in an AWS environment is to go ahead and log in to AWS console and start provisioning resources.

The following screenshot shows how to create a **Virtual Machine** (**VM**) in AWS console. We have to go through a bunch of screens and provide operating system, hardware, network, and security-related details to create it:

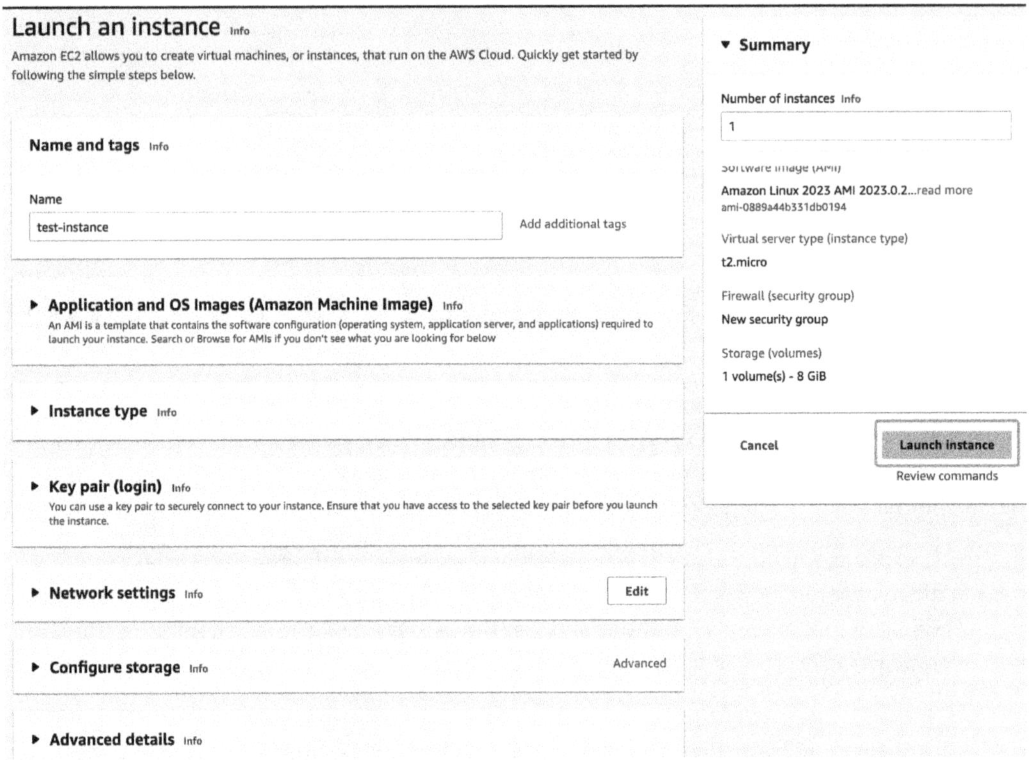

Figure 4.1 – AWS console for EC2 instance creation

This sounds easy and fun, as you can go through each screen and fill in each and every detail, showing you how easy it is to get a running VM in a matter of minutes.

Now, suppose we have to create one load balancer, two instances in different availability zones, and one security group, and configure network ACLs. Not a big deal – let's start with the AWS console, and this time, we might have to go through 10–15 screens.

Sounds alright, doesn't it?

But think about it – if you have to create the resources for thousands of microservices, you have to create exactly the same resources in different environments for redundancy. This solution is not scalable, and it is more error-prone, as someone manually has to log in, go through several screens, and key in the same information, which is time-consuming. Also, if someone makes a mistake, there is no way to review that information.

To deal with this kind of situation and provision an infrastructure in a scalable, reviewable, and repeatable way, an infrastructure as code technique is used. But what is infrastructure as code? Let's find out.

What is IaC

Infrastructure as Code (IaC) is a technique used to manage and procure hardware/software resources as a versioned code base, that can be scripted.

With the popularity of the cloud, hardware resources can be viewed as a software component, and with the help of APIs, you can procure the resources needed for your application. Cloud providers keep pool of hardware resources, and when an API call is made to provision a resource, a required capacity is made available from these pools of resources.

Cloud providers keep pools of resources in their data centers across the globe and allow you to manage these resources through their APIs, so you can request or release a resource at any point in time and pay only for what you have used. Therefore, you need automation to manage these resources for you, and this automation is done through scripts, which need configuration to configure the resources as per your need. Versioning of these infrastructure provisioning scripts is known as IaC.

In another words, IaC is the process of procuring and managing a data center through machine-readable automation files instead of physical hardware configuration or an interactive configuration tool. These scripts or files can be managed in a source control system as source code, and they can be versioned and reviewed by your peers, similar to Java or another programming language.

The following diagram shows a typical IaC process. In this diagram, you can see that an infrastructure provisioning file can be created as a script or in a declarative way. Once this file is created, it is checked into the source code repository or version control system, where it can be reviewed by peers for accuracy and any feedback. Once we have the changes ready, they can be applied in the target cloud environment and defined; the resources are created with their configuration or whatever has been defined in the file:

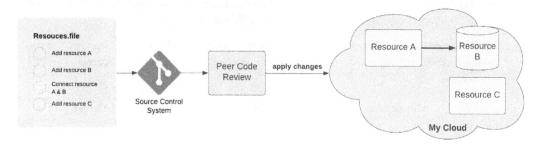

Figure 4.2 – The IaC provisioning process

This file can be applied as many times as you want and will produce similar results, and if you change anything in any commit to this file, it can be traced from the version control system. So, basically, you can compare the changes applied in each iteration.

Now, let's review the great benefits provided by IaC.

The benefits of IaC

IaC provides the following benefits through automation:

- **Speed**: With the help of an IaC tool, you can create a resource a lot faster compared to manually going through several screens. For example, if you have to recreate an environment for a critical application due to unforeseen reasons, an IaC tool can create it in a matter of minutes, while doing it manually might take hours.

- **Consistency**: With the help of an IaC tool, you can create the exact same environment as many times as you want, whereas if you have to do it manually, there are chances of making a mistake.

- **Traceability**: An IaC tool requires a configuration file, through which it creates an environment for you, so it can be easily tracked and versioned in a source code system, such as AWS CodeCommit or GitHub, and history can be used to find a change for fellow developers to review.

- **Cost reduction**: If you have to manually create resources, you'd need someone to do it manually each time, which would take human resources, so by automating using an IaC tool, you can eliminate those resources and the associated cost.

- **Customizable**: IaC tools are customizable, and parameters can be passed through at runtime to customize them for a specific environment. For example, if you want to launch 2 instances of an EC2 instance in your development environment, while in production you need 10 instances, then those values can be passed as parameters.

- **Reduced risk**: IaC tool automation produces the same results each time you run it, so it reduces the risk of human errors and, eventually, the chances of outage by misconfiguration.

- **Standardization**: With the help of an IaC tool, you can enforce and standardize security practices and other organizational standards, and then connect them with an approval process so that you know what changes are happening in your environment.

- **Self-documentation**: IaC tools are self-documented and explain how your infrastructure was created and when changes were made, so you don't need to create explicit documentation for new team members.

Now that you've got an understanding about the benefits provided by an IaC tool, let's learn about a few of these tools and how to use them in the next section.

CloudFormation

AWS CloudFormation is an IaC service provided by Amazon to help provision a required infrastructure. In CloudFormation, we create a template that describes all the AWS resources we need (such as Amazon EC2 instances, a security group, and a key pair), and CloudFormation will create those resources for us and configure as per the instructions provided in the template.

CloudFormation helps us to simplify our resource management, which can be easily repeated, and changes can be tracked and versioned.

Let's try understanding the concepts related to CloudFormation.

Template

In order to create resources in an AWS environment, you need to define them in a descriptive document, which is known as a template.

A CloudFormation template is a JSON/YAML-formatted text document, which you can save with a .json, .yaml, .template, or .txt extension. These text files are used to define the different resources and configurations you need to create them in your environment. These templates work as a blueprint to create your environment.

It doesn't matter how many times you apply these templates – they will create identical resources for you, so you can rely on them to confidently create your environment.

The following code shows a simple JSON-formatted CloudFormation template, where we specify the creation of a T2 Micro EC2 instance from an **Amazon Machine Image (AMI)** ID, ami-0ed9277fb7eb570c9:

```
{
    "AWSTemplateFormatVersion": "2010-09-09",
    "Resources": {
        "MyDemoEC2Instance": {
            "Type": "AWS::EC2::Instance",
            "Properties": {
                "InstanceType": "t2.micro",
                "ImageId": "ami-0ed9277fb7eb570c9"
            }
        }
    }
}
```

As you can see, that template is very simple, and we have provided a name, MyDemoEC2Instance, for the EC2 instance. We can also specify parameters in a template by using the Parameters tag at the same level as Resources. At runtime, you need to specify a value for these parameters. Let's recreate the preceding template using parameters:

```
{
"AWSTemplateFormatVersion": "2010-09-09",
"Resources": {
"MyDemoEC2Instance": {
"Type": "AWS::EC2::Instance",
"Properties": {
"InstanceType": "t2.micro",
```

```
"ImageId": {
"Ref": "AmiId"
}
}
}
},
"Parameters": {
"AmiId": {
"Type": "String",
"Default": "ami-0ed9277fb7eb570c9",
"Description": "Please provide an AMI ID to launch an instance"
}
}
}
```

In the preceding example, we have created a parameter named AMI ID to accept ImageId. We have defaulted AMI ID to ami-0ed9277fb7eb570c9, but the user will have the option to provide a new value. All parameters in a CloudFormation template needs to be defined inside a parameters node, and they can have a description, type, default value, and some more constraints. Once you define a parameter, you can refer back to them using a Ref tag and specify the name of the parameter as shown previously for ImageId tag, where we referred the AmiId parameter.

Stacks

In an AWS environment, all the resources created through a CloudFormation template are managed through a single unit called a stack. You can create, update and delete a collection of resources using a stack. A stack provides atomicity of AWS resources for management as a single unit. For some reason, if any resource creation fails, a stack can be entirely rolled back, depending on the configuration setting you specify.

If you decide to delete an AWS Stack, all the associated resources related to the stack will be deleted, unless delete protection is enabled on related resources. Let's suppose you defined an EC2 instance, a load balancer, and an auto-scaling group as part of your CloudFormation template. When you apply this template in your environment, all of these resources will be created for you as part of the same CloudFormation stack. You can enable delete protection on a CloudFormation stack to avoid accidental deletion of a stack. You can use the AWS UI console, an API, or the AWS CLI to create and manage a stack.

Change sets

In CloudFormation, change sets allow you to visualize the impact of changes in your live environment. Suppose you have created a stack by applying a CloudFormation template, and now you have made some changes to the template and want to apply the new version of the template. With the help of a change set, you can visualize and see a summary of changes due to happen in your environment with this new version of the template.

For example, in the new version of the template, let's suppose you renamed an EC2 instance or a DynamoDB database so that when your changes apply, CloudFormation will delete your existing resources and create a new one with a new name. If that happens, you will lose the data previously stored on the EC2 instance or Dynamo database, so a change set provides a summary of these changes, which can help us to plan better.

CloudFormation process

Now that we have understood different concepts related to CloudFormation, let's try to understand how we actually apply a CloudFormation template.

The following diagram explains the process how a CloudFormation template gets applied in an AWS environment. First, we create the JSON or YAML-formatted CloudFormation template, either by hand, by using an IDE, or by using the CloudFormation template designer service provided by AWS.

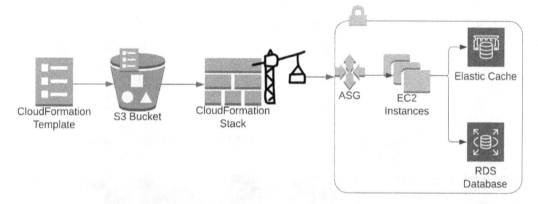

Figure 4.3 – The CloudFormation resource provisioning process

Once we have the template ready, it needs to be uploaded in an AWS bucket in order to create a stack out of it. Once we have our CloudFormation template in the S3 bucket, we can either make an API call, run an AWS CLI command, or log in to the AWS console to create a stack. In the AWS console, we can see the progress in terms of what resources are being created and what the final outcome is of the stack. Once the stack run is successful, we can go to individual resources and see that they have been launched in our account. Deleting a stack will delete all the provisioned resources if you haven't enabled termination protection on individual resources.

Let's log in to the AWS console and start creating a sample CloudFormation template using template designer:

1. Go to `https://aws.amazon.com/console/`.

2. Log in using your credentials. If you don't have an account, you can create one. We will be launching minimal resources in our environment, which should be covered in the free tier, but please make sure that you delete all the resources after completing the exercise; otherwise, AWS may charge you for the resource you use. Once you have logged into the console, you will see a home screen similar to what is shown in *Figure 4.4*:

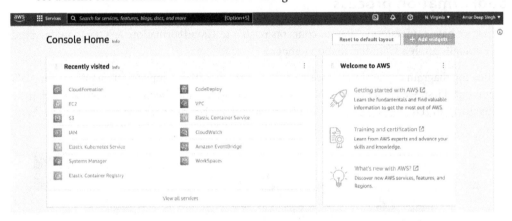

Figure 4.4 – AWS console

3. Search for the CloudFormation service in the search bar, and then click on **Designer** to go to CloudFormation designer service:

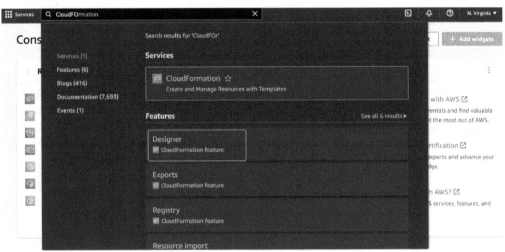

Figure 4.5 – A CloudFormation Designer search

4. Once you click on the designer service, it will take you to a screen similar to what is shown in *Figure 4.5*, where you can drag and drop resources from the **Resource types** section on the left to the main designer area in middle. Doing so will automatically generate the JSON/YAML-formatted template code in the bottom area of the screen:

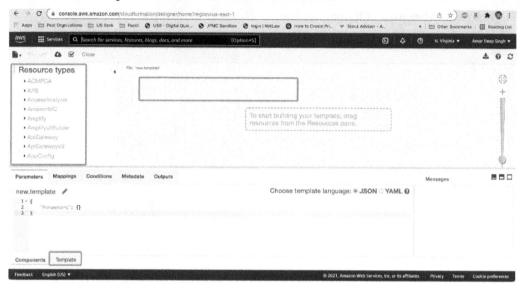

Figure 4.6 – The CloudFormation designer page

5. Let's start to drag and drop an EC2 instance from the left side of the section to the middle. Similarly, you can drag and drop other AWS services:

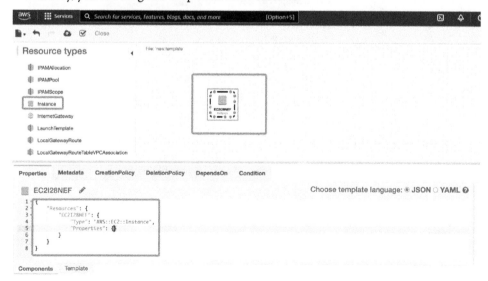

Figure 4.7 – The CloudFormation designer with an EC2 instance

Note in the preceding screenshot that as soon as we drop the EC2 instance in the designer panel, we can see the code in the components/template section automatically generate for us. You can manually update the code section, and then the designer will be refreshed. So, you can write down your code and then visualize it in the designer, or you can work the other way round.

6. Now, let's update some of the properties manually inside the EC2 resource:

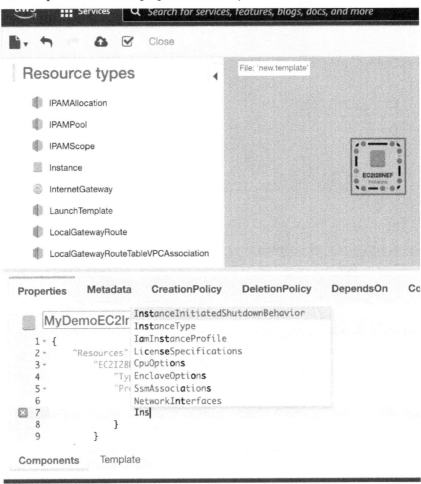

Figure 4.8 – The CloudFormation designer syntax highlighting

In order to launch an EC2 instance, you need to provide the type of instance you want to launch and what **Amazon Machine Image (AMI)** you want to use for it. In the **Properties** section, as soon as you start typing, it will provide you syntax highlighting.

7. Specify `InstanceType` as `t2.micro` to keep you under the free usage tier, use `ImageId` as a parameter, and specify the default value as `ami-0ed9277fb7eb570c9`, which is an Amazon Linux image. This parameter will let you specify a different `ImageId` at runtime.

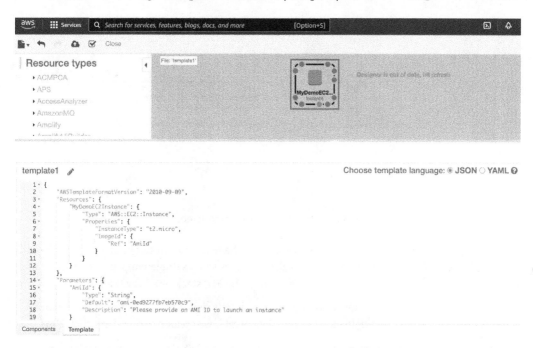

Figure 4.9 – CloudFormation template with EC2

Now, our CloudFormation template to create the EC2 instance is ready. Let's go ahead and try this template. Let's click on the cloud icon in the top-left corner to directly upload and save this template to the S3 bucket managed by AWS and create the stack. Alternatively, we can download this template from the file menu to our local system.

8. As soon as we click on the **Create stack** icon, it opens up the following screen to apply the template to our AWS environment:

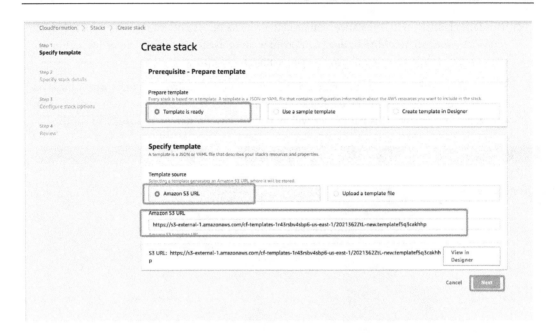

Figure 4.10 – The Create stack home screen

In the preceding screenshot, you can see that we have an option to either use our current template or upload a new template. AWS also provides some sample templates to create some pre-configured environments, such as deploying a LAMP stack or a WordPress site.

As you can see in the previous screen, the template is being referred from an Amazon S3 location, where AWS has uploaded the template, which we created earlier. Let's click **Next**.

9. Once you click next, it asks you to provide a stack name. AWS doesn't accept special symbols, except –. Whatever parameters you specify in the template will appear here; as you can see, our `AmiId` parameter appears here with the default value, and we have option to change that to provide a different `ImageId`. Let's go ahead with a default `ImageId` and click **Next**.

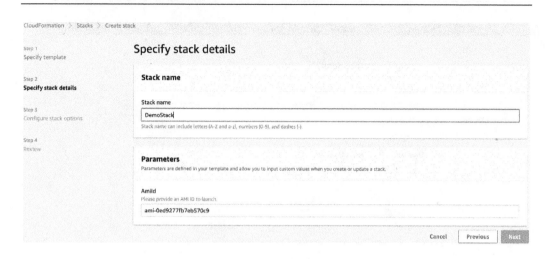

Figure 4.11 – The Create stack details screen

10. On this screen, we provide additional details for our stack creation process, such as what tags need to be added to the AWS resources created as a part of this stack. Other important details are related to how we handle the stack failovers, whether we should roll back all resources, or whether we should keep the resources we were able to create successfully.

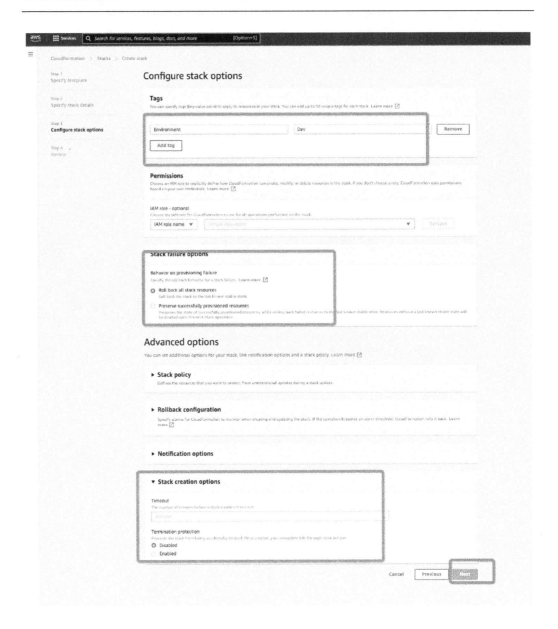

Figure 4.12 – The CloudFormation configure stack options

CloudFormation also needs a role to create these resources for us; if we don't specify anything, it will create one for us. You can also enable termination protection on this screen, which will protect you from accidentally deleting your resources. On this screen, we have provided one tag, named **Environment**, and chosen rollback for all of our resources if our stack creation fails. Let's click **Next** to go to the review screen.

11. This screen shows all the values and configurations we chose earlier. If we need to, we can edit this screen. Let's go ahead and click on the **Create stack** button, as shown in the following screenshot:

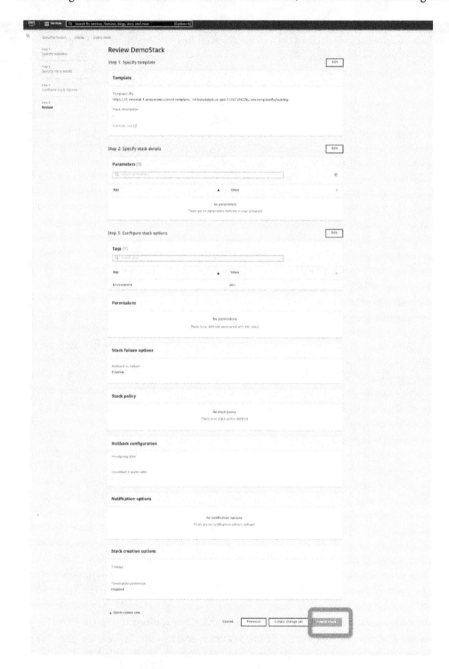

Figure 4.13 – The CloudFormation Create stack review screen

12. Once you click on the **Create stack** button, you will see your stack creation in progress. This **Events** tab will refresh with the status once the stack is created or if there is any failure:

Figure 4.14 – DemoStack creation in progress

Once the stack is created, you will see a screen similar to the following, showing that the stack has been created successfully:

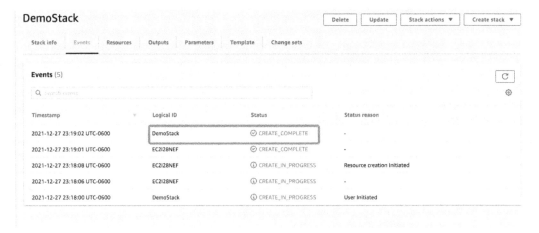

Figure 4.15 – Successful creation of the CloudFormation DemoStack

13. Now, let's click on the **Resources** tab to ensure that the required resources in our template are created. In this case, we just have an EC2 instance, and as you can see, it is listed there:

Figure 4.16 – The CloudFormation Resources EC2 instance

14. Let's click on the physical ID of the resource to take us to EC2 dashboard. On the EC2 dashboard, we can see that our EC2 instance is running and healthy, and we can see different details related to our EC2 instance:

Figure 4.17 – The CloudFormation Resources EC2 instance

15. Now, let's go back to our DemoStack, and we can see in the **Stack actions** section that different options are available to us:

Figure 4.18 – CloudFormation – Stack actions

We can see that we have different options such as **Edit termination protection**, **View drift results**, and **Delete drift**.

16. Now that we are done reviewing our stack, let's start deleting it so that we don't have to incur any additional charges from AWS. Click the **Delete** button on the stack page. It will ask you to confirm the deletion; click on the **Delete stack** button:

Figure 4.19 – CloudFormation – deleting DemoStack

17. Once you click **Delete stack**, you will see in the resources section that our resource's (EC2 instance's) state changes to **DELETE_IN_PROGRESS**:

Figure 4.20 – CloudFormation DemoStack – deletion in progress

18. Once CloudFormation stack deletion is completed, you will see that the status changes to **DELETE_COMPLETE** in the **Events** tab:

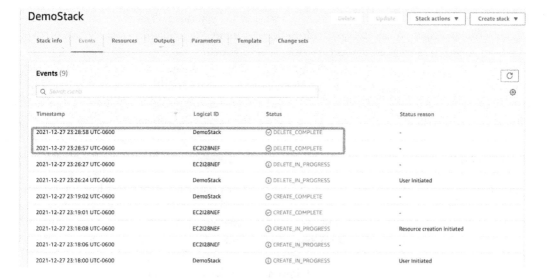

Figure 4.21 – CloudFormation – DemoStack deletion complete

19. So, once you have deleted a stack successfully, all resources associated with that stack will also be terminated. To confirm this, let's go to our EC2 dashboard and look for the EC2 instance we created earlier:

Figure 4.22 – The E2 dashboard – the terminated EC2 instance

You can see in the preceding screenshot that the instance we created has been successfully terminated.

In this section, we learned how to create an AWS CloudFormation template and apply the template using the AWS console to create a stack. Alternatively, you can apply a CloudFormation stack using the CloudFormation API or the AWS CLI. We aren't covering those options here, but it is very straightforward.

CloudFormation is Amazon's proprietary managed service, and you can easily manage your resources using this. CloudFormation has one downside in that it is limited to managing resources within the AWS cloud. If you have to create resources in any other competitor cloud provider, such as Google Cloud or Microsoft Azure, then CloudFormation won't work. In the next section, let's learn about another IaC tool, which works across the different cloud providers and is open source, so you can develop integration with your own cloud. In order to gain cross-cloud provider reusability, Terraform is preferred over CloudFormation by a large developer community.

Terraform

Terraform is an open source project and is widely used by organizations that are looking for an infrastructure provisioning tool across different cloud providers. Terraform works across multiple platforms and is not tied to a specific one, as it is platform neutral. Terraform supports all major providers, such as AWS, Microsoft Azure, Google Cloud, Oracle Cloud, and many more. The full list of providers can be found here: `https://registry.terraform.io/browse/providers`. If you are working on your private cloud and don't have a provider, Terraform allows you to extend and develop one.

We will describe our desired infrastructure using a human-readable, configuration language known as **HashiCorp Configuration Language** (HCL), a JSON-like syntax. This configuration file allows us to create a blueprint that can be versioned, shared, and reused. Once we have the configuration file ready, Terraform uses it and creates a detailed execution plan, explaining what changes it is going to make to our infrastructure. Terraform templates are declarative; we declare all resources we need. Then, Terraform finds the best way to create those for us. Terraform keeps track of the resources it creates for us, and on subsequent runs of the same template, it doesn't try to create the same resources. Instead, it identifies the changes that we are applying and makes those changes. If we remove a resource from the template, in the next run, Terraform will go ahead and remove that resource from our infrastructure. Terraform provides detailed information before applying changes in our environment. In other words, Terraform provides the necessary information before it creates, updates, or deletes resources.

Terraform generates a resource graph and creates nondependent resources in parallel, so it is optimized to quickly create an infrastructure.

Whenever we make any changes in the configuration file, Terraform determines the dependencies and creates an incremental plan to create an infrastructure without the conflict of dependencies. Terraform accepts configuration with any filename ending with a `.tf` extension. The main configuration file is usually named `main.tf`, and then you can run a set of commands to create Terraform-managed infrastructure. Before going deep into the implementation details, let's learn some of the concepts.

Terraform concepts

Terraform is an open source project built around some core concepts, such as providers, resources, and state. Let's understand these concepts before proceeding further.

Providers

In a Terraform configuration file, you need to provide a provider section, which is actually an infrastructure provider plugin service. A provider is a plugin that manages the connection, creation and configuration of the resources for that provider. For example, AWS is a provider for Terraform. On the Terraform site, there are thousands of providers available. We can configure more than one provider in a configuration file and manage resources in different cloud providers.

With the help of providers, Terraform manages the infrastructure. Each provider adds a set of resources, and these resource types are implemented by a provider. Each provides add new capabilities to Terraform, and these capabilities can be as small as adding a local utility or as large as adding a cloud capability.

The following code snippet adds an `aws provider` to a Terraform configuration file. We can provide additional configuration inside the provider section, such as a default AWS region and connection information:

```
provider "aws" {
}
```

Terraform providers are not a core part of the Terraform binary; once you add them to your configuration and initialize it, Terraform downloads them from the registry. These providers are available in the Terraform registry and have their own release schedule and versions.

Resources

A `resource` block in a configuration file identifies an infrastructure component for a provider. A resource can be a physical component or a logical component. The `resource` section starts with a `resource` keyword, and then we use a provider name such as aws for Amazon cloud followed-up by the resource type, such as `s3_bucket`. We also need to provide a name to this resource through which we can refer it in other places in the template. In the following example, we are creating an AWS S3 bucket, which we are going to refer to as `mybucket`:

```
resource "aws_s3_bucket" "mybucket" {
    bucket = "my-test-bucket"
    acl    = "private"
}
```

Once we run this template, it will create an S3 bucket with the name `my-test-bucket` and set the `acl` status of the bucket as private. Terraform will keep the remaining configuration to default, and before it finally runs it, it will ask us for confirmation.

Terraform state

When we run a Terraform template, Terraform generates a special file called `terraform.tfstate` to maintain the state of each resource it creates. Next time, when we make any change to the template, Terraform will do a difference check with this file and determine what new resources need to be created or what needs to be removed.

The `terraform.tfstate` file is the only way for Terraform to keep track of the managed resources, so when we run the same template again, Terraform avoids creating duplicate resources. This file keeps a lot more metadata than our configuration file, and sometimes, it stores sensitive information about infrastructure as well, so we need to be careful in managing this file. By default, `terraform.tfstate` is created in the same directory as the Terraform template, but in a production environment, you should back up this file, as this is the way Terraform understands your infrastructure.

Backend

As we learned in the previous section, the `terraform.tfstate` file is important, and storing this file locally on a developer's computer is a risk. Having a state file locally is also problematic when you have multiple people in a team working on changes, as they can make changes simultaneously and override infrastructure changes. To address this problem, Terraform has backends that determine where a Terraform state file can be stored centrally and shared with everyone for collaboration. Terraform locks this file and ensures that developers don't make conflicting changes simultaneously.

Terraform supports different types of backends, including Amazon S3, Azure, and JFrog Artifactory; a full list of these backends can be found here: `https://www.terraform.io/language/settings/backends`. Some of these backends support state locking and some of them doesn't, so you have to be careful what you want to choose. By default, Terraform initializes with a local backend if we don't specify anything. If you plan to use the Terraform cloud, then you don't need to define a backend, as that will automatically manage the state file and support locking.

You can define a backend as follows in a Terraform template:

```
terraform {
  backend "s3" {
    bucket = "my-template-bucket"
    key    = "bucket-key"
    region = "us-east-1"
  }
}
```

If you have an existing Terraform template and need to add a backend, then you need to reinitialize Terraform by executing the `terraform init` command. Let's now learn about input variables and how to use them in a configuration file.

Input variables

Input variables are the parameters you can pass to your template without hardcoding them into your configuration. It provides you with the flexibility to pass these values at execution time and also write code that is easier to modularize and reusable. If you use the variables in a root module, then you need to supply these values either through the Terraform CLI command options or through the environment variables.

You define a variable using a variable block in Terraform configuration, similar to the following code:

```
variable "my_ami_id" {
type = string
default = "ami-0ed9277fb7eb570c9"
}
```

In this example, `my_ami_id` is the variable name that we can refer to in another place, and the type of parameter is `string`; we have initialized a default value for this parameter, so if, at the time of execution, we didn't specify a value for the variable, it will take this value as default. Optionally, you can define the description and enforce validation on these parameters using additional configuration.

We can define the variables either in the main Terraform template file, or additionally, if we have lot of variables, we can define these variables in a separate variables file, ending with the `.tfvars` or `.tfvars.json` extension. At execution time, we can specify this variables file with the `–var-file` option, similar to the following. By default, Terraform uses `.tf file name` variables and loads variables from the variables file:

```
terraform apply -var-file="myvariables.tfvars"
```

If a variable is defined in multiple variable files, then the value defined in the last applied file takes precedence. Terraform loads the variables in the following order, so the value that gets loaded last takes precedence:

- System environment variables

- A `terraform.tfvars` file, if available

- A `terraform.tfvars.json` file, if available

- A `*.auto.tfvars` or `*.auto.tfvars.json` file, processed in filename lexical order

- `-var` and `-var-file` options supplied at runtime, in the order they are provided

Considering the file load order if a variable is defined in both the `terraform.tfvar` file and `terraform.tfvars.json`, the final value will be taken from the `terraform.tfvars.json` file.

Outputs

Outputs are the value that you want to highlight to the user once you apply a template. Let's suppose you are creating an AWS EC2 instance. Once this instance gets created, you want Terraform to display the ID of the volume attached to it. Terraform holds a lot more information related to your infrastructure, so using an output, you can request that information back from Terraform when that resource is created.

Also, sometimes, you may want to create a new resource, and if another resource depends on that resource, you can use an output to query the information from the first resource and then apply that information in the creation of a second resource. For example, you may have defined an EC2 instance and want the IP address of this newly created EC2 instance to pass to the other service; you can use the output to extract that information from the EC2 instance.

The following is an example of extracting the IP address of an EC2 instance referred to by the name `MyDemoInstance` name in Terraform template:

```
output "my_machine_ip" {
 value = "${MyDemoInstance.public_ip}"
}
```

Here, once we apply the Terraform template, we will get the IP address of the `MyDemoInstance` resource.

Terraform Commands

Terraform has commands to execute a Terraform template, and a Terraform binary is shipped with core configuration, and with the help of these commands, it downloads additional information and manages our infrastructure. Let's understand some of the basic commands.

Init

The Terraform `init` command is used to initialize a Terraform project within a working directory that contains a Terraform configuration file, also known as a template. When we run a `terraform init` command in a directory, it looks at the configuration file and prepares the backend for the Terraform state file. If there is no backend defined, then it uses a local state file. It looks at the providers and modules defined in the configuration file and download dependencies for those as well, based on the providers defined in the configuration.

Other Terraform commands such as `plan`, `apply`, and `destroy` depend on this `init` command; if you run these commands prior to executing `init`, they will fail.

Plan

The `terraform plan` command creates an execution plan before actually performing those changes in the environment. This command doesn't make any change to your infrastructure, but it reads the current state of the infrastructure, compares it with the changes you are making, and previews a snapshot of the changes.

If any new resource gets added as a part of this configuration change, `plan` will show a + symbol in front of the resource, and if it is something it is going to destroy, then it will show a – symbol in front of the resource. If there aren't any changes that need to be made to the environment, then Terraform will let you know that it has nothing to do.

You can use the `plan` command with the `-out` parameter, and that will save the plan information in a file, which you can share with your team to review the changes you are making. Later on, you can use this file with the `apply` command to make the changes in your infrastructure.

```
terraform plan -out=myexecutionplan.txt
```

Apply

The `terraform apply` command is used to execute the actions suggested in the plan. So basically, this command is responsible for making the necessary changes in the infrastructure. You can supply a previously created `plan` file to this command as a parameter, or you can directly execute the `apply` command without a saved plan file.

If you are not specifying a `plan` file, the `apply` command will first create an execution plan and then execute it:

```
terraform apply [options] path_to_plan_file
```

The preceding commands execute a previously saved plan file, and this command will apply the changes in same order, as planned. If you run an `apply` command without a plan file, it generates a new one, and the order of execution may be different.

The Terraform `apply` command asks for confirmation from a user before applying the changes to the environment, but using a command-line option, you can automatically apply the changes without a confirmation:

```
terraform apply -auto-approve
```

This command will not ask for any confirmation and start creating the required infrastructure, which is not recommended if you are not sure what you are doing. It is always recommended to review the plan before executing.

Destroy

The `terraform destroy` command is used to destroy the infrastructure managed by the Terraform. Terraform will cleanly delete all the resources you created using Terraform configuration. You need to be careful while using this command, as this command may result in an outage to your environment if you do not intend to delete a resource. Similar to the `apply` command, `destroy` also asks for your confirmation before making changes to your infrastructure, and you can run the following command to automatically approve the change:

```
terraform destroy -auto-approve
```

We covered the basic commands for Terraform that we will be using in the following sections, but a full list of Terraform commands can be found here: `https://www.terraform.io/cli/commands`. In the next section, we will go through the steps to configure our environment and create a basic template.

Setting up the environment

In order to start developing Terraform templates to provision our cloud resources, we need to set up our development environment.

Terraform installation

You need to follow the following instructions to set up Terraform:

1. Go to `https://www.terraform.io/downloads` and select the latest binary (1.1.2, in our case) and choose the appropriate environment as per your operating system. We will select Windows 64-bit. Terraform also has a **Terraform Cloud** option, where they manage and hold all your templates:

Figure 4.23 – Terraform downloads

2. Let's click on the **Windows 64-bit** link and download the Terraform binary:

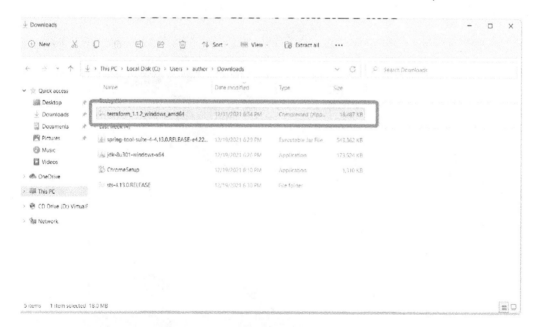

Figure 4.24 – The Terraform ZIP file

3. Right-click on the downloaded file and unzip to your preferred location. We are unzipping to `C:\Users\author\Downloads\terraform_1.1.2_windows_amd64` location:

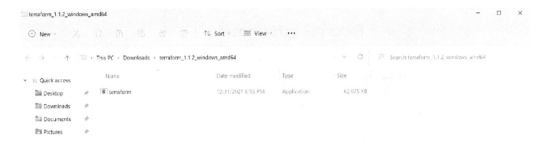

Figure 4.25 – The Terraform binary unzipped

There is only one file inside the unzipped folder, the Terraform binary. Copy this unzip location; we will add it to the Windows environment PATH variable.

4. Open the Windows search bar, type `environment`, and click on **Edit the system environment variables** link:

Figure 4.26 – The Windows environment variable search

5. In the pop-up window, click on the **Environment Variables…** button:

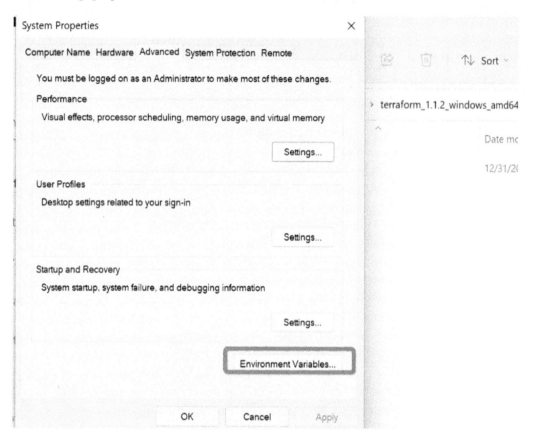

Figure 4.27 – The Windows Environment Variables… button

6.   In **Environment Variables**, click on **Path** and then the **Edit** button to modify the system path:

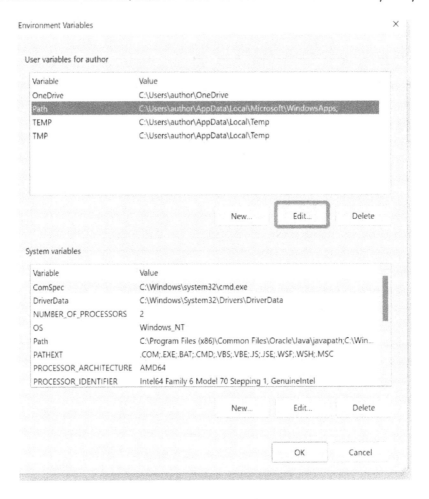

Figure 4.28 – The Windows Path variable

7. In the pop-up window, click on the **New** button:

Figure 4.29 – Editing the Windows Path variable

8. In the new line on the pop-up window, paste the Terraform binary location that we copied earlier; in our case, it is `C:\Users\author\Downloads\terraform_1.1.2_windows_amd64`.

Figure 4.30 – The Windows Path variable with Terraform

Now, click on the **OK** button on screen, which will close this popup. Click **OK** again on the next window to close the environment variables.

9.   Now, open Windows Command Prompt and type `terraform -version` to verify the Terraform installation. An output similar to the following screen will confirm that our Terraform installation is successful.

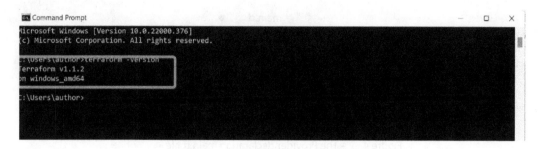

Figure 4.31 – Terraform installation verification

Now, we are done installing Terraform. In the next step, we need to start creating our templates. We will use the Visual Studio Code editor, which we installed and configured in the last chapter.

Terraform authentication

In order for Terraform to provision resources in your cloud environment, you need to provide access to Terraform with appropriate permissions. Terraform needs to authenticate with our AWS platform to create the resources that we are going to declare in our template. If you don't have the access key and secret key, then follow the instructions provided in the *Appendix* section of the book, of the book to create a pair of keys with administrative privilege. If you already have the keys, then let's proceed to the next section to start the Terraform template development.

Terraform development

We now have everything in place to get started and write a Terraform configuration file, and execute it to provision our environment. We are going to create an Amazon EC2 instance using Terraform, and we want to pass the AWS region as a parameter, using the variables section, and when our resource is created, we want to output the public IP address of the EC2 instance. Let's get started:

1. Open Visual Studio Code in an empty folder, say terraform, and let's create a configuration file in the `Terraform` directory called `main.tf`:

2. Now, let's start adding the code to the `main.tf` file; first, we will add a provider to the configuration fil, so that Terraform can download necessary settings and dependencies to connect the AWS provider to AWS services. The AWS provider needs `access_key` and `secret_key` to connect to the AWS services. There are different ways to supply `access_key` and `secret_key` using the AWS profile – using environment variables or directly specifying in the provider block, which is not a recommended way, as it can lead people to commit their credentials to public repositories.

Figure 4.32 – The Provider declaration main.tf file

3. Let's go ahead and create two environment variables in our Windows system with the names `AWS_ACCESS_KEY_ID` and `AWS_SECRET_ACCESS_KEY`, and provide the values from the programmatic user created in the previous section titled *Terraform authentication*. Terraform recognizes these variables and automatically connects to AWS services if you just specify the Terraform provider.

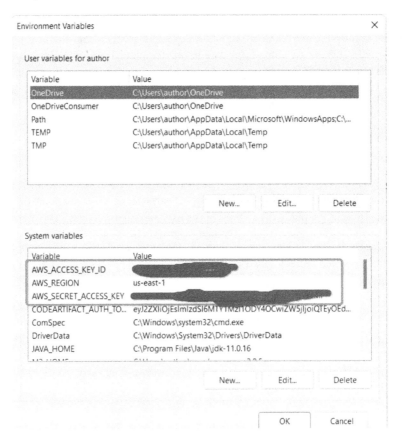

Figure 4.33 – The Terraform AWS Provider environment variables

You can also specify the default AWS region in the environment variables with the variable name AWS_REGION. In this example, we are going to supply the AWS region using parameters, which we will declare later.

4. Next, let's add a resource section to create an EC2 instance using our template. Each resource needs to be specified to what kind of provider it belongs, so the name starts with aws and then the resource type. We need to specify a name of the resource that we can refer to in other places in our template:

```
main.tf  M  X
main.tf > provider "aws"
1    provider "aws" {
2      region = var.aws_region
3    }
4
5    resource "aws_instance" "EC2DemoInstance" {
6      instance_type = "t2.micro"
7      ami = var.aws_ami_id
8    }
9
```

Figure 4.34 – AWS EC2 instance resource declaration

We have specified the type of the EC2 instance as t2.micro and the AMI as a variable, so we can specify a value at runtime while executing the Terraform template. We will declare the aws_ami_id variable in the next section.

5. Now, it is time to declare the variables we will use in provider and the resource section. Variables are declared using a variables section in the template file, followed by a name, and we can specify the type as string and a description. When we run this template, Terraform will ask us to supply this value. For aws_ami_id, we have specified a default AMI value of ami-0ed9277fb7eb570c9. If you run this code in any region other than us-east-1, make sure you change this to the appropriate AMI ID.

```
main.tf M ×
main.tf >
 1   provider "aws" {
 2     region = var.aws_region
 3   }
 4
 5   resource "aws_instance" "EC2DemoInstance" {
 6     instance_type = "t2.micro"
 7     ami = var.aws_ami_id
 8   }
 9   |
10   variable "aws_region"{
11       type = string
12       description = "Provide an aws region to apply the changes."
13   }
14
15   variable "aws_ami_id"{
16       type = string
17       description = "Provide an aws AMI id to create EC2 instance."
18       default = "ami-0ed9277fb7eb570c9"
19   }
20
```

Figure 4.35 – The variable declaration main.tf file

6. Let's go to the last section of the configuration file and add an output section to get the public IP address of the created EC2 instance after execution:

```
main.tf M ×
main.tf > output "demo_instance_ip_address"
 1   provider "aws" {
 2     region = var.aws_region
 3   }
 4
 5   resource "aws_instance" "EC2DemoInstance" {
 6     instance_type = "t2.micro"
 7     ami = var.aws_ami_id
 8   }
 9
10   variable "aws_region"{
11       type = string
12       description = "Provide an aws region to apply the changes."
13   }
14
15   variable "aws_ami_id"{
16       type = string
17       description = "Provide an aws AMI id to create EC2 instance."
18       default = "ami-0ed9277fb7eb570c9"
19   }
20
21   output "demo_instance_ip_address" {
22       value = aws_instance.EC2DemoInstance.public_ip
23       description = "Public ip of the EC2 isntance."
24   }
```

Figure 4.36 – The Terraform output declaration main.tf file

7. Now, we have everything to start, so open Command Prompt or the terminal and navigate to the Terraform directory where we saved our `main.tf` file. Let's initialize Terraform by applying the `terraform init` command to download the AWS Provider and other configurations that Terraform needs:

```
PROBLEMS   OUTPUT   DEBUG CONSOLE   TERMINAL

Copyright (C) Microsoft Corporation. All rights reserved.

Install the latest PowerShell for new features and improvements! https://aka.ms/PSWindows

PS C:\Users\author\Projects\B18011\chapter_4\terraform> terraform init

Initializing the backend...

Initializing provider plugins...
- Finding latest version of hashicorp/aws...
- Installing hashicorp/aws v4.33.0...
- Installed hashicorp/aws v4.33.0 (signed by HashiCorp)

Terraform has created a lock file .terraform.lock.hcl to record the provider
selections it made above. Include this file in your version control repository
so that Terraform can guarantee to make the same selections by default when
you run "terraform init" in the future.

Terraform has been successfully initialized!

You may now begin working with Terraform. Try running "terraform plan" to see
any changes that are required for your infrastructure. All Terraform commands
should now work.

If you ever set or change modules or backend configuration for Terraform,
rerun this command to reinitialize your working directory. If you forget, other
commands will detect it and remind you to do so if necessary.
PS C:\Users\author\Projects\B18011\chapter_4\terraform>
```

Figure 4.37 – The Terraform init command

8. Now, you can that see that Terraform has downloaded additional files in the source directory:

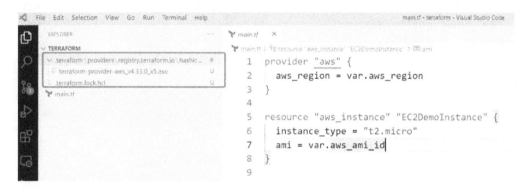

Figure 4.38 – The Terraform provider-specific setting files

9.   Now, let's execute `terraform plan` to see what changes Terraform is going to execute in our cloud environment:

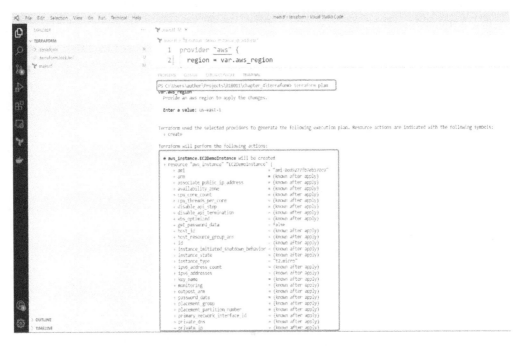

Figure 4.39 – The Terraform Plan command showing the execution plan

As you can see, we specified only two properties for the EC2 instance, but there are a lot more details needed for the EC2 instances, which are defaulted by Terraform for us. Additionally, Terraform adds a few more resources, such as the root device and the network interface, for us.

10.   Now that we know about all the changes that are going to happen in our environment, let's apply them to our environment:

Figure 4.40 – The Terraform apply command

Execute the `terraform apply` command, and it will ask for our AWS regions; provide us-east-1, as we specified in the variables. It will show us the plan regarding what it is going to apply and ask us to confirm. Type yes to confirm:

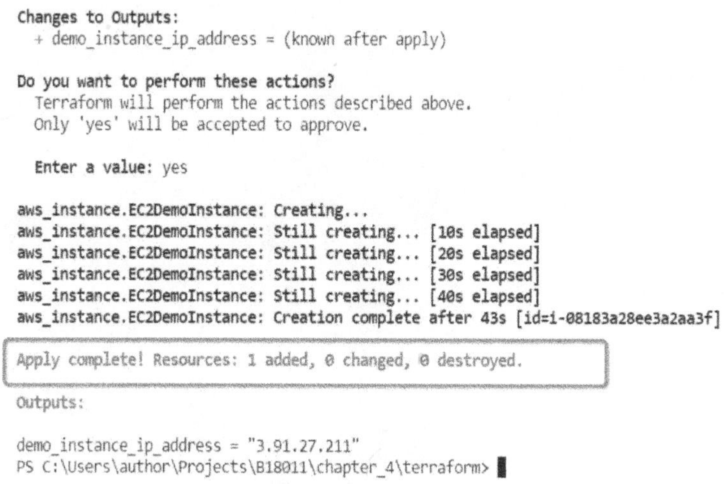

```
Changes to Outputs:
  + demo_instance_ip_address = (known after apply)

Do you want to perform these actions?
  Terraform will perform the actions described above.
  Only 'yes' will be accepted to approve.

  Enter a value: yes

aws_instance.EC2DemoInstance: Creating...
aws_instance.EC2DemoInstance: Still creating... [10s elapsed]
aws_instance.EC2DemoInstance: Still creating... [20s elapsed]
aws_instance.EC2DemoInstance: Still creating... [30s elapsed]
aws_instance.EC2DemoInstance: Still creating... [40s elapsed]
aws_instance.EC2DemoInstance: Creation complete after 43s [id=i-08183a28ee3a2aa3f]

Apply complete! Resources: 1 added, 0 changed, 0 destroyed.

Outputs:

demo_instance_ip_address = "3.91.27.211"
PS C:\Users\author\Projects\B18011\chapter_4\terraform>
```

Figure 4.41 – Terraform resource creation confirmation

As you can see, now our EC2 instance is created and has the available ID of I-0b84af8e23a421f13, and it has indicated that the public IP address of this is 54.165.204.227, as we asked it to return it to us in our output section.

11. Let's look at the Visual Studio Code editor, now that we have our terraform.tfstate created in our source directory. This is the file through which Terraform keeps the track of the environment and determines what changes needs to be applied in our infrastructure:

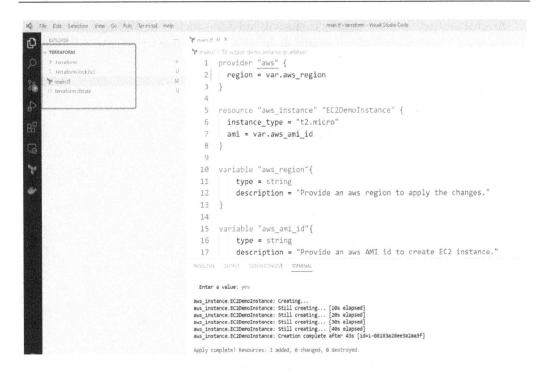

Figure 4.42 – Terraform resource creation confirmation

12. In order to validate the status of our resource, let's log in to the AWS console and go to the EC2 dashboard. As you can see in the following screenshot, our EC2 instance with the ID of i-06a7cd998ae6c9c46 is available and running with a public IP address of 34.229.124.216, similar to our Terraform output:

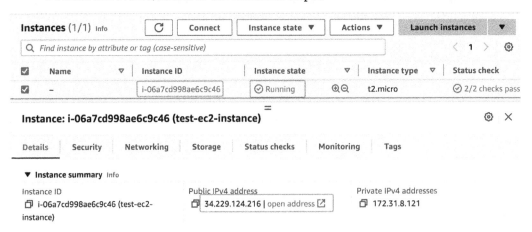

Figure 4.43 – The EC2 dashboard showing the EC2 instance running

13. We are able to successfully validate our Terraform configuration, so let's tear down the resources created by us so that there are no additional charges applied to our account. Terraform provides the `destroy` command, similar to `apply`, to gracefully clean up our environment. Type the `terraform destroy` command and press *Enter* on Command Prompt:

Figure 4.44 – The Terraform destroy command

14. Provide the AWS region value as `us-east-1` and press *Enter* for the variable value. Confirm yes for the `Do you really want to destroy all resources` question:

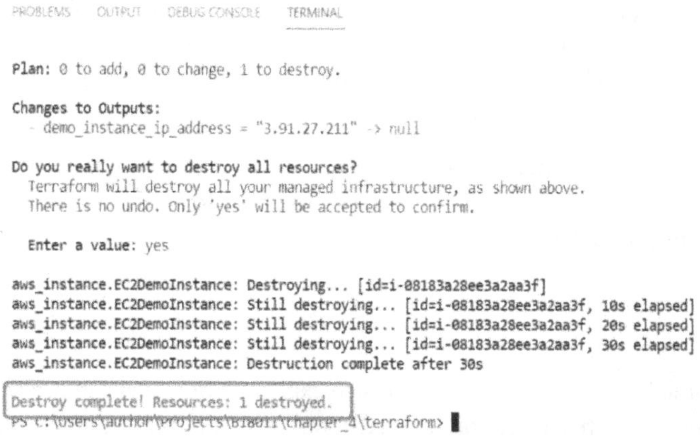

Figure 4.45 – The Terraform destroy confirmation

15. The preceding screenshot confirms that the EC2 instance that we created earlier has been destroyed. Now, let's go ahead and validate this through the EC2 dashboard:

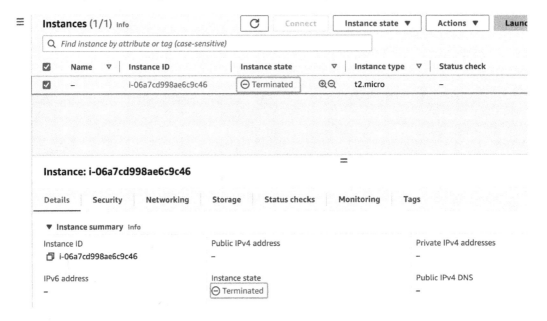

Figure 4.46 – EC2 dashboard terminated instance

The EC2 dashboard shows that the instance with the ID of i-06a7cd998ae6c9c46 is in a terminated status.

> **Note**
>
> All the source code and example code we used in this chapter can be downloaded from the GitHub repository: https://github.com/PacktPublishing/Building-and-Delivering-Microservices-on-AWS/tree/main/chapter_4.

You now have a fair understanding of the idea of Terraform and saw how easy it is to create resources using it. Creating resources in other cloud providers are also easy; you just need to specify the additional provider in the configuration/template file.

Summary

In this chapter, we explained IaC and its benefits. We also explained how you can use these tools to automate and version your infrastructure. You learned about AWS CloudFormation templates and experienced how to apply these in an AWS environment.

We covered Terraform to create an AWS cloud infrastructure and explained the different concepts. We walked you through, step by step, setting up a development environment to create Terraform templates, and then we created and destroyed the AWS resources.

In the next chapter, we will focus on the AWS CodeCommit tool provided by AWS to manage source repositories in the cloud. We will understand what a source code repository is and how AWS provides a scalable solution.

Part 2:
Build the Pipeline

This section provides in-depth knowledge about source code repositories, how to use AWS code commit, and how to set up automated code reviews using the AWS CodeGuru service. This section also explains AWS CodeArtifact and the AWS CodeBuild service.

This part has the following chapters:

- *Chapter 5, Creating Repositories with AWS CodeCommit*
- *Chapter 6, Automating Code Reviews Using CodeGuru*
- *Chapter 7, Managing Artifacts Using CodeArtifact*
- *Chapter 8, Building and Testing Using AWS CodeBuild*

5

Creating Repositories with AWS CodeCommit

In this chapter, you will learn about version control system (VCS) and how they help us to manage our code effectively. You will cover the Git VCS and examine the selection of Git-based VCS available. You will understand some of the basic commands you need to know to manage a Git repository. This chapter focuses on the AWS CodeCommit service and its benefits. As we progress, we'll be committing the sample application code we developed in *Chapter 3*, to the newly created CodeCommit repository we'll be creating in this chapter.

In this chapter, we will be covering the following topics:

- What is a VCS
- Introduction to Git
- What is CodeCommit?
- Creating and managing CodeCommit repositories
- Checking in a sample application to CodeCommit
- Migrating GitHub repositories to CodeCommit

A developer's life revolves around code. All of us who have worked as developers at some point in our careers have developed some source code. When we work alone, we don't need to worry much about sharing the code, but if we are working within a team, we need to share the code with our colleagues and integrate the changes that everyone makes. This may prompt some questions to pop up in your head, such as how do we share code with the team? What happens when someone else modifies the same area of source code that you're working on? What happens to all the hard work we did in the case of a system failure? To answer all of these questions, we use a source code to VCS. So, let's learn about VCSs and how they solve the problems just mentioned.

What is a VCS?

A VCS is software available to developers to manage and share source code. These systems are centrally hosted and provide storage, backup, and versioning of source code through a graphical user interface or a set of text commands.

A VCS keeps track of the changes made by each developer, making changes traceable, allowing them to be compared against previous versions, and making it possible to revert to those previous versions of the source code if a mistake is discovered. The following sections cover some of the benefits provided by VCS in more detail.

Traceability

Traceability provides insights into the changes made to files over a period of time by different individuals. It gives information about what has been changed since the original version and who made those changes. Traceability helps to audit the changes on a file.

History

VCSs provide the valuable documented history of a file. This helps us find out who edited the file and what date and time the changes were made. It also provides a summary of the reasons for changes having been made to files.

Branching and merging

A good benefit of version control systems is that they provide support for simultaneous updates on the same set of files without affecting each other. Each contributor can work on their independent stream of changes known as a branch. Later on, these branches are merged together and provide traceability of who made which changes.

Increased efficiency

A source version control system increases the efficiency of the team, making it easy for them to share the changes with each other and refer to previous versions of the code. Code can easily be reverted to earlier versions, and the team can integrate and merge changes without difficulty.

Easy merge conflict resolution

When two or more contributors are working on the same file, it is highly likely that the changes they make to the file will conflict. This is known as a **merge conflict**. Modern version control systems provide an easy way to manage and resolve merge conflicts, showing you exactly what changes are causing the conflict and giving you the opportunity to resolve the issue by deciding if one change is more "correct" than the other. You can also accept both changes and then resolve the conflict once the merge is done. This gives flexibility to developers to easily manage merge conflicts, providing a line-by-line comparison of the changes.

Transparent code reviews

A version control system tracks the changes made by each developer, providing a clear view of who is making each change. Each developer can see what their colleagues are committing to the repository, and can easily identify any potential problem in the code because it's easy to review what was done. As a best practice, providing a meaningful commit message along with a change ID is recommended. A bad commit message is one that doesn't provide any detail, such as `Bug fix`, whereas a good commit message would reference the bug (potentially in a bug tracking system) and explain what the update did to fix the issue.

Reduced duplication and errors

Version control systems promote the **Don't Repeat Yourself (DRY)** principle by increasing reusability of the code. A team using a version control system will produce less duplicate code as developers can look through the changes made by their colleagues and reuse code that has already been written, tested, and reviewed instead of re-inventing the wheel. This leads to fewer errors and increases teams' output.

Increased team productivity

Modern version control systems provide graphical user interfaces to easily search, navigate, branch, merge, tag, and review code, which speeds up the overall pace of development. Most version control systems also integrate seamlessly with build tools, so any code updates can easily be built and deployed to the any DEV/QA/PROD environment.

Compliance

Version control systems provide detailed tracking of change history, which can be used for auditing. An audit trail of changes can be preserved and used for regulatory compliance. You can restrict read/write-level permissions for documents based on a user's role within the organization, defining a granular level of access for users so that only those users who need to do so are able to read code, commit code, delete a branch, or merge code.

Now that you've got a fair idea about what source version control systems are and what benefits they provide, let's next understand the Git-based version control system, before deep diving into the AWS version control system, CodeCommit.

Introduction to Git

Git is a popular, free, open source, and distributed version control system that allows you to make changes to everything from small files to very large projects. Being distributed means you don't have to rely on a single server: each machine in use has a full copy of the repository along with the version history, so if a server died for some reason, you could restore the code from the copy present on another machine.

Git stores data as a stream of snapshots when you commit a change. If a file is not changed then Git doesn't store that file again, instead just using the reference to the previous identical file. When committing a changed file, Git basically captures a picture of what all changed files look like at that moment and stores a reference of that snapshot.

The following diagram shows an architectural view of a VCS repository where computers one and two have the same copy of the repository. A developer is working on a copy of the file and when they are ready, they can stage the file to the local repository using the `git commit` command. Staging means that you have marked a file's current version, to create a snapshot so file can be committed or rollback with current set of changes. Once the developer is ready to share it, they execute `git push` to send the changes to the centralized server, and the other developers can perform the `git pull` command to retrieve the code to their own local repositories:

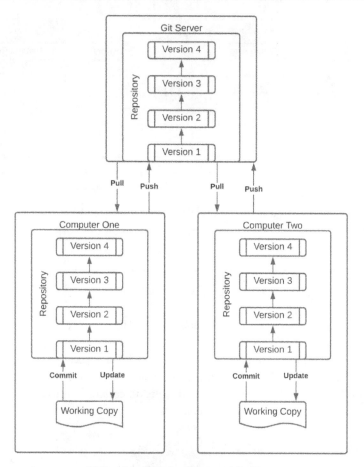

Figure 5.1 – Git repository architecture

You can see in the preceding diagram that Git maintains a local copy of the repository, allowing you to revert and stage changes as needed.

The next diagram explains the basic Git workflow and the series of commands used when a developer creates a file and maintains it using Git. Files are added to the staging area using the `git add` command and once files are staged, they are pushed to the local Git repository using `git commit`.

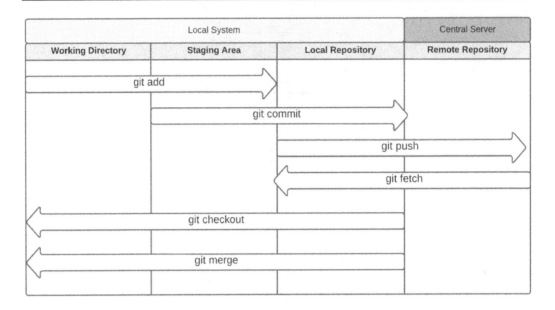

Figure 5.2 – Git repository workflow

`git commit` stores the file in the local Git repository, which is later synced to the remote repository using the `git push` command. The `git fetch` command is used to sync the remote repository changes to the local Git repository, while the `git checkout` command is used for loading the file/changes from a different branch to the working copy.

Git commands

There are several Git UI tools available, such as AWS CodeCommit, that aim to simplify interaction with the Git system, but at its core, Git is a command-line-based VCS. While it's possible to interact with Git using a GUI, you must understand the text commands as well. This section is likely to be redundant for readers who are already using Git, so if you already know Git commands, feel free to skip this section. There are a lot of commands, so let's try first to understand some of the most popular ones that see frequent use:

- Creating repositories:

  We can either create a new Git repository locally or clone an existing remote repository. The following commands are used to start working with Git:

  - `git init`

  This command is used to initialize a Git repository in an existing directory. Later on, you can push this repository to the remote server.

  - `git remote add origin URL`

This command is used to attach an existing local Git repository to a remote repository specified using the remote URL. Using this technique, you can change the remote server for an existing Git repository.

- `git clone URL`

This command is used to fetch an existing Git repository to your local system. This command connects to the remote URL provided and downloads the repository contents into your local system, allowing you to work without an internet connection. Later on, you can sync this local copy to the remote server.

- Changing files:

As a developer, you have to make changes to files and sync your changes with your peers. This section explains the different commands you use to make changes to repositories:

- `git add`

This command adds a file to the staging area and snapshots the file in order to prepare it for versioning. You need to specify the filename or filename pattern as part of this command, such as `git add *.java`. This command will add all files with the `.java` extension.

- `git commit`

This command takes your changes from the staging area and adds them to the local Git repository. This command records the snapshot created in Git system adds to the version history permanently. This command takes the -m parameter to specify a commit message, where we provide the reason for the change for easier reference in the commit log later. An example is `git commit -m "This is a commit message"`.

- `git log`

This command gets the latest version of the history for the working branch. You can use this command to see what changes have been made and by whom.

- `git diff`

This command shows the difference between two branches of the repository. If you don't supply any other parameters, it will show you changes between your local repository and what's been committed. If you want to compare the differences between specific branches, you need to specify the branch names as parameters to this command, such as `git diff branch_one branch_two`.

- `git show`

This command shows the metadata about a particular commit and what changes have been made as part of that commit. You need to provide the commit ID as follows: `git show commit_id`.

- Repository synchronization:

    The following commands are used to synchronize your local repository with the remote server repository, so you can push your changes or take updates from your fellow developers:

    - `git fetch` – This command downloads all the history from the remote tracking server branches, downloading all commits, files, and references from the remote server to your local repository without altering your local copy. The `fetch` command is a safe way to review commits and see what has been changed before integrating changes with your local repository.

    - `git merge` – This command lets you merge a remote branch into your current local repository.

    - `git push` – This command lets you push your current local branch changes to the remote branch. This command uploads the changes from your local computer to the remote centralized Git server.

    - `git pull` – This command is a combination of `git fetch` and `git merge`. This command updates the current local working branch from the remote branch, downloading and merging all new commits and history from that remote branch.

Now that you've got an understanding of the basic `git` commands used to check code in to Git repositories, let's next focus on the AWS CodeCommit service offered by Amazon for Git-based repositories.

What is CodeCommit?

Git is an open source version control system. Multiple products have been built using Git and provide enterprise-grade support, including GitHub Enterprise, Bitbucket, GitLab, and AWS CodeCommit.

AWS CodeCommit is a fully functional Git version control system that makes it easy to collaborate. It is a fully managed version control system that hosts secure Git-based repositories. It is a highly scalable and secure source control service, allowing us to create secure private repositories in the cloud without the need to provision the hardware to host the Git software or worry about maintenance of the system.

Let's learn some of the benefits AWS CodeCommit provides:

- **Fully managed**: AWS CodeCommit service is fully managed, so you don't need to purchase hardware or launch an EC2 instance to run the Git software to host your source code repositories. AWS will automatically back up the repositories and scale disk and CPU resources as required based on the number of users accessing the repository and the load being generated. This automatic scaling by design allows the service to meet the growing needs of your project.

- **Data security**: Your Git repositories in AWS CodeCommit are fully secured and no one gets access to the repository until and unless you provide it explicitly. CodeCommit is fully integrated with AWS **Identity and Access Management (IAM)**, so you can easily define who will have what level of access to your source code repositories. AWS CodeCommit provides data encryption at rest and in transit, so your data is always secure and easily accessible to you. CodeCommit supports both the HTTPS and SSH protocols for communication with your repository.

- **High availability**: AWS CodeCommit is a highly available service; it is deployed on redundant hardware and is built with fault-tolerant architecture to allow it to survive hardware failures. AWS CodeCommit is built on top of existing proven AWS services including S3 and DynamoDB. The CodeCommit service is designed to be accessible all the time, so you can work freely and access your code with high speed whenever you need.

- **Affordable pricing**: AWS CodeCommit uses a pay-as-you-go model, so you only pay for the period and resources you use. You don't have to pay for the hardware or compute power being used to run the service, only for the active users and the storage cost of the repository data. AWS CodeCommit offers 5 active users for free, with defined limits for storage.

- **Collaborate on code**: AWS CodeCommit makes it easy for you to collaborate with your team using branching, merging, and code reviews. CodeCommit also makes it simple to integrate with the AWS CloudGuru service for automated code reviews of Java and Python code.

- **Faster development lifecycle**: AWS CodeCommit gives you choice to store your code within the same region that you use to deploy your production workload, reducing network latency and allowing faster deployment, in turn creating faster development cycles.

- **Workflow support**: AWS CodeCommit generates events that can be used to trigger SNS notifications or AWS Lambda functions, allowing you to design workflows to perform complex activities on these events or simply send notifications.

- **Easy to learn**: Git is a widely used tool and the developer community is well versed in it. AWS CodeCommit has been designed using Git, just deployed in the cloud, so you don't have to learn any new skills to use it. All Git commands are supported by CodeCommit and you can use any Git-based client to connect to your AWS CodeCommit repository. Because CodeCommit implements all Git commands, it can be used with any IDE that supports the Git protocol.

Now that we have understood the benefits provided by AWS CodeCommit, let's look at some of the service's limitations. Every great product comes with limitations, and these might not be a concern for most users. The following are some of the limits you have with the AWS CodeCommit service:

- By default, you can have a maximum of 1,000 repositories per account, but this can be increased to 25,000 by making a service limit increase request to AWS Support.

- While using the CodeCommit console or API or the AWS CLI, any individual file can be a maximum of 6 MB in size. In the case of an individual blob file, the size shouldn't be more than 2 GB.

- There is a maximum of 100 files allowed in a single commit.

- There is a maximum of 1,000 open pull requests allowed at a time.

- There is a maximum of 10 triggers allowed in a single CodeCommit repository.

A full list of limitations can be found at `https://docs.aws.amazon.com/codecommit/latest/userguide/limits.html`.

Creating repositories

Now that we have understood the benefits and limitations of the AWS CodeCommit service, let's start playing with it and get some hands-on experience. We're going to create a repository and then push our sample microservice code to it. Follow these steps to create a new repository:

1. Log in to the AWS console and search for `CodeCommit`:

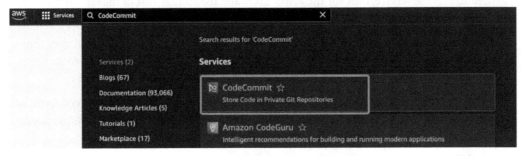

Figure 5.3 – AWS console CodeCommit search

2. This will take you to the CodeCommit home page, where you can create new repositories or configure rules for those repositories that have already been created:

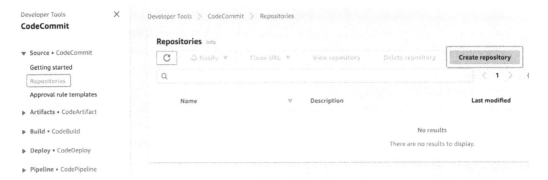

Figure 5.4 – CodeCommit home page

3. Click on the **Create repository** button to create a new repository for our `aws-code-pipeline` microservice. The screen that appears will ask you to provide a repository name and a description; please use the values shown in the following screenshot:

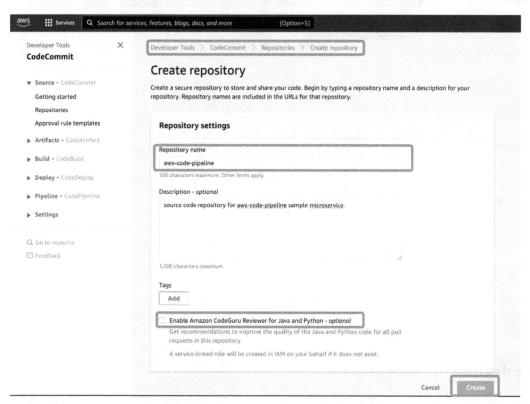

Figure 5.5 – CodeCommit create repository

This screen also gives you the option to add tags to your repository, which you can use for billing purposes or to easily find resources. Optionally, you can enable CodeGuru Reviewer for this source code repository. We will be discussing the CodeGuru service in later chapters. Click on the **Create** button to create the repository.

4. Once you click the **Create** button, a Git repository will be created in the AWS CodeCommit service:

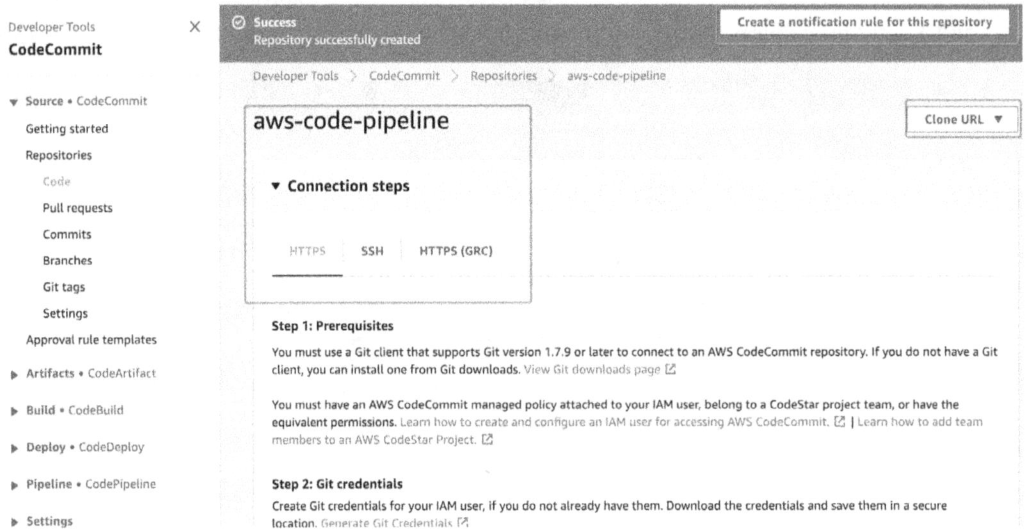

Figure 5.6 – CodeCommit repository create confirmation

As you can see in the screenshot, there are multiple connection options to interact with our CodeCommit repository. In the next section, we will obtain the Git credentials in order to push our sample microservice code to the CodeCommit repository.

CodeCommit credentials

Now that you have your first AWS CodeCommit repository, it is time to push some code to your repository. We need to first create a CodeCommit user and provide them access using CodeCommit permissions. By default, a root user is not allowed to push code to a CodeCommit repository using Git.

You can follow the steps mentioned in *Appendix, Creating an IAM Console User*, to create a user for CodeCommit. Once you've created the `console_user` user with admin access, let's obtain the Git credentials for this user using the following steps:

1. Log in to your AWS Console using `console_user` and the password you set as referred to in the preceding paragraph. Click on the username dropdown and select **Security Credentials**:

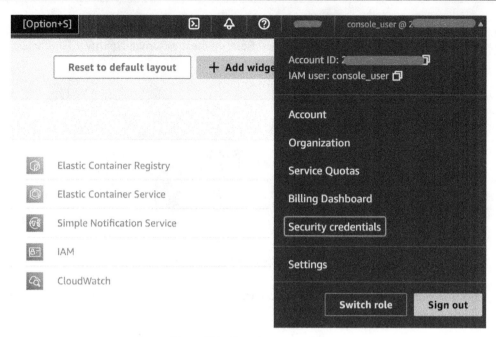

Figure 5.7 – Security credentials

2. Now, you should have landed on the security credentials screen. Click on the **AWS CodeCommit credentials** tab to get the Git username/password for CodeCommit:

Figure 5.8 – AWS console – My security credentials

3. You can either upload the SSH public key to authenticate or you can generate a username/password for CodeCommit Git repositories. Here, we use the username/password mechanism. Let's click on the **Generate credentials** button:

Figure 5.9 – AWS CodeCommit credentials

4. Now, as you can see, we have our Git credentials, which we can use to push code to the CodeCommit Git repository:

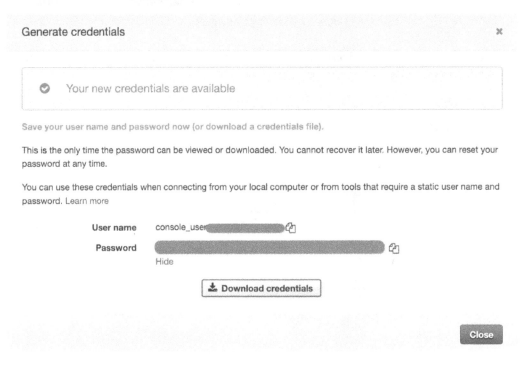

Figure 5.10 – CodeCommit generated credentials

Note down your username and password – we will use this to push our code to the Git repository.

Pushing code to a CodeCommit repository

We now have our CodeCommit repository created, and the credentials set up. Next, let's push the source code of our `aws-code-pipeline` microservice we created in *Chapter 3*, to this new repository.

Before we can push the source code, we need a Git client on our development machine. Follow the instructions provided in *the Git Installation section in the Appendix*, to install a Git client. Once this is done, let's work through the following instructions to push the code to our CodeCommit repository:

1. Open a terminal/Command Prompt and go to your microservice source code folder, which we created in *Chapter 3*.

2. We need to initialize the Git repository in our source code directory. Type `git init` and press *Enter*:

▣ Command Prompt

```
C:\Users\author\Projects\aws-code-pipeline>git init
Initialized empty Git repository in C:/Users/author/Projects/aws-code-pipeline/.git/

C:\Users\author\Projects\aws-code-pipeline>_
```

Figure 5.11 – Git repository initialization

3. In order to connect our remote CodeCommit repository to our local source code directory, we need to obtain the CodeCommit repository URL. So, let's log in to the AWS console with the `console_user` username we created earlier in this chapter.

4. Go to the CodeCommit repository and copy the URL by clicking the **HTTPS** link. Our CodeCommit Git repository URL is `https://git-codecommit.us-east-1.amazonaws.com/v1/repos/aws-code-pipeline`:

Figure 5.12 – aws-code-pipeline repository URL

5. Now go to terminal/Command Prompt and set the remote URL for our local code repository. Run the following command:

```
git remote add origin https://git-codecommit.us-east-1.amazonaws.com/v1/repos/aws-code-pipeline
```

```
C:\Users\author\Projects\aws-code-pipeline>git remote add origin https://git-codecommit.us-east-1.amazonaws.com/v1/repos
/aws-code-pipeline

C:\Users\author\Projects\aws-code-pipeline>
```

Figure 5.13 – Git add remote URL

6. Now we need to add our source code files to the staging area of the local Git repository. We use the `git add` command to do this. We are using a filename pattern that will add all files in our source code directory to the Git repository staging area except the one declared in the `.gitignore` file:

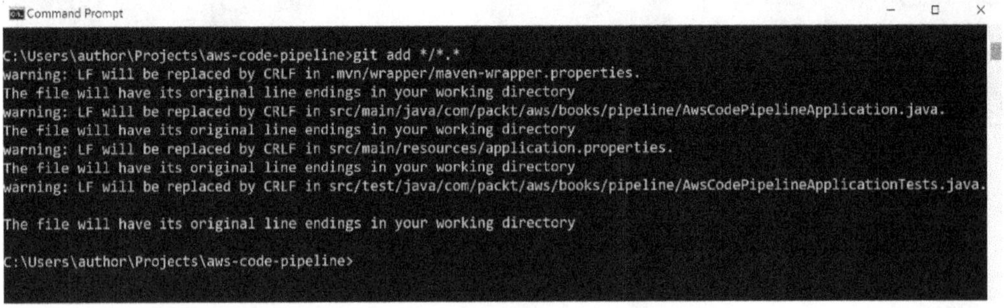

```
Command Prompt                                                                                    –    □    ×

C:\Users\author\Projects\aws-code-pipeline>git add */*.*
warning: LF will be replaced by CRLF in .mvn/wrapper/maven-wrapper.properties.
The file will have its original line endings in your working directory
warning: LF will be replaced by CRLF in src/main/java/com/packt/aws/books/pipeline/AwsCodePipelineApplication.java.
The file will have its original line endings in your working directory
warning: LF will be replaced by CRLF in src/main/resources/application.properties.
The file will have its original line endings in your working directory
warning: LF will be replaced by CRLF in src/test/java/com/packt/aws/books/pipeline/AwsCodePipelineApplicationTests.java.

The file will have its original line endings in your working directory

C:\Users\author\Projects\aws-code-pipeline>
```

Figure 5.14 – Git add files command

As you can see in the preceding screenshot, Windows systems use **Carriage Return plus Line Feed (CRLF)**, while Linux and Mac systems use the **Line Feed (LF)** character for line end in files. Now, let's commit these added files to the local Git repository from the staging area. We use the `git commit` command to stage the changes to the local repository and add a message with the –m parameter:

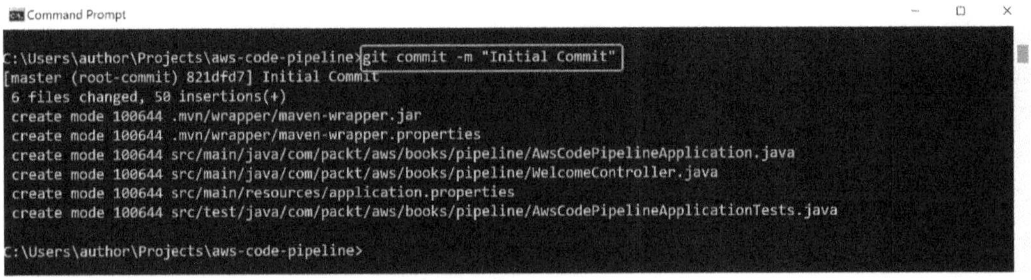

```
Command Prompt                                                                                    –    □    ×

C:\Users\author\Projects\aws-code-pipeline>git commit -m "Initial Commit"
[master (root-commit) 821dfd7] Initial Commit
6 files changed, 50 insertions(+)
create mode 100644 .mvn/wrapper/maven-wrapper.jar
create mode 100644 .mvn/wrapper/maven-wrapper.properties
create mode 100644 src/main/java/com/packt/aws/books/pipeline/AwsCodePipelineApplication.java
create mode 100644 src/main/java/com/packt/aws/books/pipeline/WelcomeController.java
create mode 100644 src/main/resources/application.properties
create mode 100644 src/test/java/com/packt/aws/books/pipeline/AwsCodePipelineApplicationTests.java

C:\Users\author\Projects\aws-code-pipeline>
```

Figure 5.15 – Git repository workflow

7. It is now time to push the source code to the CodeCommit repository we created from the local repository. Let's use the `git push -u origin master` command to push the code. With this, we create a remote master branch and push the code to it.

> **Note**
>
> If you are using the Git repository on your system for the first time, `git push` may ask you to provide Git credentials. If so, use the Git credentials we created earlier in this chapter.

```
Command Prompt
C:\Users\author\Projects\aws-code-pipeline>git push -u origin master
Enumerating objects: 28, done.
Counting objects: 100% (28/28), done.
Delta compression using up to 2 threads
Compressing objects: 100% (16/16), done.
Writing objects: 100% (28/28), 53.59 KiB | 6.70 MiB/s, done.
Total 28 (delta 0), reused 0 (delta 0), pack-reused 0
To https://git-codecommit.us-east-1.amazonaws.com/v1/repos/aws-code-pipeline
 * [new branch]      master -> master
branch 'master' set up to track 'origin/master'.

C:\Users\author\Projects\aws-code-pipeline>
```

Figure 5.16 – Git repository push command

As you can see now, the source code is committed to the repository and our local Git repository is now tracking the remote master branch.

8. Let's now go to the AWS console and look at our repository to confirm that our files are available in AWS CodeCommit:

Developer Tools > CodeCommit > Repositories > aws-code-pipeline

aws-code-pipeline 🔔 Notify ▼ | main ▼ | Create pull request   Clone URL ▼

aws-code-pipeline Info   Add file ▼

Name

📁 src

📄 .gitignore

📄 pom.xml

Figure 5.17 – aws-code-pipeline CodeCommit repository

You should see that all of your source code is available in the CodeCommit repository. CodeCommit supports a lot more functionality of Git beyond just pushing code to it. Let's explore these other features of AWS CodeCommit next.

Beyond the basics of CodeCommit

We now have our microservice source code available in the CodeCommit repository. Now, let's learn about more features provided by the CodeCommit Git repositories.

Creating branches

A branch in a Git system is a pointer to a snapshot of the changes. The following are the steps to create a new branch in a CodeCommit repository:

1. In order to create a new branch in CodeCommit, click on the **Branches** link in the left panel. This will take you to a page displaying all of your existing Git branches. On this screen, click on the **Create branch** button:

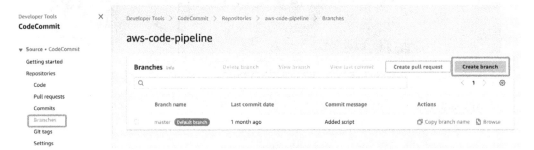

Figure 5.18 – CodeCommit create branch repository

2. In the pop-up window that appears, enter a new branch name and specify a source to create the branch from. Then, click the **Create branch** button to create the branch:

Figure 5.19 – CodeCommit create new branch popup

3. If successful, you should see your newly created branch listed on the following screen.

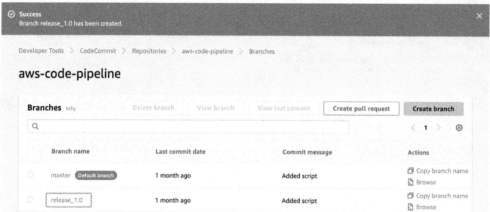

Figure 5.20 – aws-code-pipeline CodeCommit branches

After creating a branch, it is time to add some files to it so we can later create a pull request and merge our changes to the master branch.

Adding files to the repository

In order to add a file to the CodeCommit Git repository, you can use either the `git add` command and then push the file using a Git client, or alternatively, you can use the CodeCommit user interface to do this. Follow these steps to add a file to the newly created `release_1.0` branch using the AWS console:

1. Select **release_1.0** from the branches on the following screen. Click on the **Add file** button to add a new `ReadMe.md` file to the `release_1.0` branch:

Figure 5..21 – CodeCommit – Add files through console

2. Add the file contents and provide the filename as `ReadMe.md`, along with the author name and email address. You also need to provide a commit message before clicking on the **Commit changes** button:

Figure 5.22 – ReadMe.md file commit through the CodeCommit console

3. Now, as you can see, the file is committed and the changes are available in the `release_1.0` branch of our repository:

Figure 5.23 – ReadMe.md file commit confirmation

Now we need to create a pull request from our `release_1.0` branch to the master branch. We'll learn how to do that in the next section.

Pull requests and code merges

A pull request is the process of initiating a code merge request from one branch to another branch. In a pull request, a developer/contributor takes their code changes and tries to add them to the main repository branch. You can set up an approver list, and those approvers can then approve any changes before they are merged with the target branch. Let's create a pull request to merge the changes to the master branch using the following steps:

1. In order to create a pull request, you can either click on the branch and then click on the **Create pull request** button, or alternatively, you can click the **Pull requests** link in the left panel, as can be seen in the following figure:

Figure 5.24 – CodeCommit pull request

The page shown in the preceding screenshot lists any pending pull requests from you or others related to your repository. You can filter these pull requests using the **Open pull requests** drop-down menu.

2. Click on the **Create pull request** button as shown in the preceding screenshot, then specify the **Destination** and **Source** branches from the relevant dropdowns and click the **Compare** button:

Figure 5.25 – CodeCommit create new pull request

3. Once you click the **Compare** button, it will show you whether the pull request is mergeable or if it has some conflicts. This screen also displays what changes have been made to the files, highlighting all added or deleted contents by comparison with the file version available in the **Destination** branch master:

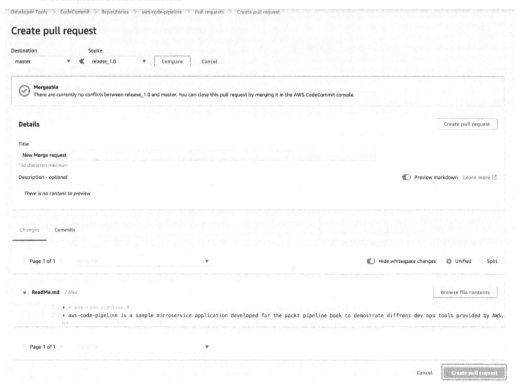

Figure 5.26 – CodeCommit pull request comparison

Click on the **Create pull request** button to create the pull request for merging.

4. The following screen shows the opened pull request and displays the option to merge the request. This screen also gives a summary of the approvals required to merge this request, how many commits are part of this pull request, and what the changes are:

Figure 5.27 – CodeCommit pull request pending screen

5. Once you click on the **Merge** button, the next screen shows the different merge options and explains what they mean. By default, the **Fast forward merge** option is selected. Click on the **Merge pull request** button:

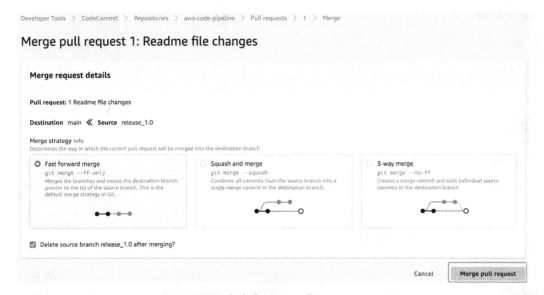

Figure 5.28 – CodeCommit pull request merge

6. Now you can see that our `release_1.0` source branch has been merged with the target master branch:

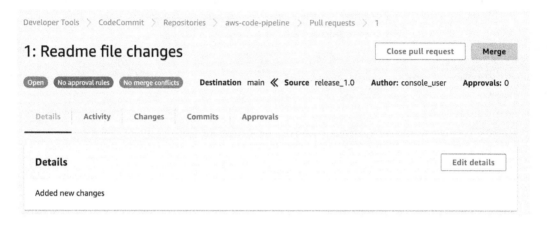

Figure 5.29 – CodeCommit pull request merge confirmation

7. Now, let's go to our repository and select the master branch. As you can see, the `ReadMe.md` file we added is now available in the master branch:

Figure 5.30 – CodeCommit merged changes in master branch

Now all the code changes are available in the master branch; if we have to look at the history of all changes made so far in our repository, we can see that under `Commits`. Let's see what a commit is.

Commits

A commit is a set of changes you stage as a single unit; you can think of a commit as a milestone in your repository timeline. To see the history of changes within the repository, follow these steps:

1. Click on the **Commits** link in the left panel to view a list of the commits merged in your selected branch:

Figure 5.31 – List of commits

2. You can click on the links in the **Commit ID** column to see the details of the changes made as part of this commit:

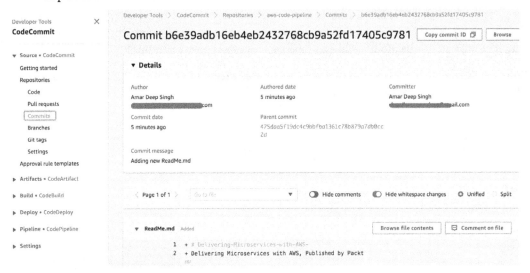

Figure 5.32 – Commit change details

3. In order to visualize the commits in a repository, click on the **Commit visualizer** tab under **Commits**. This screen shows the visualization of the commits and how these commits got merged into the repository over time:

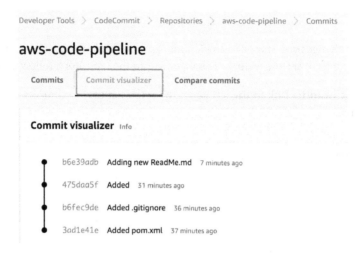

Figure 5.33 – Commit change visualizer

You can compare different commits and see what has been changed using the **Compare commits** tabs. Next, let's learn about Git tags and see how they are different than branches.

Git tags

A Git tag is a specific point in the history of a Git repository that can be referred back to create an exact copy of the changes in the past. For example, a tag can be created for a production release. Once a tag is created, it can't be modified like a branch and does not have a commit history.

CodeCommit supports tagging repositories. You can see existing tags by clicking on **Git tags** in the left panel, as shown in the following figure:

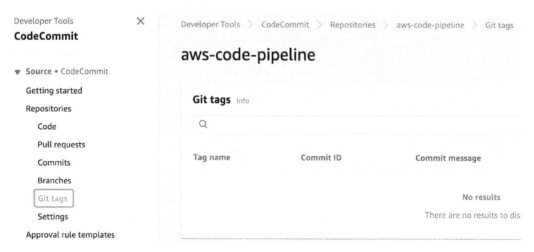

Figure 5.34 – CodeCommit Git tags

To create a new tag for the repository, you can use the `git tag` command from a Git client.

Repository settings

The repository settings in CodeCommit allow a user to change the details of a repository. You can change the name and description, set the default branch, delete the repository, create triggers, and add new resource tags. You can also enable Amazon CodeGuru Reviewer from this screen:

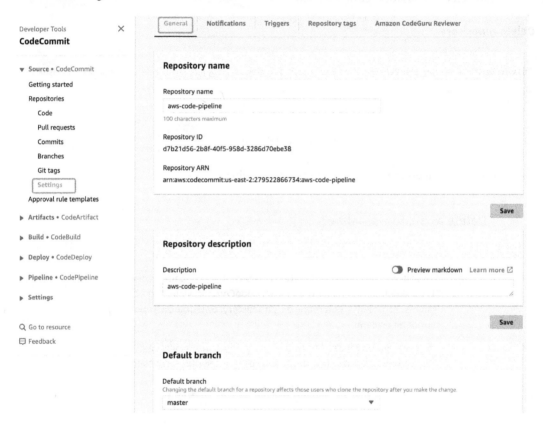

Figure 5.35 – CodeCommit repository settings

The settings screen has different tabs for setting up different things. Let's learn how to delete an existing repository.

Deleting a repository

The **Delete repository** option allows a user to delete their repository data from the CodeCommit service. Deleting a repository is a destructive action and will remove all branches, tags, commits, and history related to the repository, so only do this if you really need to irretrievably delete the repository:

1. To delete a repository completely from AWS CodeCommit, click on the **Delete repository** button on the repository settings screen:

Delete repository

This will delete the repository in AWS CodeCommit, including all branches, triggers, comments, pull requests, and history.

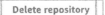

Figure 5.36 – Delete repository

2. This action will prompt you for confirmation as follows:

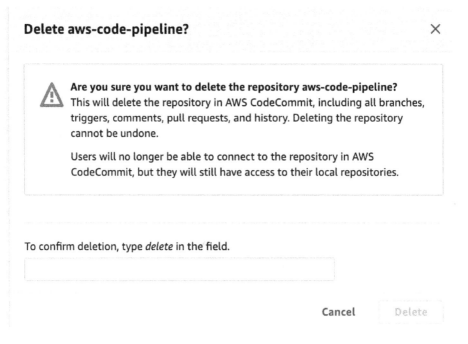

Figure 5.37 – CodeCommit delete repository confirmation prompt

3. Here, we do not intend to delete the repository we created, so click on the **Cancel** button. If you did want to delete it, however, then you would type the confirmation text `delete` into the relevant field and click the **Delete** button.

The repository settings also allow you to enable notifications for different Git events. Click on the **Notifications** tab on the settings screen to enable notifications.

Repository notifications

A CodeCommit repository notification is an alert you can set up to receive upon a specific CodeCommit event. CodeCommit uses AWS **Simple Notification Service** (**SNS**) to deliver these notifications. Follow these instructions to create a notification for your CodeCommit repository:

1. Click on the **Create notification rule** button on the **Notifications** tab:

Figure 5.38 – CodeCommit repository notification rules

2. Provide a name for the notification. Choose the type of details you are interested in receiving in the notification. All Git events supported by the CodeCommit are listed for you to choose from, so choose the events you are interested in receiving notifications for. You need to select the target type for these notifications and specify the target:

Figure 5.39 – CodeCommit notification rule creation

CodeCommit supports **AWS Chatbot (Slack)** and **SNS topic** as the target types. Here, we select **SNS topic** and choose the **NotifyMeNow** SNS topic. Follow the instructions provided in the *Creating an SNS Topic* section in the *Appendix* to create an SNS topic to receive notifications. Lastly, click on the **Submit** button.

3. The following screen shows the confirmation that `Repo_Notification` has been created and is active:

Developer Tools > CodeCommit > Repositories > aws-code-pipeline > Settings
aws-code-pipeline
General
Notifications
Triggers
Repository tags
Amazon CodeGuru Reviewer
Notification rules
Notifications rules for aws-code-pipeline
View details
Edit
Delete
Create notification rule
1
Notification rule
Created
Sending notifications
Notifying events
RepoNotification
1 minute ago
Enabled
11 events

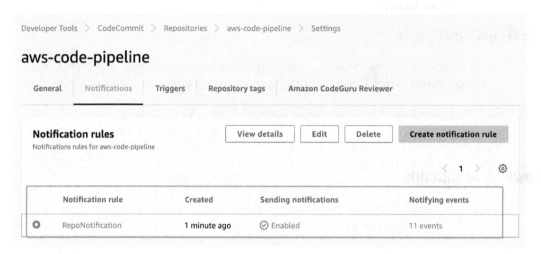

Figure 5.40 – Notification rule creation confirmation

4. The following is an SNS notification message for a CodeCommit push event:

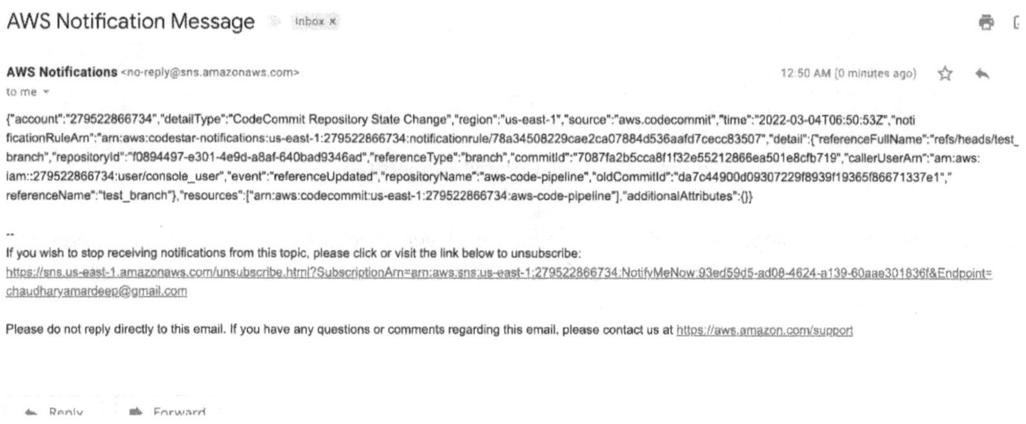

AWS Notification Message
Inbox ×
AWS Notifications <no-reply@sns.amazonaws.com>
12:50 AM (0 minutes ago)
to me
{"account":"279522866734","detailType":"CodeCommit Repository State Change","region":"us-east-1","source":"aws.codecommit","time":"2022-03-04T06:50:53Z","notificationRuleArn":"arn:aws:codestar-notifications:us-east-1:279522866734:notificationrule/78a34508229cae2ca07884d536aafd7cecc83507","detail":{"referenceFullName":"refs/heads/test_branch","repositoryId":"f0894497-e301-4e9d-a8af-640bad9346ad","referenceType":"branch","commitId":"7087fa2b5cca8f1f32e55212866ea501e8cfb719","callerUserArn":"arn:aws:iam::279522866734:user/console_user","event":"referenceUpdated","repositoryName":"aws-code-pipeline","oldCommitId":"da7c44900d09307229f8939f19365f86671337e1","referenceName":"test_branch"},"resources":["arn:aws:codecommit:us-east-1:279522866734:aws-code-pipeline"],"additionalAttributes":{}}
--
If you wish to stop receiving notifications from this topic, please click or visit the link below to unsubscribe:
https://sns.us-east-1.amazonaws.com/unsubscribe.html?SubscriptionArn=arn:aws:sns:us-east-1:279522866734:NotifyMeNow:93ed59d5-ad08-4624-a139-60aae301836f&Endpoint=chaudharyamardeep@gmail.com
Please do not reply directly to this email. If you have any questions or comments regarding this email, please contact us at https://aws.amazon.com/support
Reply
Forward

Figure 5.41 – SNS code push notification

Now, whenever any activity takes place on your `aws-code-pipeline` repository, you will receive an SNS notification similar to that shown in the preceding screenshot. In the next section, let's take a look at CodeCommit triggers.

CodeCommit Triggers

CodeCommit triggers are commonly used for sending emails to subscribers or notifying external build systems on CodeCommit events. CodeCommit triggers use AWS SNS and the AWS Lambda service:

1. Click the **Create trigger** button on the **Triggers** tab.

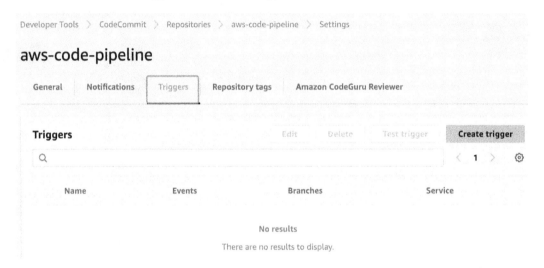

Figure 5.42 – CodeCommit Triggers tab

2. On the screen that appears, provide a name for the trigger, select the event type you are interested in, and choose the branch name. We choose **AWS Lambda** to process the trigger. Click the **Create trigger** button:

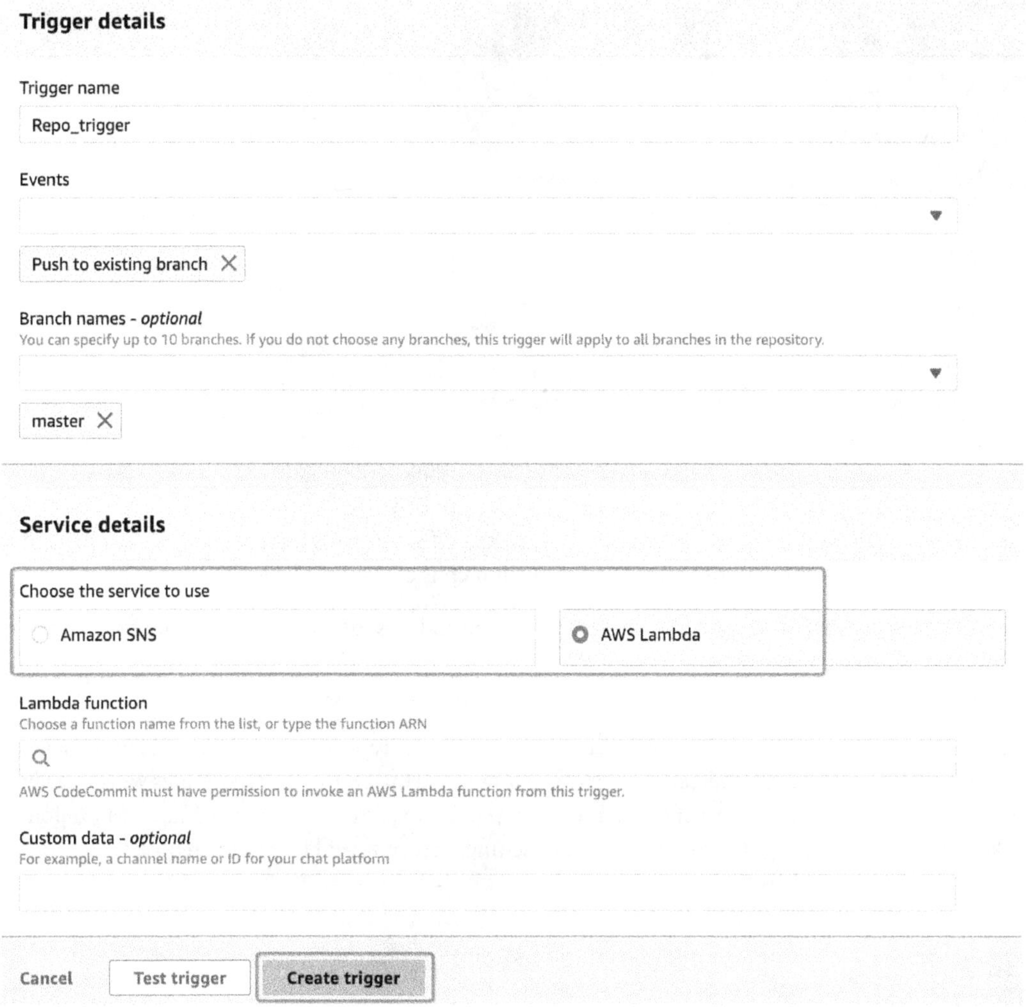

Trigger details

Trigger name

Repo_trigger

Events

▼

Push to existing branch ✕

Branch names - *optional*
You can specify up to 10 branches. If you do not choose any branches, this trigger will apply to all branches in the repository.

▼

master ✕

Service details

Choose the service to use

○ Amazon SNS   ● AWS Lambda

Lambda function
Choose a function name from the list, or type the function ARN

🔍

AWS CodeCommit must have permission to invoke an AWS Lambda function from this trigger.

Custom data - *optional*
For example, a channel name or ID for your chat platform

Cancel   Test trigger   **Create trigger**

Figure 5.43 – CodeCommit Trigger details screen

3. The following screen shows the confirmation that a trigger is now associated with the master branch of our repository:

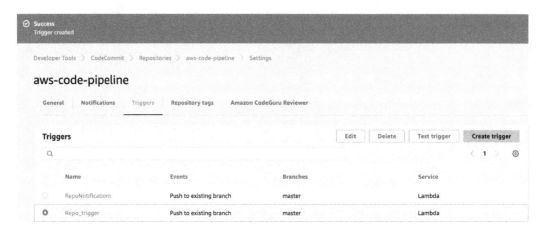

Figure 5.44 – CodeCommit create trigger confirmation

Based on the events we configured, a trigger will be kicked off whenever a suitable CodeCommit event occurs such as code is checked in to a branch , new branch is created or deleted.

Every AWS resource supports tagging, so let's see how we can add tags to an AWS CodeCommit repository.

Repository tags

Repository tags allow us to associate key-pair metadata with the repository. Tags allow us to manage, identify, organize, and search resources in the AWS environment. By tagging our repositories we can easily categorize them by product or department and cost can be billed to correct cost center within organization. Click on the **Repository tags** tab on the repository settings screen to add tags to the repository:

Developer Tools > CodeCommit > Repositories > aws-code-pipeline > Settings > Repository tags

Edit Repository tags

Repository tags Info

A tag is a label that you assign to an AWS resource. Each tag consists of a key and an optional value. You can use tags to help manage and secure your resources or to help track costs.

Key	Value - *optional*	
cost center	2332423	Remove tag

Add tag

Cancel   **Submit**

Figure 5.45 – CodeCommit resource tagging

Once you have created the tags here, you can click on the **Submit** button and your tags will be added to the repository. You can then use these tags for budgeting and other purposes.

Amazon CodeGuru Reviewer

AWS CodeGuru is an **Artificial Intelligence (AI)**-powered service that integrates with CodeCommit and performs code analysis, code reviews, and vulnerability assessments. We can associate a CodeCommit repository with CodeGuru by clicking the associate repository. We will be covering more details of the CodeGuru service in later chapters. To associate the `aws-code-pipeline` repository with **Amazon CodeGuru Reviewer**, click on the **Associate repository** button:

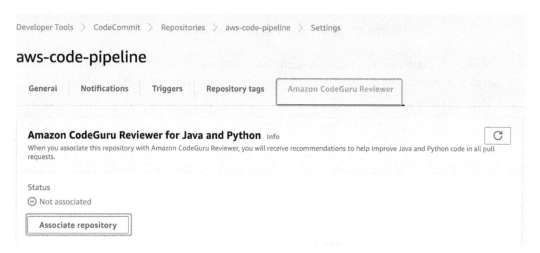

Figure 5.46 – CodeGuru Reviewer

Once you associate your repository with the CodeGuru service, the service will scan your repository and present the results for review. You will learn more about the CodeGuru service in the coming chapters. In the next section, let's learn how to configure approval rules.

Approval rule templates

As the name suggests, approval rules allow us to control the pull request merge process, define how many approvals are required, and who can make approvals before a pull request can be merged.

CodeCommit allows us to create and manage approval rule templates, which can be associated or disassociated with one or more repositories. Let's create a new approval rule template for our CodeCommit repository, requiring a minimum of two approvers before allowing users to merge a request with the master branch. Follow these instructions:

1. Click on the **Approval rule templates** link in the left panel and click on the **Create template** button:

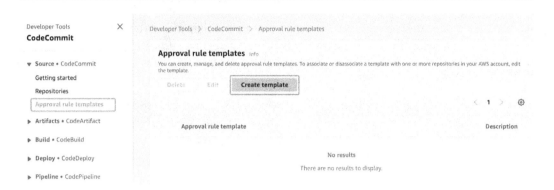

Figure 5.47 – CodeCommit approval rule templates

2. Now fill out details for the name, description, and number of approvals needed before merging. Approval pool member types can be an IAM user name, a role name, or the ARN of a role/user name:

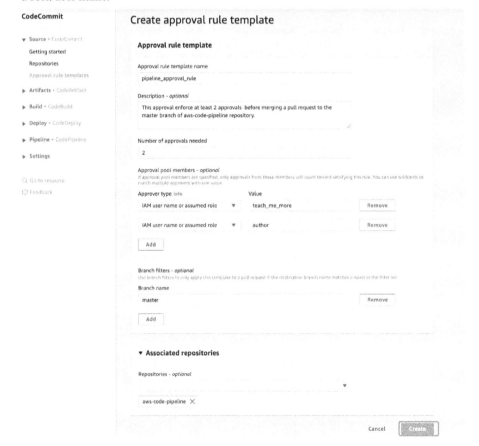

Figure 5.48 – CodeCommit create approval template

3. Enter `master` under the **Branch name** field, associate our `aws-code-pipeline` repository, and click on the **Create** button. The following screen shows a confirmation that our approval rule template has been created successfully. The same approval rule template can be applied to multiple repositories:

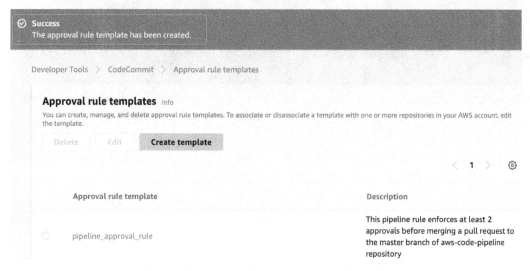

Figure 5.49 – Approval template create confirmation

Approval rule templates can be applied to all the repositories, so you don't have to define them each time, you can just associate new repositories. We have covered pretty much all aspects of CodeCommit, so now let's learn how you can migrate an existing GitHub repository to the AWS CodeCommit service.

Repository migration to CodeCommit

There are several public and private Git repository providers, including GitHub, Bitbucket, and GitLab, but AWS CodeCommit provides a lot of additional benefits as discussed earlier in this chapter. Migrating an existing Git-based repository to AWS CodeCommit is really easy, as we'll now demonstrate:

1. We have a sample GitHub repository hosted at `https://github.com/teach-me-more/sample-github-repo.git`. In order to migrate this repository to AWS CodeCommit, we first need to clone the existing GitHub repository to our local workstation using the `git clone` command and navigate to the cloned directory:

```
Command Prompt

C:\Users\author\Projects>git clone https://github.com/teach-me-more/sample-github-repo.git
Cloning into 'sample-github-repo'...
remote: Enumerating objects: 8, done.
remote: Counting objects: 100% (8/8), done.
remote: Compressing objects: 100% (7/7), done.
remote: Total 8 (delta 1), reused 0 (delta 0), pack-reused 0
Receiving objects: 100% (8/8), 5.46 KiB | 310.00 KiB/s, done.
Resolving deltas: 100% (1/1), done.

C:\Users\author\Projects>cd sample-github-repo

C:\Users\author\Projects\sample-github-repo>_
```

Figure 5.50 – Sample GitHub repository cloning

2. As the next step, we need to create a new CodeCommit repository using the steps provided earlier in this chapter. We are using the same name as our GitHub repository:

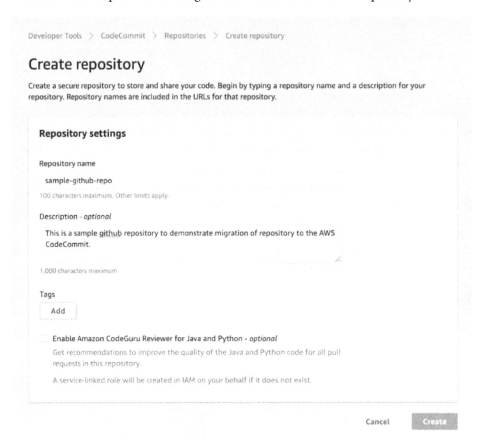

Figure 5.51 – CodeCommit create example migration repository

3. Now copy the newly created repository URL from CodeCommit. In our case, it is `https://git-codecommit.us-east-1.amazonaws.com/v1/repos/sample-github-repo`. Then, run the `git push` command using the CodeCommit URL from your local workstation repository:

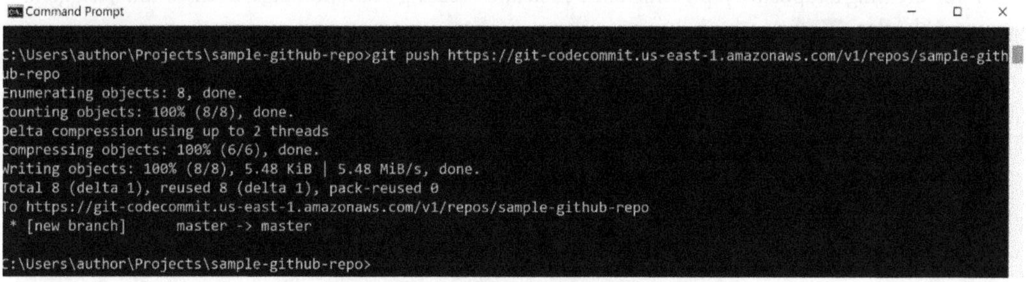

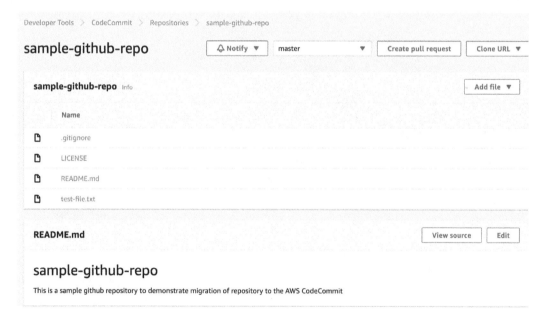

Figure 5.52 – GitHub repository push to the CodeCommit repository

4. As we can see from the preceding screenshot, the code has been pushed successfully. Let's go to the AWS CodeCommit console and verify the successful migration of the new repository:

Figure 5.53 – CodeCommit migrated repository

As we can see, our repository from GitHub has successfully been migrated to AWS CodeCommit. Using the same steps we used here, you can migrate any Git-based repository to AWS CodeCommit.

Summary

In this chapter, we learned about version control systems, Git-based repositories, and their basic usage.

We learned about the AWS CodeCommit service and the advantages it provides over other providers. We covered creating a repository, pushing code to it, and creating branches, tags, notifications, triggers, and approval rules.

In the next chapter, we will focus on the AWS CodeGuru service and learn how to perform automated code reviews on a CodeCommit repository.

6

Automating Code Reviews Using CodeGuru

In this chapter, you will learn about the AWS CodeGuru service and how this service helps you to develop high-quality, highly performant code. You will learn how CodeGuru Reviewer can help you to find issues in your code and how CodeGuru Profiler helps you to run code in a production environment and find performance issues. This chapter will walk you through enabling code reviews for the sample application repository. In this chapter, we will be covering the following topics:

- What is AWS CodeGuru?

- Understanding CodeGuru Reviewer

- The benefits and limitations of CodeGuru Reviewer

- Integrating CodeGuru with CodeCommit

- Integrating CodeGuru with GitHub

- Understanding CodeGuru Profiler

- The benefits and limitations of CodeGuru Profiler

- Sample application monitoring with CodeGuru Profiler

In the previous chapter, we learned that once we develop code, we can create pull requests for our fellow developers, and they can review it and provide feedback. This code review process is manual and time-consuming, and often, people make mistakes in finding issues in code. Catching performance and security issues is tricky. The AWS CodeGuru service makes it possible to automatically detect issues in your code and speed up the overall development.

What is AWS CodeGuru?

AWS CodeGuru is a service provided by Amazon to automatically perform code reviews and application profiling. CodeGuru uses machine learning and artificial intelligence to detect code quality and security-related issues and provide recommendations to fix those issues. The CodeGuru service also provides runtime monitoring of the application in production and detects any anomalies in application behavior.

> **Application profiling**
>
> Application profiling is a mechanism to understand the performance of an application at runtime. In application profiling, you collect different metrics related to CPU usage, memory usage, and the time taken to execute a piece of code. In application profiling, you get a view of the application's behavior and see how your code is performing in certain conditions and then you use that data to improve the quality of the code and **mean time to restore** (**MTTR**) metrics.

The AWS CodeGuru service has two components: CodeGuru Reviewer and CodeGuru Profiler. *Figure 6.1* shows how you can use both of these components to improve the code quality and the application's performance.

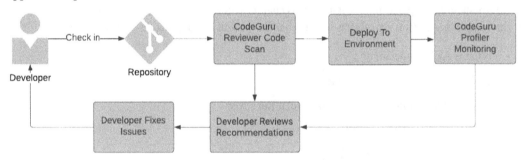

Figure 6.1 – CodeGuru workflow

You can set up your repository to perform CodeGuru Reviewer scans, and once the developer checks code or creates a pull request, the CodeGuru Reviewer scans will perform the analysis, thus providing security recommendations and finding bugs or quality issues. The developer can fix those issues and once your team is comfortable with the code quality, you can deploy your code and set up CodeGuru Profiler to monitor your environment and learn about the runtime performance. Based on the runtime analysis, CodeGuru Profiler provides recommendations that developers can apply to improve the application's performance. Let's learn more about CodeGuru Reviewer and CodeGuru Profiler next.

CodeGuru Reviewer

CodeGuru Reviewer allows you to scan your source code and find vulnerabilities and performance optimization opportunities. It uses program analysis and machine learning techniques to detect any code defects and provide recommendations.

At a high level, these are the types of checks AWS CodeGuru Reviewer provides:

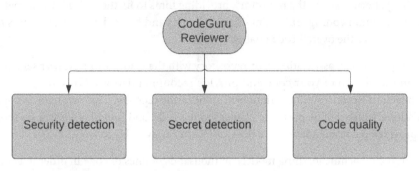

Figure 6.2 – CodeGuru Reviewer recommendation types

Let's look at all of these categories one by one and understand the capabilities provided by each CodeGuru category.

Security detection

CodeGuru Reviewer performs automated reasoning analysis on code and detects security flaws between different method calls in the request chain. CodeGuru Reviewer helps to detect and fix the **Open Web Application Security Project (OWASP)** top 10 common code vulnerabilities.

CodeGuru helps to detect and correct the usage of Java and Python crypto libraries and ensures that correct hashing algorithms are used to protect information. CodeGuru scans the code and makes sure that the user input is validated properly from untrusted sources and that no sensitive information is leaked and logged in logs. CodeGuru also helps to protect against application security vulnerabilities such as JSON injection, LDAP injection, log injection, and API-related security issues using the AWS security best practices.

Secret detection

CodeGuru Reviewer performs secret detection in configuration files or source code. It uses machine learning to find hardcoded passwords, API keys, SSH keys, JSON web tokens, access tokens, and database keys. CodeGuru Reviewer finds these secrets in your repository and provides instructions on how you can configure these credentials using the AWS Secrets Manager service.

Code quality

CodeGuru Reviewer helps to improve code quality by identifying concurrency, resource leaks, code duplication, and code reusability issues. It provides code quality improvement suggestions based on a number of sources, including AWS best practices, language-specific recommendations for Java, and Python best practices.

CodeGuru performs human-like code reviews using machine learning and identifies any deviations from the coding patterns used in the repository, providing hints to fix those. CodeGuru analyzes the code and finds common coding and maintainability issues and helps development teams to solve those issues to improve the overall code quality.

As a developer, you start by associating your repository with the CodeGuru Reviewer service, which will trigger an initial scan of your code and provide recommendations. Once the initial scan is complete, CodeGuru Reviewer will continuously monitor the repository for commits and update its recommendations, which will be posted to changes to the code, and provide feedback on those. You can see the provided recommendations on the CodeGuru reviewer UI console.

Amazon has trained machine learning models on thousands of lines of code in order to generate its recommendations. You can tell CodeGuru Reviewer whether the provided recommendations are helpful or not, and this feedback is used to improve the machine learning algorithms.

The benefits of CodeGuru Reviewer

CodeGuru Reviewer provides the following benefits:

- CodeGuru Reviewer enables development teams to find issues before they reach production.

- CodeGuru Reviewer generates code comments line by line, wherever it finds an issue, so it is easy for the developer to fix those issues.

- CodeGuru Reviewer not only finds defects in code but also provides potential fix suggestions, so developers can quickly make changes, thus reducing the lead time it takes to research and fix any issues.

- CodeGuru Reviewer provides easy integration with several Git-based code repositories, named CodeCommit, Bitbucket, GitHub, and GitHub Enterprise.

- You can enable CodeGuru Reviewer as part of your CI/CD pipeline and ensure that each release is free of security flaws using CodeGuru Security Detector.

- CodeGuru Security Detector ensures that you are using the best practices for other AWS services such as **AWS Key Management Service (AWS KMS)** and **Amazon Elastic Cloud Compute (Amazon EC2)**, **application programming interfaces (APIs)**, common Java or Python crypto, and **Transport Layer Security (TLS)/Secure Socket Layer (SSL)** libraries and provides recommendations for using them and preventing security holes in your code.

- CodeGuru Reviewer helps development teams to focus on reviewing business logic implementation while common programming errors and bugs can be identified by the CodeGuru Reviewer service. It can be configured to automatically perform code reviews as part of pull requests raised by developers.

The limitations of CodeGuru Reviewer

CodeGuru Reviewer is an evolving product, and it comes with a set of limitations. As of 2022, CodeGuru Reviewer has the following shortcomings:

- CodeGuru Reviewer is limited to only the Java and Python programming languages
- CodeGuru Reviewer can only integrate with Git-based repositories such as GitHub, GitHub Enterprise, Bitbucket, and AWS CodeCommit
- CodeGuru Reviewer Security Detector only supports Python 3.x and Java 8 through Java 11 to identify issues

Now that we have understood the benefits and limitations of CodeGuru Reviewer, let's focus on understanding how to use the CodeGuru Reviewer service.

> **Note**
>
> CodeGuru Reviewer is not free, but a free tier is available for a limited period of time for smaller repositories. CodeGuru Reviewer charges a fixed price per month for each repository plus a fee for every 100,000 lines of code, so be aware that executing the following examples may result in a charge against your account. There are several other code scanning tools available on the market for providing insights into your code and helping you fix code quality and security-related issues. SonarQube, Fortify, Snyk, and BlackDuck are some of the code-scanning tools that cover different aspects of code. We will be learning more about Snyk in upcoming chapters and integrating it into our AWS CodePipeline.

CodeGuru Reviewer in action

There are multiple ways that you can interact with CodeGuru Reviewer. You can access it in the following ways:

- The **AWS Command Line Interface (AWS CLI)**
- The CodeGuru Reviewer API
- The AWS **Software Development Kit (SDK)**
- The Amazon CodeGuru Reviewer UI console

In this section, you are going to learn how to associate a repository with CodeGuru Reviewer using the AWS UI console.

Associating a repository

Follow these steps to configure a GitHub repository to integrate with CodeGuru Reviewer:

1.   Log in to your AWS UI console and search for the Amazon CodeGuru service.

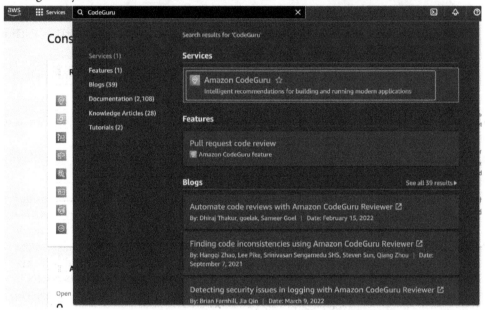

Figure 6.3 – AWS console CodeGuru service search

2.   Now in the **Getting started** section, select CodeGuru Reviewer and click the **Get started** button.

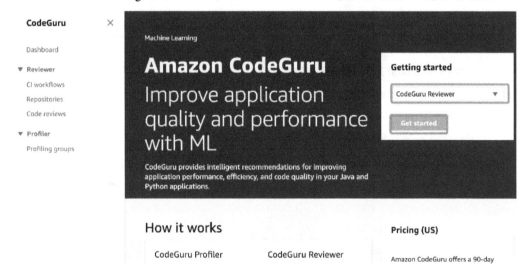

Figure 6.4 – AWS console CodeGuru service Getting started page

3. This takes you to the **Associate repository** section. On this screen, you can choose your repository type. CodeGuru Reviewer provides integration with four types of Git-based repositories: AWS CodeCommit, Bitbucket, GitHub or GitHub Enterprise Cloud, and Git hosted by GitHub Enterprise Server. As you can see, as soon as we select AWS CodeCommit as the repository type, all our CodeCommit repositories available in that region are listed for selection.

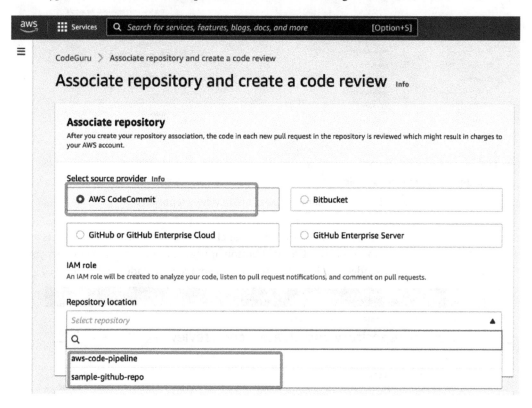

Figure 6.5 – CodeGuru Reviewer CodeCommit repository selection

We can simply click on the **Associate repository and run analysis** button and move on. But our available repositories don't have any vulnerabilities or code issues, so we are going to run an analysis on a GitHub repository that we developed to showcase how CodeGuru Reviewer works.

4. You can find the GitHub repository at this URL: `https://github.com/teach-me-more/codeguru-reviewer-sample.git`. We have left a few vulnerabilities related to the unverified input in the sample repository code.

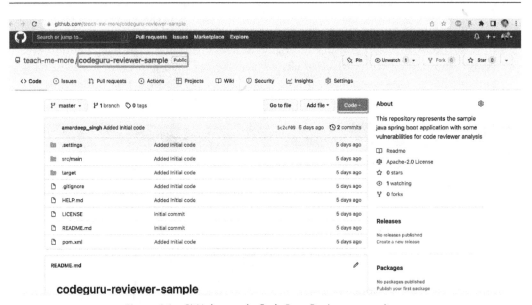

Figure 6.6 – GitHub sample CodeGuru Reviewer repository

5. Let's select GitHub as a source provider on the **Associate repository** screen. As soon as you select GitHub or GitHub Enterprise Cloud, a button appears on the screen to connect your GitHub account to the CodeGuru Reviewer service, so your source code can be analyzed.

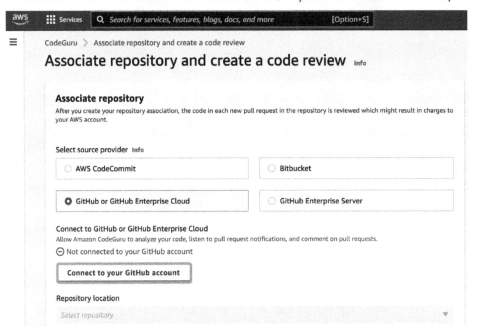

Figure 6.7 – CodeGuru Reviewer GitHub connection

Once you click on the **Connect to your GitHub account** button, a pop-up window will open, and GitHub will ask you to enter your credentials.

6. In the pop-up window, provide your GitHub credentials and click the **Sign in** button.

Figure 6.8 – GitHub login pop-up window

7. Once CodeGuru is able to connect to GitHub, the list of repositories will update and you will be able to see a list of your repositories. Select the codeguru-reviewer-sample repository. You can fork this repository from this URL: https://github.com/teach-me-more/codeguru-reviewer-sample.git.

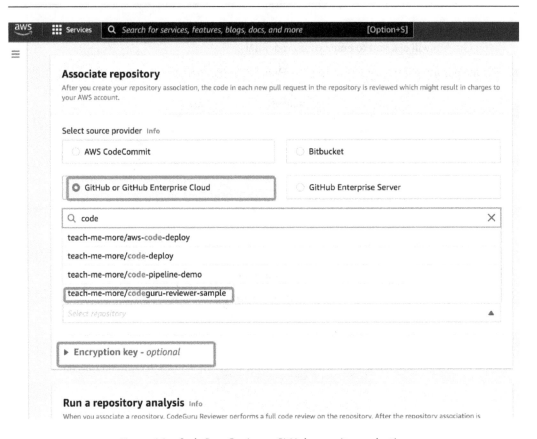

Figure 6.9 – CodeGuru Reviewer GitHub repository selection

Once you select the repository, you can expand the **Encryption key** section and customize the encryption for the source code. We are leaving the default option. CodeGuru automatically encrypts the data using an AWS-managed key.

8. When running Reviewer, you need to select the source code branch on which you want to perform the initial code analysis report. For demonstration, we are selecting the master branch. You can give your scan a custom name, or if you leave it blank, AWS CodeGuru Reviewer will generate a name for your scan. Now, click on the **Associate repository and run analysis** button to start scanning your source code. This action is going to cost you some money, which depends on the base price, the number of scans you perform, and the number of lines of code you scan:

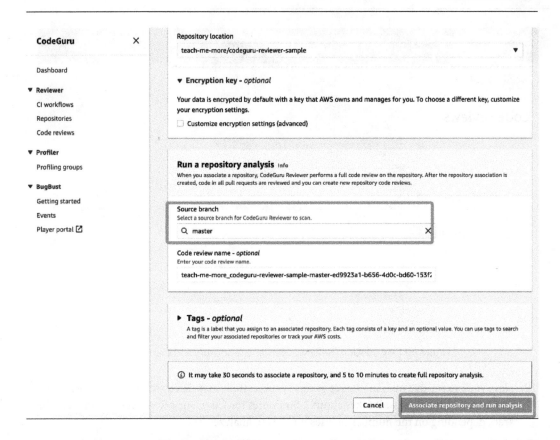

Figure 6.10 – CodeGuru Reviewer branch selection

9.   Now you can see that your repository is being associated with CodeGuru Reviewer. It takes about 30 seconds for it to be associated. Until then, the status will be **Pending**:

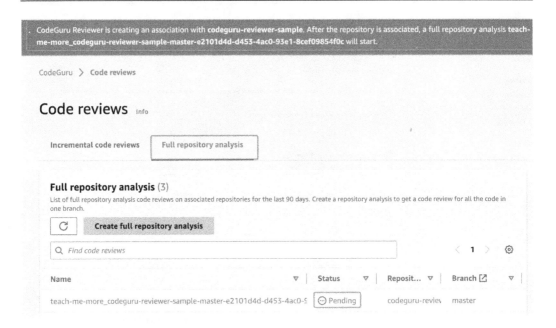

Figure 6.11 – CodeGuru Reviewer pending repository association

10. Now, you can see that our GitHub repository is successfully associated with CodeGuru Reviewer and a scan is in progress. It takes about 3-5 minutes for CodeGuru Reviewer to complete the scan, depending on the number of lines it needs to analyze:

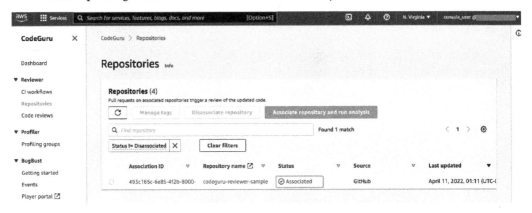

Figure 6.12 – CodeGuru Reviewer associated repository

11. In order to view the CodeGuru scan report, we need to navigate to the code reviews using the link in the left panel. Click **Code reviews** and click on the **Full repository analysis** tab. Here, you can see all the scans run on the different repositories. The **Incremental code reviews** tab is for the scan reports related to a particular pull request. For our sample repository, `codeguru-reviewer-sample`, you can see that the scan is complete and we can see that there are three recommendations provided by CodeGuru Reviewer:

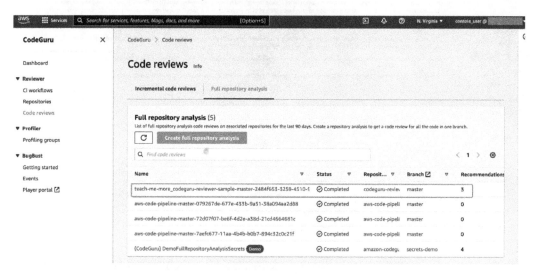

Figure 6.13 – CodeGuru Reviewer completed full analysis

12. Once you click on recomendation link, it will take you to the detailed report page, where you can see the scan details and list of findings with recommendations:

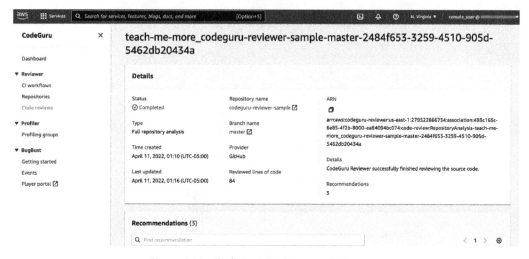

Figure 6.14 – CodeGuru Reviewer analysis report

As you can see in *Figure 6.15*, CodeGuru Reviewer explains the issue with our Java code in `HomeController.java` on line number 36. In summary, it explains that we are performing an operation on a non-validated user input field, received in our REST controller:

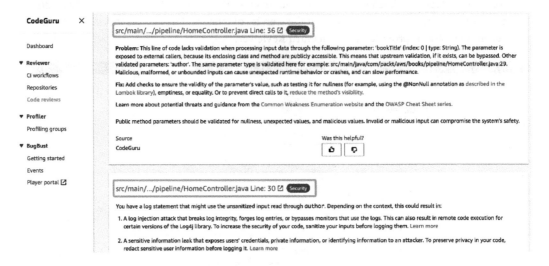

Figure 6.15 – CodeGuru Reviewer scan recommendations

13. CodeGuru categorized this code issue as a security issue as per OWASP standards. If you click on the link, it will take you to GitHub and point you to the exact line of the code where this problem exists:

```
 github.com/teach-me-more/codeguru-reviewer-sample/blob/master/src/main/java/com/packt/aws/books/pipeline/HomeController.java#L36-L36
22
23        @GetMapping("/book")
24        public ResponseEntity<Void> createBook(@RequestParam String bookTitle,
25                        @RequestParam(required = false) String author) {
26            Random random = new Random();
27            Integer id = random.nextInt();
28            BookEntity book = new BookEntity();
29            if (author != null) {
30                LOGGER.info("Author name is {}", author);
31                book.setAuthor(author);
32            } else {
33                LOGGER.info("Missing author name for id {}", id);
34            }
35            if (author != null) {
36                bookTitle = bookTitle.toLowerCase();
37                LOGGER.info("Author name is {}", bookTitle);
38                book.setTitle(bookTitle);
39            } else {
40                LOGGER.error("Missing book title for book id {}", id);
41            }
42            book.setId(id);
43            bookRepositiry.save(book);
44
45            return new ResponseEntity<>(HttpStatus.CREATED);
46
47        }
48
49    }
```

Figure 6.16 – HomeController.java code issue indicated by CodeGuru Reviewer

You can use the recommendation provided by CodeGuru Reviewer to fix this issue, by validating the user input field book title before performing any operations.

14. To look at the summary of all repositories, issues, and lines of code metrics, you can click on the **Dashboard** link and that will show you a summary and history of all your scans:

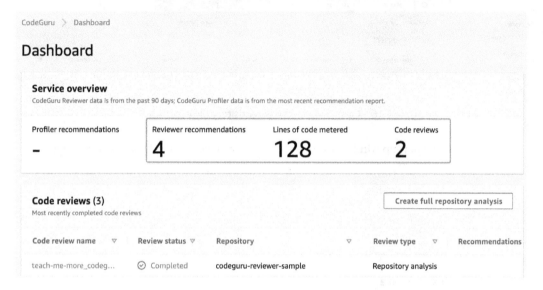

Figure 6.17 – CodeGuru Reviewer dashboard

Now that we have learned how to associate a repository, run a scan, and review the recommendations, let's continue and learn how to remove repositories from CodeGuru Reviewer.

Disassociating a repository

Code reviewer scans are important, and you can continue scanning your code as part of your development practice. But if for some reason you no longer need CodeGuru to scan your code (perhaps you've onboarded an alternate tool so you want to stop CodeGuru Reviewer from scanning your code), you can take the following steps:

1. To disassociate a repository from the code scans, you need to click on the **Repositories** link in the left panel. Next, select the code repository and then click on the **Disassociate repository** button:

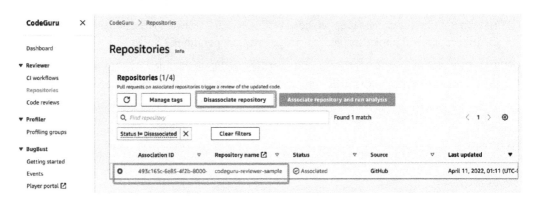

Figure 6.18 – CodeGuru Reviewer repository disassociation

2.   This opens up a pop-up window to confirm the action. Click on the **Disassociate** button:

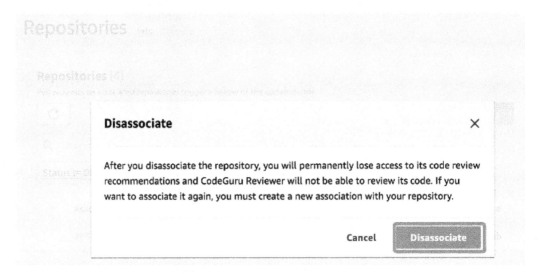

Figure 6.19 – CodeGuru Reviewer Disassociate confirmation

3.   This action will start the repository's disassociation from CodeGuru Reviewer and will display a confirmation once the task is completed:

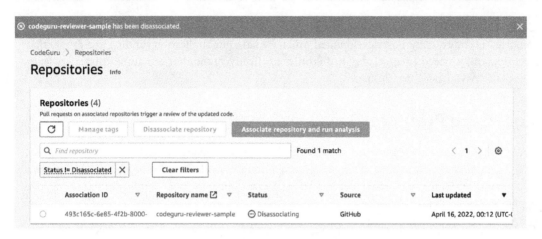

Figure 6.20 – CodeGuru Reviewer Repositories screen

4. To reassociate a repository, you can go to the **Repositories** screen again, click on the **Associate repository and run analysis** button, and follow the same steps as mentioned in the previous subsection of this chapter:

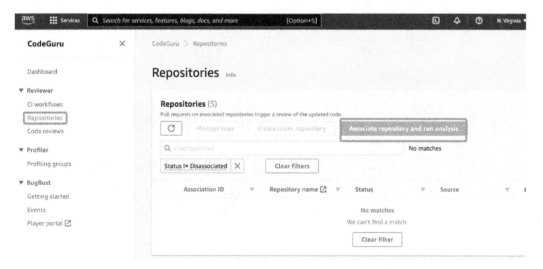

Figure 6.21 – CodeGuru Reviewer repository reassociation

CodeGuru Reviewer helps us in the development phase of an application where we can use it to improve the code quality. Post development, when we have our application running in a production environment, we need to use CodeGuru Profiler to effectively monitor our application. Let's learn how to use AWS CodeGuru Profiler.

CodeGuru Profiler

AWS CodeGuru Profiler continuously monitors application performance, collects performance metrics, and provides recommendations to fine-tune the application. CodeGuru Profiler helps development and operations teams to better understand the runtime behavior of an application and improve application performance.

CodeGuru Profiler uses machine learning algorithms to analyze the metrics collected from the application and find issues. It helps to detect memory leaks, expensive lines of code, and CPU bottlenecks. CodeGuru Profiler also provides different graphical presentations of the profiling data, which provide significant insights into running code on the CPU and the time it takes, detects any memory leaks, and so on.

Profiler requires you to have an agent running with your application, which collects the metrics from your application and pushes them to the CodeGuru Profiler service to perform analysis and visualization.

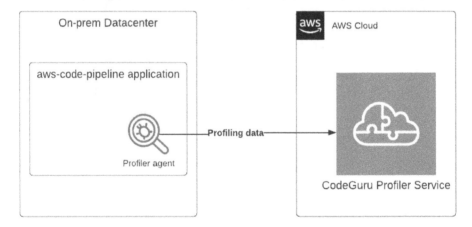

Figure 6.22 – CodeGuru Profiler working process

The preceding figure shows how a CodeGuru Profiler agent is running inside the sample aws-code-pipeline application in a local datacenter and pushes metrics to the CodeGuru Profiler service.

The benefits of CodeGuru Profiler

CodeGuru Profiler provides the following benefits:

- It collects your performance metrics in one place and provides you with insights into your application.

- It helps you to detect high CPU usage in your application and provides recommendations to improve the application's performance.

- It helps you to learn where expensive object creation is happening and whether the same expensive objects are being created multiple times. For example, creating a `Logger` or `ObjectMapper` object multiple times in a class.

- With the help of Profiler, you can understand your application better, and that helps in reducing the cost of ownership of the hardware on which your application is running.

- You can understand the heap memory utilization of the application over time with the help of CodeGuru Profiler.

- You can use CodeGuru Profiler with AWS Lambda functions to identify performance issues.

- CodeGuru Profiler is not limited to only AWS-hosted services such as EC2, Lambda, Amazon ECS, Amazon EKS, and AWS Fargate. It works well with on-premises applications as well.

The limitations of CodeGuru Profiler

As of writing this book, CodeGuru Profiler has the following shortcomings:

- CodeGuru Profiler is limited to only Java, Python, and JVM-based languages such as Scala and Kotlin

- You have to package the CodeGuru Profiler agent with your application or install the agent on the host machine to collect metrics data

- Heap summary visualizations are only available for JVM-based languages

Now that we understand the benefits and limitations of CodeGuru Profiler, let's understand how to set up our sample application, aws-code-pipeline, which we created in *Chapter 3*, with CodeGuru Profiler.

Setting up CodeGuru Profiler

Before we start collecting performance metrics, we need to set up a profiling group in AWS CodeGuru Profiler to start pushing metrics to it. For the demonstration, we are going to run the `aws-code-pipeline` application as a standalone application from our local machine, which you can consider as an on-premises computer.

Following are the instructions to create the profiling group:

1. Click on the **Profiling groups** link under **CodeGuru** in the left panel and click on the **Create profiling group** button:

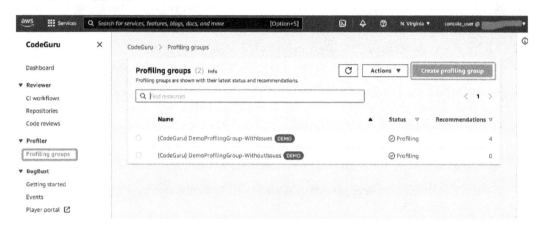

Figure 6.23 – CodeGuru Profiler dashboard

2. Now, provide a name for this profiling group, and you can select the type of compute platform. If you have a Lambda function, you can choose the **Lambda** option here to monitor performance. However, here, we are going to choose the **Other** option, which is good for monitoring any JVM-based or Python application running on ECS, EKS, Fargate, EC2, or on-prem infrastructure. Select **Other** and click the **Create** button to create the profiling group.

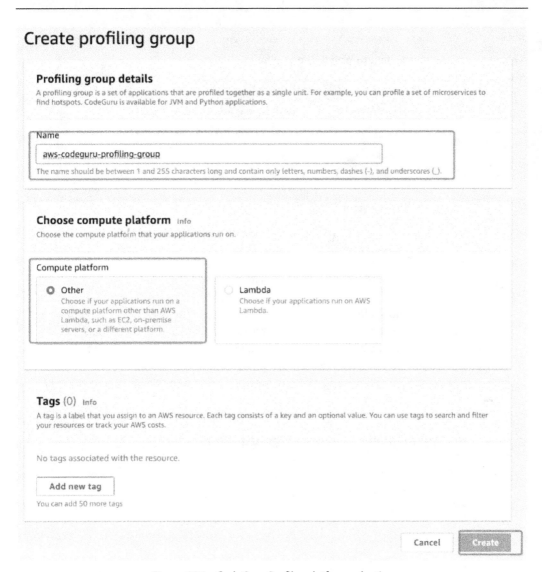

Figure 6.24 – CodeGuru Profiler platform selection

3. Now a profiling group is created, but we need to assign permissions to the user account to start pushing metrics to the CodeGuru profiling service. Click on **Give access to users and roles**.

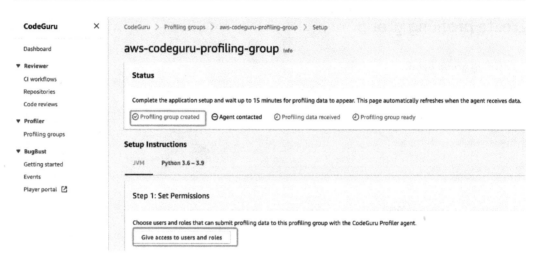

Figure 6.25 – CodeGuru Profiler group creation

4. In the pop-up window, choose users and the corresponding role for those who will be pushing JVM metrics to CodeGuru Profiler. We have created a programmatic user named COMMAND_USER with an access key and a secret key. Select COMMAND_USER to be able to push the metrics. The CodeGuru Profiler agent will use these credentials to send the data to the CodeGuru service. In the production environment, selecting a role is recommended. Click the **Save** button to continue:

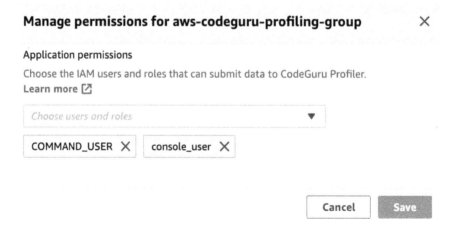

Figure 6.26 – CodeGuru Profiler permission selection

5. For the next step, we have to set up a code profiler Java agent to start pushing the metrics from the client application. In order to do that, we have two options: either using the JAR file direct download or including that in your code using `pom.xml`. For the first method, we need to download the code profiler agent JAR file from `https://repo1.maven.org/maven2/software/amazon/codeguruprofiler/codeguru-profiler-java-agent-standalone/1.2.1/codeguru-profiler-java-agent-standalone-1.2.1.jar` to the host machine and start with the command displayed as follows:

```
java -javaagent:codeguru-profiler-java-agent-standalone-
1.2.1.jar="profilingGroupName:aws-codeguru-profiling-
group,heapSummaryEnabled:true"
```

The following screenshot shows the different options available to start the CodeGuru Profiler agent to push the data:

Step 2: Start agent

There are two ways to do this:

Simple (recommended)

1. Download the JAR file ⊕

2. Restart your JVM with the `-javaagent` parameter:

```
java -javaagent:codeguru-profiler-java-agent-standalone-1.2.2.jar="profilingGroupName:aws-codeguru-profiling-group,heapSummaryEnabled:tr      📋 Copy
```

Your application will now start as normal with the CodeGuru Profiler agent running in the background. After you complete the application setup, it will take up to 15 minutes for a profile to be received Learn more ☑

Custom (Requires recompiling and re-deploying your app)

Choose your language:

▼ Java

To update your Java application to collect and send data to CodeGuru Profiler:

1. Add a dependency on the Profiler agent JAR. Learn more ☑

2. Open the Java file containing your application main function

3. Add the following import:

```
import software.amazon.codeguruprofilerjavaagent.Profiler;      📋 Copy
```

4. Add the following code to your main function:

```
new Profiler.Builder()                                          📋 Copy
    .profilingGroupName("aws-codeguru-profiling-group")
    .awsCredentialsProvider(myAwsCredentialsProvider) // optional
    .withHeapSummary(true) // optional - to start without heap profiling set to false or remove line
    .build().start();
```

Figure 6.27 – CodeGuru Profiler agent download screen

For our aws-code-pipeline application, we are going to take a different approach. We are going to add a new Maven dependency to include the code profiler agent in our application.

6.   Open `pom.xml` in the aws-code-pipeline project and add the dependencies highlighted in *Figure 6.28*, along with the CodeGuru Profiler repository:

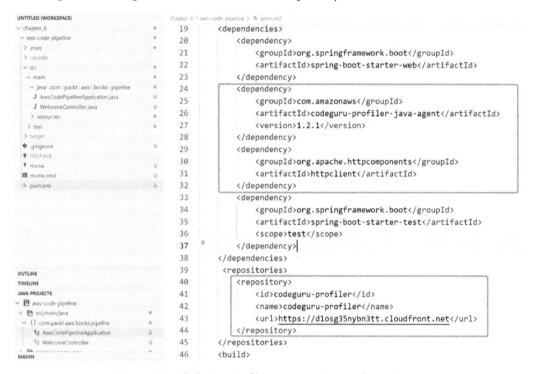

Figure 6.28 – CodeGuru Profiler Java agent Maven dependency

7.   Now open up the `AwsCodePipelineApplication.java` file and add the following lines of code to initialize the Profiler agent. Make sure you update the correct profiler group name that you provided in the AWS console while creating the profile group:

```
pom.xml 9+, U        J AwsCodePipelineApplication.java U ✕

chapter_6 > aws-code-pipeline > src > main > java > com > packt > aws > books > pipeline > J AwsCodePipelineApplication.java > ...
  1    package com.packt.aws.books.pipeline;
  2
  3    import org.springframework.boot.SpringApplication;
  4    import org.springframework.boot.autoconfigure.SpringBootApplication;
  5
  6    import software.amazon.codeguruprofilerjavaagent.Profiler;
  7
  8    @SpringBootApplication
  9    public class AwsCodePipelineApplication {
 10
       Run | Debug
 11    public static void main(String[] args) {
 12            new Profiler.Builder()
 13                    .profilingGroupName("aws-codeguru-profiling-group")
 14                    .withHeapSummary(true)
 15                    .build()
 16                    .start();
 17            SpringApplication.run(AwsCodePipelineApplication.class, args);
 18    }
 19
 20  }
```

Figure 6.29 – CodeGuru Profiler Java initialization code

8. Now, before we build and start our aws-code-pipeline application, we need to set up the AWS credentials in the `environment` variable, so that when our application starts, our local code profiler agent is able to connect with our AWS CodeGuru Profiler service and can push the data. Create three environment variables in Windows **System variables** manager: `AWS_ACCESS_KEY_ID`, `AWS_SECRET_ACCESS_KEY`, and a default `AWS_REGION`, where we created the profiler group:

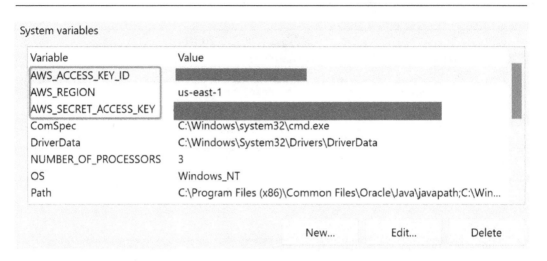

Figure 6.30 – AWS credential environment variables

9. Right now, if you look at the profiler screen, the agent is not activated and the **Agent contacted** status is grayed out:

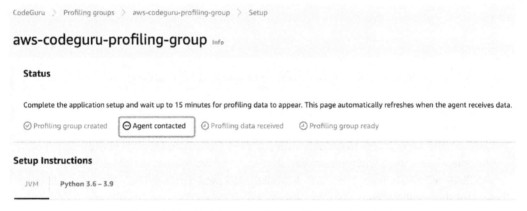

Figure 6.31 – CodeGuru profiler pending agent contact

10. Let's start our local aws-code-pipeline application. This should have the code profiler agent running on our host computer and pushing data to the AWS CodeGuru service. Go to the **Run** menu and click on **Run Without Debugging**:

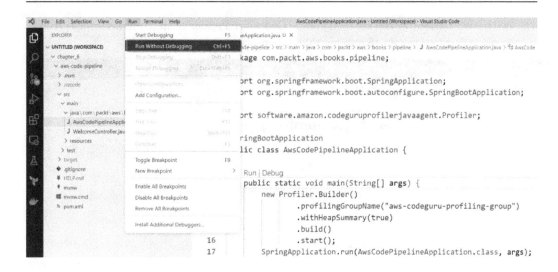

Figure 6.32 – Spring Boot aws-code-pipeline application startup

11. From the logs, we can see that the sample application has started successfully and the agent has already started sending information to CodeGuru Profiler:

Figure 6.33 – Spring Boot Profiler Java agent startup

12. In the AWS console, we can see that now the CodeGuru Profiler agent status has turned green. Profiling data takes about 15 minutes to start appearing in the console:

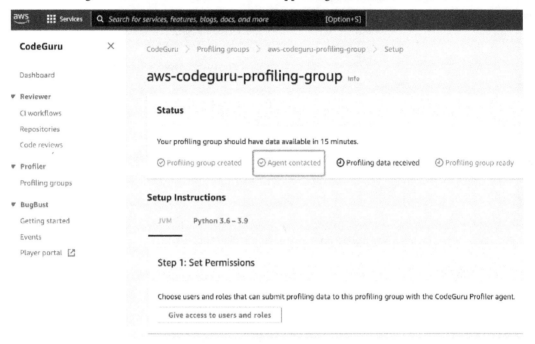

Figure 6.34 – Profiler Java agent successful contact with the CodeGuru service

13. Once Profiler is ready, it will start showing the different performance indicators:

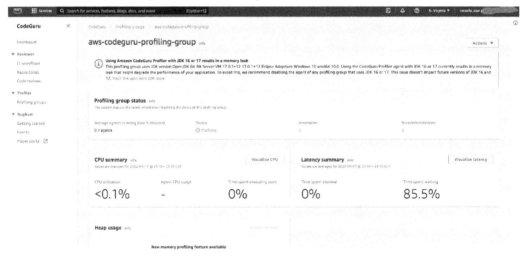

Figure 6.35 – CodeGuru Profiler dashboard for profiling group

14. You can click on the **Visualize CPU** link to see the metrics related to the CPU such as how much time the CPU is spending on processing a task, and also how much time is ideal.

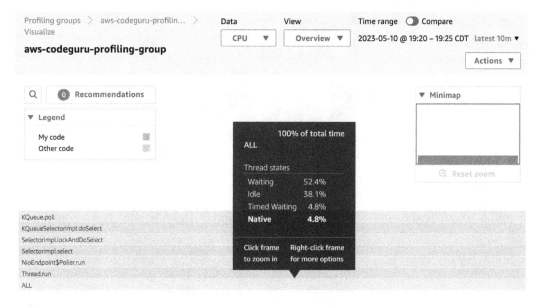

Figure 6.36 – CodeGuru Profiler CPU metrics visualization

15. In a similar way, you can see the latency metrics and hotspots related to the application profile:

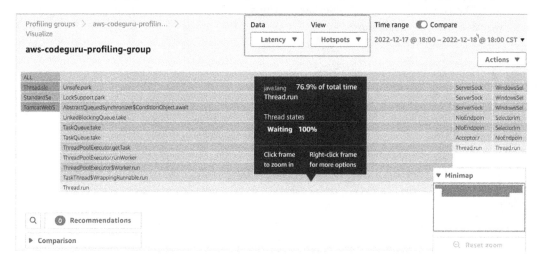

Figure 6.37 – CodeGuru Profiler latency metrics visualization

16. You can check out the sample app provided by AWS CodeGuru Profiler to view the heap summary report:

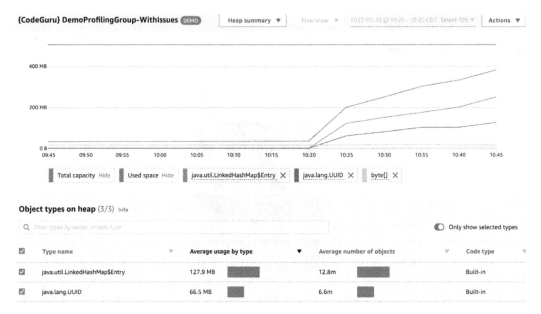

Figure 6.38 – CodeGuru Demo app heap memory visualization

17. CodeGuru Profiler provides recommendations if any issues are found in the application profiling data. The sample application profile data provided by AWS shows application issues, for which we can see recommendations:

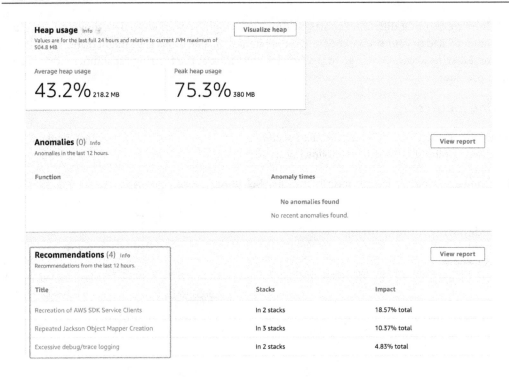

Figure 6.39 – Demo app CodeGuru Profiler recommendations list

18. You can click on the **View all reports** link to see the detailed recommendations report:

Figure 6.40 – CodeGuru demo app recommendations report

While looking through different reports, at times you might've wanted to drill down on certain information related to your source code and see what lines of code or package might display a certain behavior. Filtering allows you to go to that level, so let's look at the filtering capabilities CodeGuru Profiler provides.

Filtering metrics

You can filter through the different metrics specific to your code or any other packages available as part of your application by using the following steps:

1. To filter the data, click on **Actions** and click **Select my code**:

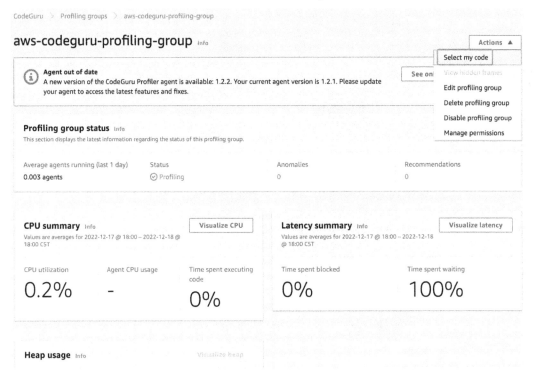

Figure 6.41 – CodeGuru profiling group filtering

2.   Select the appropriate package and click **Save**:

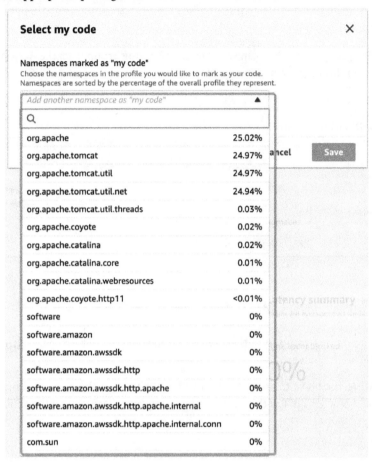

Figure 6.42 – CodeGuru code filter selection

This will filter the profiling group data and display data corresponding to the selected code.

Deleting a profiling group

Deleting a profiling group is easy. You can select **Delete profiling group** from the **Actions** menu and this will delete the profiling group and its reports:

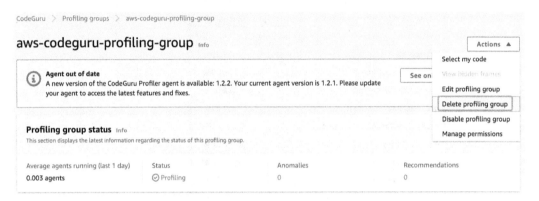

Figure 6.43 – CodeGuru Delete profiling group

Once you delete a profiling group from the AWS account, you will not be able to collect metrics by using the CodeGuru Profiler agent.

In this section, we learned how to utilize CodeGuru Profiler to learn about the runtime behavior of our application and how we can use this information to improve the quality of the application.

Summary

In this chapter, we learned about CodeGuru and the different features that the CodeGuru service provides: CodeGuru Reviewer and CodeGuru Profiler. We enabled CodeGuru Reviewer to scan our repository and then later utilized CodeGuru Profiler to start monitoring our sample application.

In the next chapter, we will focus on the AWS CodeArtifact service and how it supports managing code artifacts for CodePipeline.

7

Managing Artifacts Using CodeArtifact

In this chapter, you will learn about code artifact repositories and how the CodeArtifact service from AWS helps you to manage artifacts. You will understand how you can push your code artifacts to the AWS CodeArtifact service. In this chapter, we will cover the following topics:

- What is an artifact?
- Artifact repository
- AWS CodeArtifact
- The benefits of CodeArtifact
- The limitations of CodeArtifact
- CodeArtifact domains
- CodeArtifact repositories

In the previous chapters, we learned about microservice development using Java and Maven projects. We learned that to compile and run our application, to build the application we need, a set of dependencies needs to be available to us; these dependencies are pulled from a centralized public Maven artifact repository. CodeArtifact is the managed artifact repository provided by AWS where we can push our generated artifacts and download different application dependencies. Before we dig more into AWS CodeArtifact, let us first learn what an artifact is.

What is an artifact?

An artifact is a software binary or code that you package, which can be distributed or shipped as a single unit to be executed or can be used as a dependency for other software programs. For example, in our `aws-code-pipeline` sample application, which we developed in *Chapter 3*, we produced a JAR file (`aws-code-pipeline-xxx.jar`) as an artifact that can be executed as a Spring Boot application.

Similarly, in our application, we are using a bunch of Maven dependencies as we import those in our `pom.xml` files; those dependencies are the artifacts that get downloaded by Maven to build our application.

An artifact is a term used for the software package, and it can have different packaging types or extensions. For example, an Android mobile app will have an APK file as an artifact.

Now that we understand what an artifact is, we need to understand how to manage and store these artifacts effectively.

Artifact repository

An artifact repository is a type of software that stores the different software packages or artifacts generated as an outcome of the software build process and makes them available for easy download and automated deployment. A typical artifact repository provides easy integration with the different build and deployment tools and allows you to store and manage the different versions of the software artifacts and dependencies.

There are lots of commercial and open source artifact repositories available on the market. Some of the popular ones are Sonatype Nexus, JFrog Artifactory, AWS CodeArtifact, Azure Artifacts, Docker Hub, and GitHub Packages. Sonatype Nexus and JFrog Artifactory are especially popular and widely used in the industry.

As a developer, you build and use several open source libraries in your build environment and produce different libraries as dependencies for other projects within an organization. An artifact repository helps you to control and manage open source libraries and versions that you want to allow your internal teams to use, and you can also share your private artifacts within the organization. An artifact repository works as a proxy and locally caches the public libraries so your builds are faster and not dependent on public artifact repositories.

Now that we have an idea of what an artifact is and what an artifact repository does, let's learn about the AWS CodeArtifact service provided by Amazon.

AWS CodeArtifact

AWS CodeArtifact is a service provided by AWS for managing artifacts, in which you can create repositories, upload your artifacts, and share artifacts easily and securely. AWS CodeArtifact is a fully managed service and integrates well with commonly used package managers and build tools, and you can secure it using AWSKey Management Service (KMS).

AWS CodeArtifact uses highly available and durable S3 storage to store the artifacts' data, while metadata about the repositories is stored in Amazon DynamoDB. Artifacts are stored in an encrypted form so that they are secured at rest and stored in multiple zones.

The following diagram shows how the AWS CodeArtifact service works. A CodeArtifact repository connects to the public repositories to download artifacts. Developers connect to the CodeArtifact repository to download dependencies, and AWS CodeBuild (or any other build tool) can be used to connect to the CodeArtifact repository to publish the artifacts after the build process:

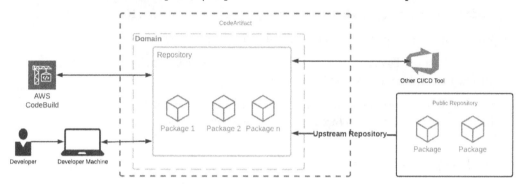

Figure 7.1 – CodeArtifact repository process

Now that we understand what the CodeArtifact service is, let's explore some of the benefits it provides.

The benefits of CodeArtifact

The following are some of the benefits that the AWS CodeArtifact service provides:

- CodeArtifact is a fully managed service, so you don't have to worry about adding storage or compute capacity or patching the servers

- CodeArtifact supports the majority of popular build tools, such as Maven, Gradle, pip, Twine, npm, and Yarn

- CodeArtifact is a *pay-as-you-go* service model, so you don't have to make any upfront investment and you only pay for the storage cost, the number of requests you make, and the data transferred out of the AWS regions

- CodeArtifact utilizes S3 infrastructure and stores the data across multiple availability zones, making it highly available and durable for access

- CodeArtifact provides easy integration with **KMS**, IAM, and the CloudTrail service, which provides encryption access control and auditing over which user has access to which artifacts

- CodeArtifact provides integration with public repositories such as Maven Central and also allows you to manage the approval of the packages that can be used

- CodeArtifact supports easy integration into existing development workflows, such as CodeBuild and CodePipeline, so that you can easily publish and use the artifacts

The limitations of CodeArtifact

AWS CodeArtifact is a highly scalable and performant artifact repository, but at the time of writing, CodeArtifact has the following limitations:

- CodeArtifact only supports four major package types – Maven, npm, NuGet, and Python

- CodeArtifact can search a maximum of 25 upstream repositories while resolving a package

- CodeArtifact can have a maximum of 1,000 repositories per domain, but this limit can be changed through an AWS quota increase request

- CodeArtifact supports a maximum of 100 write requests from one AWS account per second, but this limit can be changed through an AWS quota increase request

CodeArtifact domains

In AWS CodeArtifact, before you create a repository to store your artifacts, you need to create a domain. A domain is a group of repositories, so within a domain, you can create more than one repository. A domain provides a level of abstraction on top of a repository, so you can apply policies directly to the domain and they will be applicable to all repositories within the domain. The following diagram shows how you can have multiple domains within the CodeArtifact service and in a single domain you can create multiple repositories:

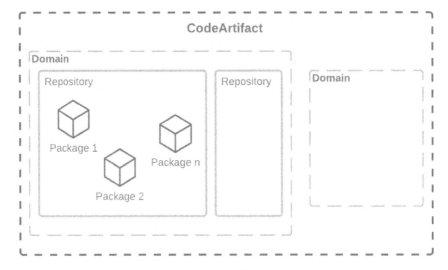

Figure 7.2 – CodeArtifact domain

Several repositories contain your packages. AWS CodeArtifact can be shared within organizations across AWS accounts. Even if an artifact is available in multiple repositories, a domain ensures that it is stored only once, so it saves on the storage cost. Having a domain allows you to create repositories with your choice of name, as repository names need to be unique within that domain only.

AWS recommends having only one domain for one AWS account, but you can create more than one domain if need be.

Let's follow these instructions to create a new domain within our AWS account:

1. Log in to your AWS account and go to the CodeArtifact service:

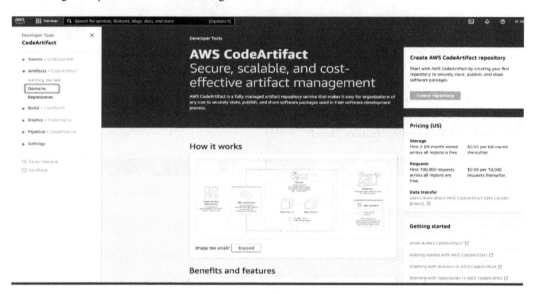

Figure 7.3 – CodeArtifact | Getting started

2. Click on **Domains** in the left panel and click on the **Create domain** button, as shown:

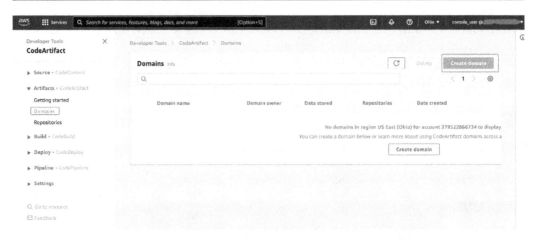

Figure 7.4 – CodeArtifact Create domain

3. Provide a domain name and select the **AWS KMS key** type. AWS domains are encrypted and all repositories under these domains will be protected automatically, so you don't have to manage the encryption at the repository level:

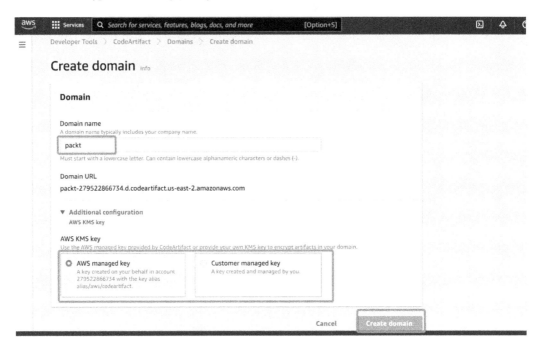

Figure 7.5 – CodeArtifact domain name details

AWS CodeArtifact supports **AWS managed key** encryption or you can upload your key using AWS KMS. Click on the **Create domain** button.

> **Note**
>
> KMS is a secure, highly available, and scalable key management service provided by AWS, which lets you create, manage, and rotate cryptographic keys for the encryption/decryption of data. KMS supports the encryption of data in your application and several AWS services, such as **Amazon Simple Storage Service (S3)**, **Amazon Relational Database Service (RDS)**, **Amazon Elastic File System(EFS)**, **Amazon Elastic Block Store (EBS)**, and DynamoDB. KMS allows you to either use AWS-managed keys or you can bring your own material for securing the data. KMS supports data encryption, decryption, signing, the verification of digital signatures, validating JSON web tokens using **Hash-based Message Authentication Code (HMAC)**, and auto key rotation for AWS-managed keys.

4. Once a domain is created, you can apply a domain-level policy to it. Click on **Apply domain policy** to create a domain policy that will be applicable to all repositories created under the domain:

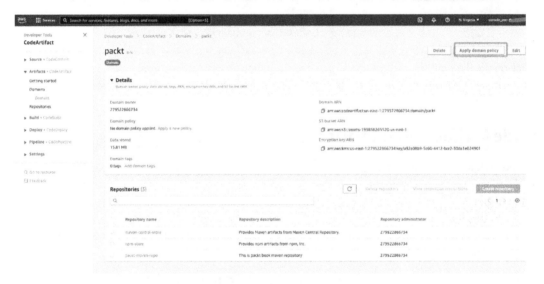

Figure 7.6 – CodeArtifact domain details

5. You can copy, paste, or modify the existing policy, as shown in the following screenshot, or you can choose some preexisting example policies by clicking on the **Example policies** button:

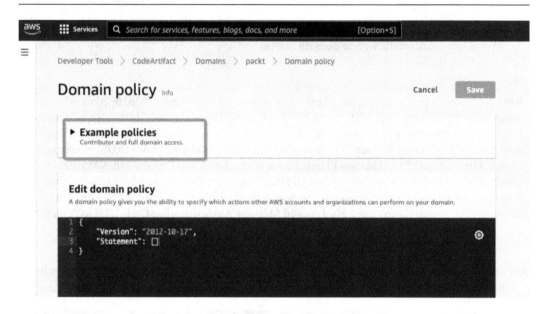

Figure 7.7 – CodeArtifact domain policy

Click on the **Save** button once you have changed the policy. In our case, we aren't making any changes. Now that we have our domain ready, let's move on to understand what a CodeArtifact repository is and how we can create one.

CodeArtifact repositories

A CodeArtifact repository is a collection of artifacts or packages. A repository maintains the different versions of an artifact, where each version maps to a set of assets. A CodeArtifact repository is exposed using an endpoint that, by default, goes through the internet, but you can obtain VPC-specific endpoints so that traffic doesn't flow through the public internet and only goes through your private network if needed. You can download and publish packages to this CodeArtifact repository endpoint using existing tools such as Maven, npm, pip, and NuGet.

Now that we have an understanding of what a CodeArtifact repository is, let's continue to create a new repository:

1.   Once you have a domain, you can create repositories under this domain:

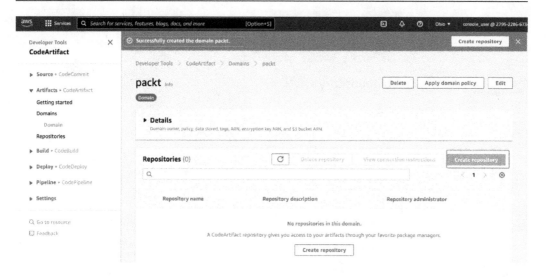

Figure 7.8 – CodeArtifact | Create repository

2. Provide the repository name details and description and select the upstream repository. An upstream repository connects the CodeArtifact repository to the official artifact distribution authority, such as npm and Maven Central, from where you can download original artifacts:

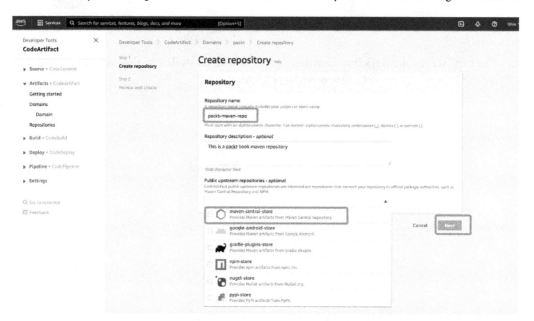

Figure 7.9 – CodeArtifact upstream repository selection

3. You can select multiple upstream repositories; then, click **Next**:

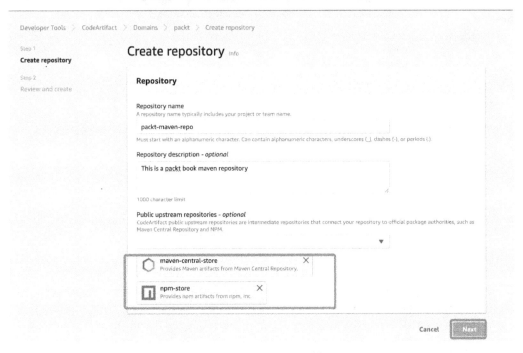

Figure 7.10 – CodeArtifact repository creation

4. The next screen displays how our dependencies will be pulled in through the upstream repositories. Click on the **Create repository** button:

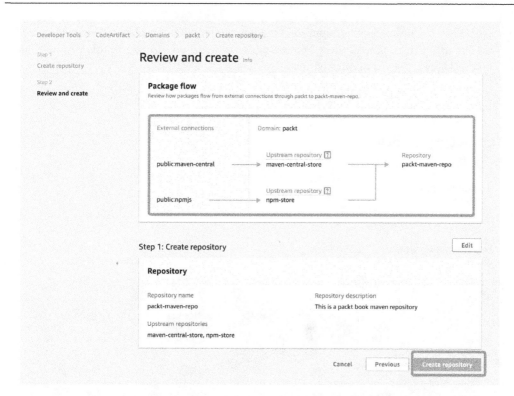

Figure 7.11 – CodeArtifact repository creation review

5. Once you have a repository created, you can click on the **Apply repository policy** button to apply a policy that will be applicable to the current repository:

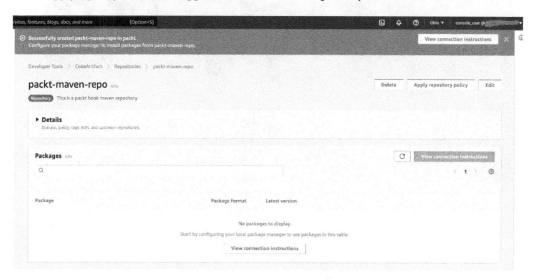

Figure 7.12 – CodeArtifact | Create repository confirmation

6. By default, the resource policy applied at the domain level is applicable to the newly created `packt-maven-repo` repository. You can apply additional policies to a repository by clicking on the **Apply repository policy** button:

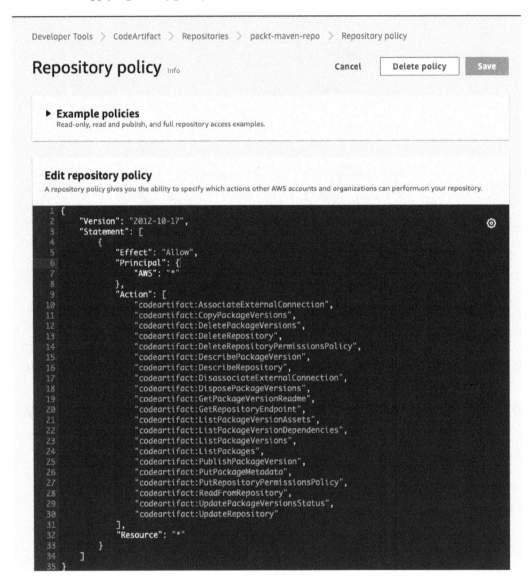

Developer Tools > CodeArtifact > Repositories > packt-maven-repo > Repository policy

Repository policy Info

Cancel | Delete policy | Save

▶ **Example policies**
Read-only, read and publish, and full repository access examples.

Edit repository policy
A repository policy gives you the ability to specify which actions other AWS accounts and organizations can perform on your repository.

```
1  {
2      "Version": "2012-10-17",
3      "Statement": [
4          {
5              "Effect": "Allow",
6              "Principal": {
7                  "AWS": "*"
8              },
9              "Action": [
10                 "codeartifact:AssociateExternalConnection",
11                 "codeartifact:CopyPackageVersions",
12                 "codeartifact:DeletePackageVersions",
13                 "codeartifact:DeleteRepository",
14                 "codeartifact:DeleteRepositoryPermissionsPolicy",
15                 "codeartifact:DescribePackageVersion",
16                 "codeartifact:DescribeRepository",
17                 "codeartifact:DisassociateExternalConnection",
18                 "codeartifact:DisposePackageVersions",
19                 "codeartifact:GetPackageVersionReadme",
20                 "codeartifact:GetRepositoryEndpoint",
21                 "codeartifact:ListPackageVersionAssets",
22                 "codeartifact:ListPackageVersionDependencies",
23                 "codeartifact:ListPackageVersions",
24                 "codeartifact:ListPackages",
25                 "codeartifact:PublishPackageVersion",
26                 "codeartifact:PutPackageMetadata",
27                 "codeartifact:PutRepositoryPermissionsPolicy",
28                 "codeartifact:ReadFromRepository",
29                 "codeartifact:UpdatePackageVersionsStatus",
30                 "codeartifact:UpdateRepository"
31             ],
32             "Resource": "*"
33         }
34     ]
35  }
```

Figure 7.13 – CodeArtifact repository policy

You can make necessary changes to the repository policy and click **Save**. Similar to the domain policy screen here, you can also use the example policies as needed. We have our repository ready; now, let's try connecting to this repository so that our `aws-code-pipeline` sample application, which we developed in *Chapter 3*, can publish and download artifacts from our AWS CodeArtifact repository.

Connecting to the repository

In order for our application to be able to connect to the CodeArtifact repository, we need to have an authentication token, which will make a connection to our repository and allow us to publish and download artifacts. Since we are using Maven to build our `aws-code-pipeline` sample application, we need to set up our Maven configuration so that Maven can communicate with our repository and can download and publish the required packages:

1. Click on the **View connection instructions** button under our new repository:

Figure 7.14 – CodeArtifact repository connection

2. This will open the pop-up screen as shown in the following screenshot, asking you to select a package type, for which we are looking for connection instructions:

Figure 7.15 – CodeArtifact repository connection details

3. Select **mvn** and a screen will be shown with steps to configure Maven. As a first step, we need to install the AWS CLI on our development machine. Follow the instructions provided at `https://docs.aws.amazon.com/cli/latest/userguide/getting-started-install.html` to download and install the AWS CLI. Once the CLI is configured using the `aws configure` command and `AWS_ACCESS_KEY_ID`, `AWS_SECRET_ACCESS_KEY` and an appropriate AWS region are set, proceed to the next step.

4. We need to create an `auth` token so that our system can make a connection to the CodeArtifact repository. Type the following command to generate the `auth` token and set that as an environment variable, `CODEARTIFACT_AUTH_TOKEN`. Linux/macOS users can use the `export` command instead of `set`:

    ```
    set CODEARTIFACT_AUTH_TOKEN=`aws codeartifact get-authorization-
    token --domain packt --domain-owner 279522866734 --query
    authorizationToken --output text`
    ```

    This command will generate a token and export that as an environment variable, which is valid for 12 hours.

5. Now, go to the Maven `.m2` directory and create/edit the Maven `settings.xml` file and add the following code. The following `settings` file is used for configuring Maven:

    ```xml
    <settings xmlns="http://maven.apache.org/SETTINGS/1.0.0"
    xmlns:xsi="http://www.w3.org/2001/XMLSchema-instance"
    xsi:schemaLocation="http://maven.apache.org/SETTINGS/1.0.0
    https://maven.apache.org/xsd/settings-1.0.0.xsd">
      <profiles>
      <profile>
        <id>packt-packt-maven-repo</id>
        <activation>
          <activeByDefault>true</activeByDefault>
        </activation>
        <repositories>
          <repository>
            <id>packt-packt-maven-repo</id>
            <url>https://packt-xxxxxxxxxxxx.d.codeartifact.
    us-east-1.amazonaws.com/maven/packt-maven-repo/</url>
          </repository>
        </repositories>
      </profile>
    </profiles>
    <servers>
      <server>
        <id>packt-packt-maven-repo</id>
        <username>aws</username>
        <password>${env.CODEARTIFACT_AUTH_TOKEN}</password>
      </server>
    ```

```
      </servers>
   </settings>
```

Here, in this `settings.xml` file, we have created a default profile for our all Maven projects and defined how to download required dependencies in our internal repository. The `servers` section defines how to connect to our internal remote repository. If you noticed, the `server` ID in the preceding code is a combination of the domain name provided earlier and a hyphen (`-`) followed by the repository name.

6. Add the following code to the `pom.xml` file of the `aws-code-pipeline` project, so that we can publish the generated artifact to our CodeArtifact repository:

```
<distributionManagement>
   <repository>
      <id>packt-packt-maven-repo</id>
      <name>packt-packt-maven-repo</name>
      <url>https://packt-279522866734.d.codeartifact.us-east-1.
amazonaws.com/maven/packt-maven-repo/</url>
   </repository>
</distributionManagement>
```

7. Now, run the `mvn clean install` command to build the project to make sure that we have our repository set up correctly and that we can build our `aws-code-pipeline` project without issues.

8. Once the project is built successfully, let's push this new `aws-code-pipeline.xx.jar` artifact to our CodeArtifact repository using Maven's native `mvn deploy` command:

Figure 7.16 – CodeArtifact Maven artifact deployment

9. The successful execution of the `mvn deploy` command confirms that the artifact has been published successfully to `packt-packt-maven-repo`. In the CodeArtifact console, we can see that our artifact is now available in the repository:

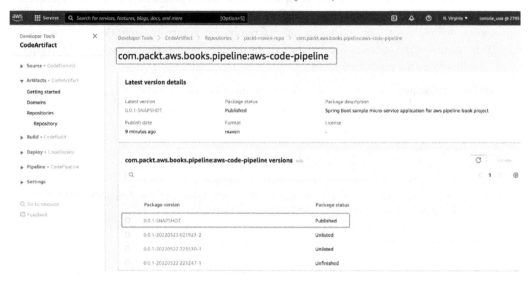

Figure 7.17 – CodeArtifact uploaded package details

10. Once an artifact is available in the repository, it can be shared for deployment or can be used as a dependency in other projects. The repository stores the different versions of each artifact, and you can download the specific version of the artifact. You can see the different dependencies associated with a package by clicking on the **Dependencies** tab. The installation details of the package can be found on the right side of the following screen:

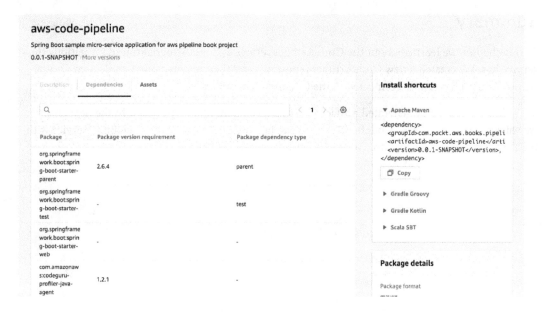

Figure 7.18 – CodeArtifact dependencies and installation details

11. Different assets associated with a package can be viewed under the **Assets** tab:

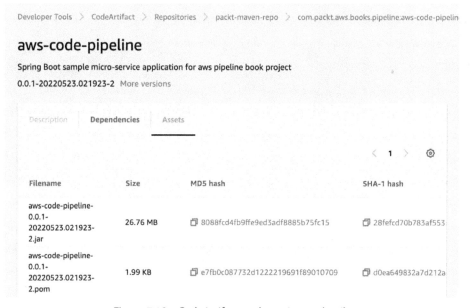

Figure 7.19 – CodeArtifact package Assets details

At this point, you have a good understanding of the CodeArtifact service and how to use this service.

Summary

In this chapter, we learned about the CodeArtifact service and all the benefits and limitations this service has. We created a new CodeArtifact repository and then updated our sample `aws-code-pipeline` application to connect with the CodeArtifact repository.

We published the newly generated artifact to the AWS CodeArtifact repository. In the next chapter, we will learn about the AWS CodeBuild service and how we can integrate the CodeBuild service with the CodeArtifact service to automatically publish generated artifacts.

8

Building and Testing Using AWS CodeBuild

In this chapter, you will learn about the AWS CodeBuild service and how you can use this service to build your source code. In this chapter, we will be covering the following topics:

- What is AWS CodeBuild?
- The benefits of using CodeBuild
- The limitations of CodeBuild
- Understanding buildspec files
- Integration with CodeCommit
- Testing using CodeBuild
- Build metrics
- Build notifications
- Build triggers
- Local build support

In the previous chapter, we learned about the CodeArtifact service and how we generate the artifacts as part of the build process. Developers write code using the programming language of their choice. Once they are done, they need to merge the code to a central repository such as CodeCommit for integration. Once the changes have been merged, code needs to be built as well as tested to make sure that there are no compilation errors and all of the changes pass the unit tests. AWS CodeBuild provides a platform for building and testing the source code. In this chapter, we will learn how to build software using the CodeBuild service.

What is AWS CodeBuild?

The process by which we generate an artifact is known as the build process. The AWS CodeBuild service provides a highly available and scalable build platform to compile source code, run test cases, and produce ready-to-deploy software artifacts.

AWS CodeBuild is a **continuous integration (CI)** service, allowing you to build source code at scale without provisioning any hardware – you only pay for what you use. CodeBuild allows you to build your code using prepackaged build environments. Alternatively, you can create your own custom build environments by uploading Docker images. AWS CodeBuild can run multiple builds simultaneously and you don't have to worry about patching or scaling the infrastructure.

When you submit a CodeBuild job, the CodeBuild service runs it in a temporary container and discards the container after the job is completed. These containers are freshly created every time you submit a job. You can run multiple build jobs in parallel and execute them immediately.

The following diagram shows how the CodeBuild service allows us to integrate with different source code repositories and build software packages for distribution or deployment:

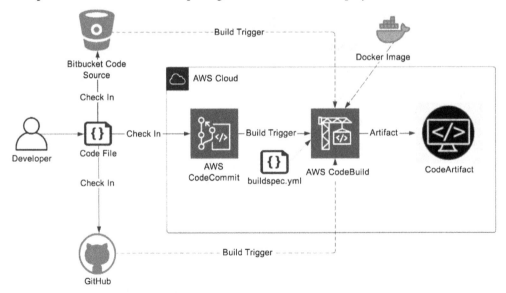

Figure 8.1 – AWS CodeBuild process

Based on the configuration provided, the CodeBuild service provisions the build environment for building your source code. CodeBuild uses Docker images to quickly set up the build environment. You can upload your own custom Docker images for CodeBuild to provision environments using your provided images. CodeBuild uses a metafile called `buildspec.yml` to execute instructions during the build process. We will learn more about the `buildspec.yml` file later in this chapter. Now that we have understood what AWS CodeBuild service is, let's learn about the benefits this service provides.

The benefits of using CodeBuild

Jenkins, Circle CI, GitHub Actions, and CloudBees CI are some other popular CI tools that help to build software packages from source code. Some of these are open source and can be hosted on your own infrastructure or on the AWS cloud. AWS CodeBuild provides the following benefits over those products from its competitors:

- AWS CodeBuild is a fully managed service so you don't have to worry about purchasing hardware or software, or doing installations, updates, and patching.

- AWS CodeBuild integrates well with your existing source code repositories such as GitHub, GitHub Enterprise, CodeCommit, Bit Bucket, S3, and so on, so you don't have to learn about any new integration.

- AWS CodeBuild is fully integrated with the AWS IAM service, so you can control who has access to the build process and access can be restricted on the CodeBuild project level. AWS also provides encryption of the generated artifacts and integration with the CodeArtifact and CodeCommit services, so your source code and binaries are always secure and don't leave the AWS infrastructure.

- AWS CodeBuild auto-scales to meet the requirements of the number of builds you need. It can also run builds concurrently, instead of your builds waiting in a queue, which increases developer productivity.

- AWS CodeBuild provides preconfigured build environments for Java, Python, Node.js, Ruby, Go, Android, .NET Core for Linux, and Docker, but it also allows you to extend the environment by uploading your own Docker images.

- AWS CodeBuild is a full CI tool, which means it can trigger builds automatically when developers check in code using repository hooks. CodeBuild can easily be integrated into existing CI/CD pipelines such as Jenkins as a worker node.

- AWS CodeBuild allows you to define specific commands to be executed during the different phases of the build process using the `buildspec.yaml` file, providing greater flexibility to customize the build process.

- AWS CodeBuild gives you the flexibility to choose different compute types to run your build. You can choose from three different levels of compute resources based on your needs.

- AWS CodeBuild charges you only for the number of minutes your build is running; there is no additional cost for provisioning the required infrastructure. You don't have to pay for idle time when your builds are not running.

- AWS CodeBuild integrates well with **Simple Notification Service** (**SNS**) to trigger notifications about the build process.

- AWS CodeBuild publishes detailed logs and metrics about the builds to the CloudWatch service, based on which you can create custom dashboards to monitor your build processes.

Now that we have covered the benefits of CodeBuild, let's also look into its limitations.

The limitations of CodeBuild

The AWS CodeBuild service is a highly available and scalable service, but it also has the following limitations as of writing this book:

- AWS CodeBuild can run a maximum of 60 concurrent builds in each region, so issues arise when you need more than 60. Although this limit can be adjusted by requesting a quota increase, this is an extra step users are required to take.

- AWS CodeBuild allows you to query a maximum of 100 requests at a time to get build information using the AWS CLI and AWS SDK.

- AWS CodeBuild allows a maximum of 5,000 concurrent build projects per Region, but this limit can be adjusted by requesting a quota increase.

- AWS CodeBuild has the limitation of allowing build timeouts between a range of 5 and 480 minutes only.

- AWS CodeBuild can maintain the build history for a maximum of 1 year only.

- You can only create a maximum of 5 reports per project. Usually, however, you won't need more than this unless there is a special requirement in a particular project.

Now that we have a good understanding of the CodeBuild service and its limitations, let's try creating a CodeBuild project and building some code.

Creating an AWS CodeBuild project

In order to understand the AWS CodeBuild service better, we need to create a CodeBuild project and then connect it to our source code repository to build some code. The instructions to create a CodeBuild project are as follows:

1. Log in to the AWS console, search for CodeBuild, open it, and click on the **Create project** button:

Figure 8.2 – AWS CodeBuild getting started page

2. Provide the project name and description. You can also specify any resource tag if needed:

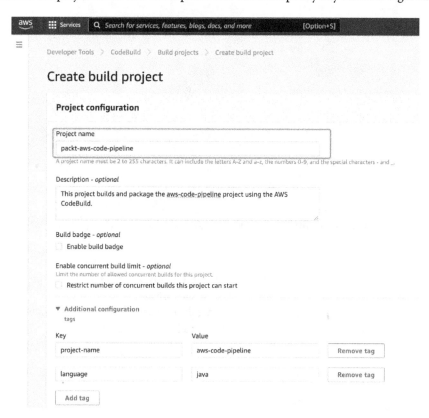

Figure 8.3 – Creating the build project in AWS CodeBuild

3. The **Source** section allows you to specify what type of source code repository you want to integrate and download source code to build from. You can integrate CodeBuild with any of the following source code providers:

    - Amazon S3
    - AWS CodeCommit
    - GitHub
    - BitBucket
    - GitHubEnterprise

    The steps to connect with any of these are the same: you have to select the given provider, then authenticate, and select the appropriate repository and branch to use. Here, we are using **AWS CodeCommit** as our source code provider, as the code we want to build is located on an AWS CodeCommit repository.

Figure 8.4 – AWS CodeBuild source code provider selection

4. In our case, we want CodeBuild to connect to our existing `aws-code-pipeline` CodeCommit repository, so as soon as you select **AWS CodeCommit** from the **Source provider** dropdown, your existing source code repositories will be listed. You can select the source code reference type and depth of the repository to clone. Then, select the appropriate branch that you want to build as part of the build process – in our case, we choose the **master** branch:

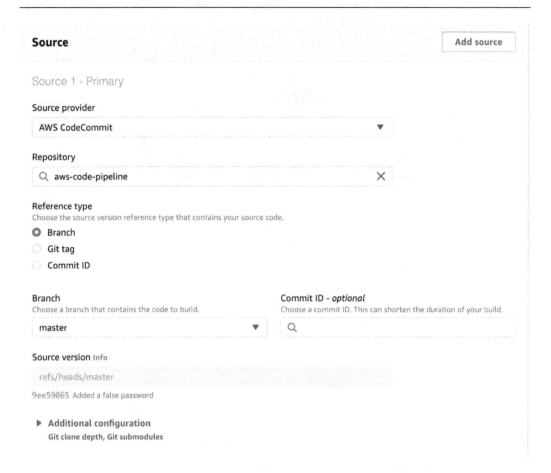

Figure 8.5 – AWS CodeBuild source code repository selection

Note that AWS CodeBuild allows you to select more than one source code location.

5.   The **Environment** section allows us to choose from an existing environment for our build. Alternatively, we can upload a custom Docker image to prepare the CodeBuild environment when we build our code. Here, we are selecting a managed Linux environment with the latest available image:

Figure 8.6 – AWS CodeBuild runtime environment configuration

CodeBuild needs a service role in order to run the builds. This **Environment** screen is where we can select an existing role, or if not, this service role will be created automatically. Lastly, click on the **Additional configuration** link.

6. The **Additional configuration** section provides us the option to customize different environment-related details and timeout configurations. You can define the environment variables that you want your build environment to use during the build execution. You can choose different compute types and filesystem configurations and attach storage to your build servers. However, for our current exercise, we won't change anything here from the defaults:

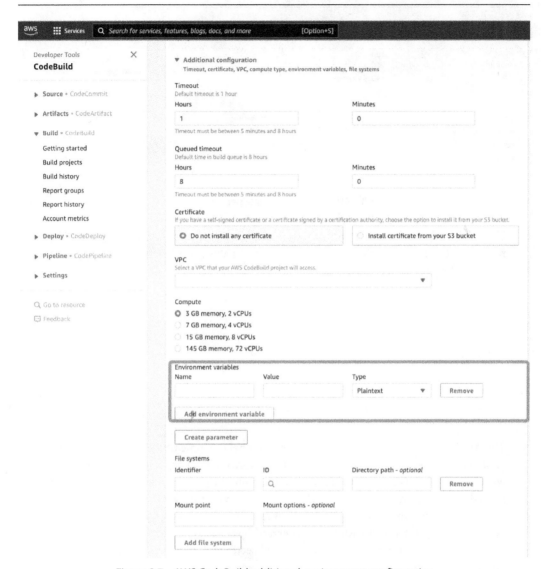

Figure 8.7 – AWS CodeBuild additional environment configuration

7. The **Buildspec** section allows us to specify a build configuration file which must be named
    `buildspec.yml`. This file contains detailed instructions on how our project should be built.
    We will learn about `buildspec.yml` files in detail later in this chapter. In this section, you
    can either specify the location of your `buildspec.yml` file or you can choose the **Insert
    build commands** option, which will provide an inline editor to create the `buildspec.yml`
    file. As we will create a `buildspec.yaml` file later in this chapter, let's leave the default setting
    for now, so that CodeBuild can pick up `buildspec.yaml` file from the root of our project:

Figure 8.8 – AWS CodeBuild buildspec file selection

8.   The **Batch configuration** section allows us to configure our build as a batch job, where we can run multiple jobs as a single execution. We are not enabling the batch option for this project as we don't need this in our use case:

Figure 8.9 – AWS CodeBuild batch configuration

9.   The **Artifacts** section is where we configure the final outcome of the build process. We can configure it depending on whether we want to store the final generated artifact in S3 or we want to discard it. For this exercise, we want to store the generated artifacts in an S3 bucket named packt-aws-code-pipeline. **Artifacts packging** section allows us to configure the packaging type to be in .zip format, but for now we won't change anything and keep the default packing type as **None** which will generate the artifact in .jar format using the Maven command. In the **Additional configuration** subsection of the **Artifacts** section, allows us to enable encryption and caching, but we are not changing anything at this moment:

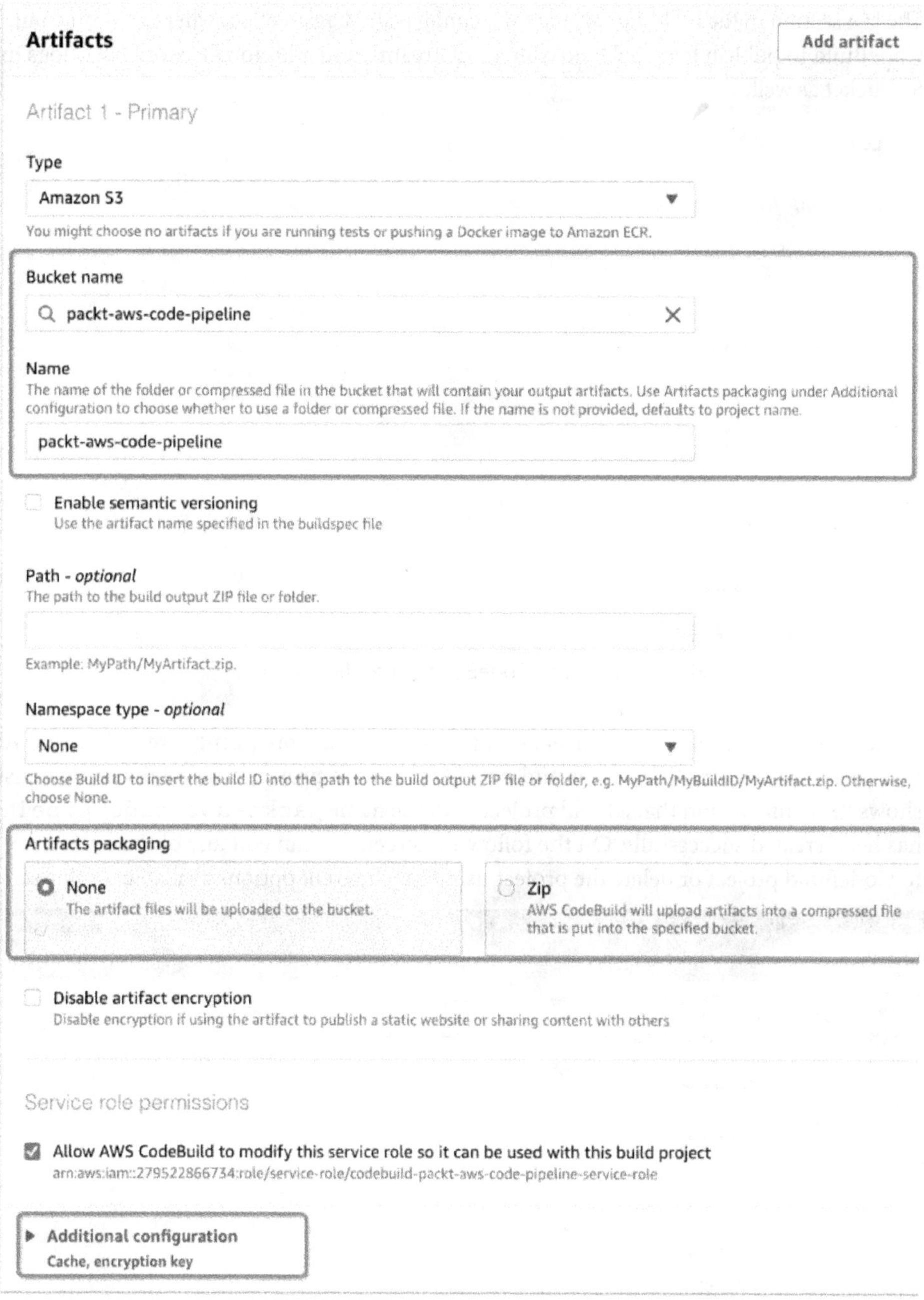

Artifacts

Add artifact

Artifact 1 - Primary

Type

Amazon S3   ▼

You might choose no artifacts if you are running tests or pushing a Docker image to Amazon ECR.

Bucket name

🔍 packt-aws-code-pipeline   ✕

Name
The name of the folder or compressed file in the bucket that will contain your output artifacts. Use Artifacts packaging under Additional configuration to choose whether to use a folder or compressed file. If the name is not provided, defaults to project name.

packt-aws-code-pipeline

☐ **Enable semantic versioning**
Use the artifact name specified in the buildspec file

Path - *optional*
The path to the build output ZIP file or folder.

Example: MyPath/MyArtifact.zip.

Namespace type - *optional*

None   ▼

Choose Build ID to insert the build ID into the path to the build output ZIP file or folder, e.g. MyPath/MyBuildID/MyArtifact.zip. Otherwise, choose None.

Artifacts packaging

🔘 **None**
The artifact files will be uploaded to the bucket.

⚪ **Zip**
AWS CodeBuild will upload artifacts into a compressed file that is put into the specified bucket.

☐ **Disable artifact encryption**
Disable encryption if using the artifact to publish a static website or sharing content with others

Service role permissions

☑ **Allow AWS CodeBuild to modify this service role so it can be used with this build project**
arn:aws:iam::279522866734:role/service-role/codebuild-packt-aws-code-pipeline-service-role

▶ **Additional configuration**
Cache, encryption key

Figure 8.10 – AWS CodeBuild Artifacts configuration

10. The last section in the build project creation config is the **Logs** section, where you can configure CodeBuild to publish logs to CloudWatch log streams, and additionally store these logs in an S3 bucket as well:

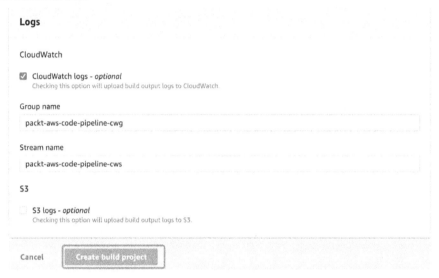

Figure 8.11 – AWS CodeBuild log configuration

11. Now, click on the **Create build project** button shown in the preceding screenshot, and AWS will create the required resources and provision a build project. The following screenshot shows the confirmation that a build project with the name `packt-aws-code-pipeline` has been created successfully. On the following screen, we can edit any of the setting related to CodeBuild project or delete the project using the different options available on this screen:

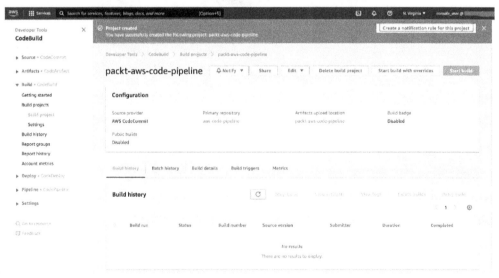

Figure 8.12 – AWS CodeBuild project creation confirmation

We have now created our first build project. However, before we start executing builds, we need to create a report group in the CodeBuild service, from where we can see our code coverage and test case execution reports.

Testing using CodeBuild

Unit testing is an integral part of any software development, it helps you to evaluate the quality of the code before it is deployed to production. Developers write test cases against the code to find flaws in the code and make sure no scenario is left untested. Modern test frameworks such as JUnit for Java, JTest for JavaScript, pytest for Python, and NUnit for .NET benefit from easy integration with the CodeBuild service.

CodeBuild allows developers to create test reports based on the metadata created by the different testing frameworks for unit testing, configuration testing, and functional testing. CodeBuild reads the test metadata files created by the different testing frameworks, and based on that, helps us to visualize the test reports and create a time series view for the test cases.

The following are the file formats supported by CodeBuild as of writing this book:

- JUnit XML (.xml)
- TestNG XML (.xml)
- Cucumber JSON (.json)
- NUnit XML (.xml)
- NUnit3 XML (.xml)
- Visual Studio TRX (.trx)

In CodeBuild, you need to create a report group in order to see the test case reports. Work through the following section to learn how to create a report group to see the code coverage and test case reports.

Creating a report group

A report group allows the AWS CodeBuild service to push test cases and code coverage reports during the build process. Reports groups allow us to visualize the data from test cases, see the reports generated during different build executions, and analyze the trends of test cases over time. Let's continue to create a report group and use it to publish our microservice test case data. The steps to do this are as follows:

1. Click on **Report groups** in the left panel and click the **Create report group** button:

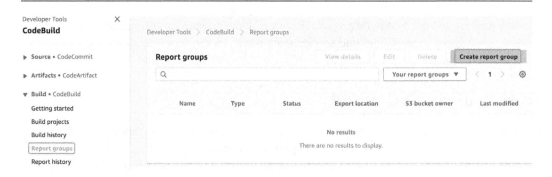

Figure 8.13 – AWS CodeBuild Report groups screen

2. Fill in the **Report group name** field and select **Test** from the **Report type** dropdown. Report groups allow us to store data related to test case results in S3 buckets for additional analysis or long-term storage. In our case, we don't want to store the data, so uncheck the **Export to Amazon S3** option. However, if you did want to store the data, then you would specify the S3 bucket details here. Additionally, you can specify any resource tags if needed:

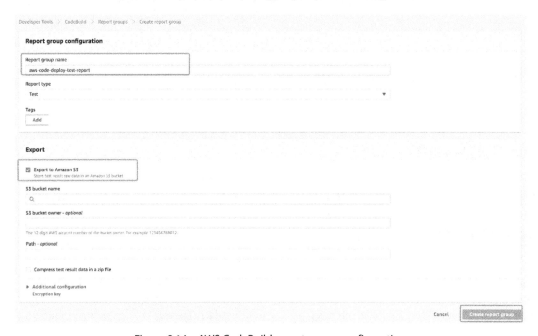

Figure 8.14 – AWS CodeBuild report group configuration

3. Clicking the **Create report group** button will take you to the following screen and confirm that the report group was created successfully. Note the **Report group ARN** value shown on this screen – in our case, it is `arn:aws:codebuild:us-east-1:279522866734:report-group/aws-code-deploy-test-report:`

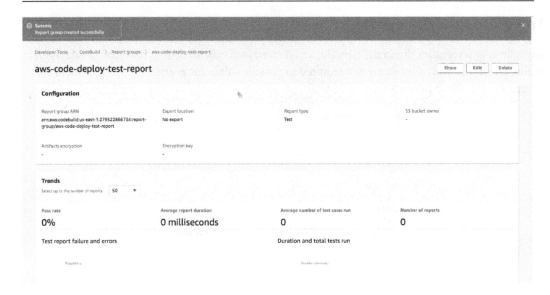

Figure 8.15 – AWS CodeBuild create report group confirmation

We are now done creating our first report group and our build project. However, before we start executing builds, we need to understand the `buildspec.yml` file so that we can add this to our microservice `aws-code-pipeline` project.

Understanding buildspec files

A `buildspec.yml` file is a YAML file that contains a set of commands and additional information to be used at build time. You can specify a `buildspec.yml` file during the build project creation or you can include it as part of the source code.

> **Note**
>
> **YAML Ain't Markup Language (YAML)** is a configuration file syntax. YAML is a human-friendly data serialization language for all programming languages. YAML files are mostly used for configuring applications where data is being transmitted or stored. You can read more about YAML and its syntax at `https://yaml.org/`.

By default, the `buildspec.yml` file should be created in the root directory of the source code and specifically named `buildspec.yml`. However, you can override the name and location of the `buildspec.yml` file and specify your own values in the build project settings if required. In order to successfully create and work with our `buildspec.yaml` file in our project, we need to understand the syntax of the file. Let's review the different configuration options available in the `buildspec.yml` file.

The buildspec file syntax

The following screenshot shows the typical `buildspec.yml` file structure. Let's look at each node in this YAML file. Note that we are not required to provide values for every node; most of them can be skipped if we don't need them:

```
1 version: 0.2
2 run-as: Linux-user-name
3 env:
4   shell: shell-tag
5   variables:
6     key: "value"
7   parameter-store:
8     key: "value"
9   exported-variables:
10     - variable
11   secrets-manager:
12     key: secret-id:json-key:version-stage:version-id
13 proxy:
14 batch:
15 phases:
16   install:
17     run-as: Linux-user-name
18     on-failure: ABORT | CONTINUE
19     runtime-versions:
20       runtime: version
21     commands:
22       - command
23     finally:
24       - command
25   pre_build:
26   build:
27   post_build:
28 reports:
29   report-group-name-or-arn:
30     files:
31       - location
32 artifacts:
33   files:
34     - location
35   name: artifact-name
36 cache:
```

Figure 8.16 – buildspec.yml file syntax

Let's look at some of the important `buildspec.yml` nodes in detail to understand their usage:

- `version`: This is a required key and represents the version of the buildspec you are using. As of writing this book, the latest version is `0.2`.

- `run-as`: This is an optional sequence available to Linux users only. It represents the specified user that the subsequent commands will be run as. If you don't specify a `run-as` sequence, it uses the root user by default. You can specify the `run-as` element at the global level or at different phase blocks of the `buildspec.yaml` file as shown in the preceding file. If you define it at the global level, then it will apply to all phases and all commands will be executed as the user you specify in the `run-as` element.

- env: This is an optional key containing information about the environment variables available to the build project. env keys have child elements such as shell, variables, parameter-store, and so on. Let's examine some of these in more detail:

  - shell: This is an optional key that specifies the shell used to execute the commands. As of writing this book, the supported shell tag types for Linux are bash and bin/sh, and for Windows, we have PowerShell.exe and cmd.exe.

  - variables: This defines a list of environment variables you want to supply to CodeBuild during build execution. You specify each key/value on a separate line using the following syntax:

    ```
    env:
        variables:
            key: "value"
            key: "value"
    ```

  - parameter-store: This contains key/value mappings, each provided on a separate line. This is mandatory if you want to retrieve the variables stored in the Amazon EC2 Systems Manager Parameter Store. The keys specify the variable names used in your build process, while the values are the name of the parameter stored in the EC2 Systems Manager Parameter Store.

- secrets-manager: This contains key mappings with the syntax secret-id:json-key:version-stage:version-id, each specified on a separate line. This is mandatory if you want to retrieve the variables stored in the AWS Secrets Manager. Here, the key specifies the variable name you will use in your build process, while secret-id refers to the ARN of the secret, and json-key is the name of the Secrets Manager's key-value pair.

- exported-variables: This is an optional key; it is used to export the environment variables to other pipeline stages. The variables you specify in the exported section must be available in the container during the build and it can be an environment variable itself.

- phases: This is a required sequence in the buildspec file. Build processes go through a set of lifecycle events and you can add specific commands to be executed in each phase. There are four different phases supported by buildspec files: install, pre_build, build, and post_build. Let's learn about each of these phases one by one:

  - install: This is an optional block. Any command that needs to be run during the installation phase can be specified in this section. This section should be used to install any package or framework needed for the build. See the following example:

    ```
    phases:
      install:
        runtime-versions:
          java: corretto11
    ```

```
    commands:
      - echo started install phase
```

- `pre_build`: This is an optional block. Any command that needs to be run before the actual build phase can be configured here. This section can be used for loading dependencies, logging in to any required server, or configuring environment before build phase. See the following example:

```
phases:
  pre-build:
    commands:
      - echo Build started on `date`
      - cd aws-code-deploy/test
```

- `build`: This is an optional block, used to specify the commands you need to run during the build:

```
phases:
  build:
    commands:
      - echo Build started on `date`
      - mvn compile
```

- `post_build`: This is an optional block that allows you to specify any commands that need to be run after the build step; for example, packaging the build artifacts, pushing a Docker image to the Docker repository or Amazon ECR, or sending a build notification:

```
phases:
  post_build:,
    commands:
      - echo Build completed on `date`
      - mvn package
```

Most of the phases discussed in the preceding list have the following sections. These sections define what is being performed in that particular phase. Let's now examine all the sections each phase can have:

- `run-as`: This is an optional sequence and available to Linux users only. This specifies the user under which the subsequent commands will be run. If you have already specified a `run-as` value at the top level of the `buildspec.yml` file, then you can override it for a specific phase here as required.

- `on-failure`: This is an optional sequence that specifies what action needs to be taken when an error is encountered during a given phase. An `ABORT` value means that the build will fail in the current phase itself, while a `CONTINUE` value indicates that even if there is an error in the current phase, the build should continue to the next phase.

- `runtime-versions`: This is an optional block that specifies the runtime environment in which to run the commands. If `runtime-versions` is specified, it should include at least one runtime. You can use more than one runtime by providing a version number or environment variable. See the following example:

```
phases:
  install:
    runtime-versions:
      java: corretto11
      python: 3.x
      ruby: "$LATEST_RUBY_VAR"
```

- `commands`: This block contains one or more commands to be executed during the current phase. For some reason, if one or more commands fail in the current phase, then that phase will be considered as a failure.

- `finally`: This is an optional block that indicates the commands that need to be executed after the commands specified in the `commands` section are completed. In case of an error in the `commands` block, the `finally` block will still be executed.

`reports`: The `reports` section signifies the test and code coverage reports being generated as part of the build process. The `reports` section has the following subsections:

- `report-group-name-or-arn`: This is the optional section; you specify the name or the **Amazon Resource Name** (**ARN**) for the report group to which you want reports to be sent. You can specify an existing report group name or a new report group; CodeBuild will automatically create a report group if one doesn't already exist under a given name. You can specify a maximum of five report groups for each project.

- `files`: If you have specified a report group, then this section is required. This section lets you specify the location of the raw test report data generated in the build process. You can specify a single file or a filename pattern.

- `file-format`: This is an optional section that allows you to specify the type of the test/ coverage reports generated. The following are the different test file formats supported by CodeBuild:

  - `CUCUMBERJSON`

  - `JUNITXML`

  - `NUNITXML`

  - `NUNIT3XML`

  - `TESTNGXML`

  - `VISUALSTUDIOTRX`

- The following are the supported code coverage file formats:

  - CLOVERXML

  - COBERTURAXML

  - JACOCOXML

  - SIMPLECOV

If you don't specify a file format, CodeBuild will use JUNITXML by default.

base-directory

This is an optional mapping and represents one or more top-level directories related to the original build directory. This directory is used by CodeBuild to find the stored raw test report files.

discard-paths

This is an optional section used to specify whether test report files are flattened in the output or not. If you specify no or don't provide a value at all, then each report file will be generated in the same output directory where the given test is located. If you specify yes, then the test result file is generated at the root of the directory level. For example, if the path to a test result is com/packt/demoapp/CodeDeployTestResult.xml, then answering yes to discard-path option will generate the test result file at /CodeDeployTestResult.xml.

artifacts

This is an optional section used to prepare the final outcome of the build process. In this section, you specify the files CodeBuild needs to upload to the S3 bucket as the final result. This section is not required if you have already specified a post-build phase in your buildspec.yml file that will generate a final build package, such as a .jar or .war file for a Java web application.

files

This is a required section inside the artifacts section. This lets you specify one or more files that are part of your artifact.

name

This is an optional section where you specify a name for the build artifact. You can specify a name using a combination of static and dynamic values such as date and time to give a unique name to each artifact. See the following example:

```
artifacts:
  name: code-pipeline-demo-$(date +%Y-%m-%d)
  files:
```

```
      - target/code-pipeline-demo.jar
      - appspec.yml
      - scripts/*
```

At this point, we have a good understanding of the buildspec.yml file. Next, let's create one for the aws-code-pipeline project and check this file in to the CodeCommit repository so the CodeBuild service can use this file to execute the instructions to build our project.

Creating a buildspec file

To create a buildspec.yml file for our aws-code-pipeline microservice, perform the following instructions:

1. Open the VS Code IDE, click **File | New File**, name the file buildspec.yml, and save the file:

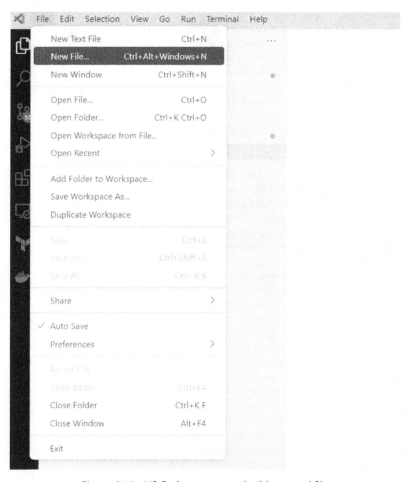

Figure 8.17 – VS Code create new buildspec.yml file

2. Add the following code to the buildspec.yml file and click **File | Save**. In this file, we are installing corretto11 Java for compiling our source code in the install phase, and in the pre_build phase we just print the Java version, although in a more complex application, we could do many more things. In the build phase, we run our test cases, while in the post_build phase we package the Spring Boot JAR file, which we publish as an artifact in the artifact section:

```
version: 0.2
phases:
  install:
    runtime-versions:
      java: corretto11
  pre_build:
    commands:
      - echo Starting pre-build phase
      - java -version
  build:
    commands:
      - echo Build started on `date`
      - mvn test
  post_build:
    commands:
      - echo Build completed on `date`
      - mvn package
reports:
  arn:aws:codebuild:us-east-1:279522866734:report-group/
aws-code-deploy-test-report:
    base-directory: 'target/surefire-reports'
    files:
      - TEST-com.packt.aws.books.pipeline.
AwsCodePipelineApplicationTests.xml
    discard-paths: yes
    file-format: JUNITXML
artifacts:
  files:
    - target/aws-code-pipeline*.jar
    - appspec.yml
    - scripts/*
  discard-paths: yes
```

Executing test cases in the build phase generates the surefire test report XML files, which we upload to the test group we created earlier with the arn:aws:codebuild:us-east-1:xxxxxxxxxxxx:report-group/aws-code-deploy-test-report ARN. You can download the full source code for this chapter from https://github.com/PacktPublishing/Building-and-Delivering-Microservices-on-AWS/tree/main/chapter_8/aws-code-pipeline.

3. Now we have the `buildspec.yml` file ready, so let's check this file in to our CodeCommit repository using the `git` commands we learned in earlier chapters. As you can see in the following screenshot, we now have our `buildspec.yaml` file available in the CodeCommit repository:

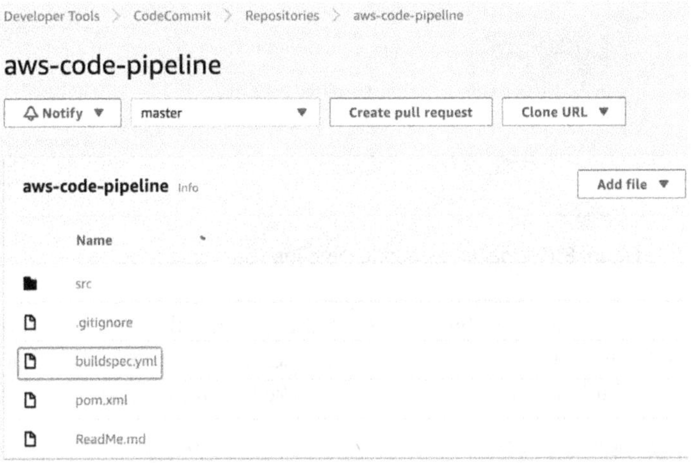

Figure 8.18 – AWS CodeComit aws-code-pipeline repository buildspec.yml file

Now that we have everything ready, let's go ahead and start the build for our CodeBuild project to see everything in action.

Starting a build

When we start a build, the CodeBuild service is responsible for preparing the required infrastructure based on the configuration and environment type we specified during the build project creation. Once the build environment is ready, CodeBuild will execute the instructions we provided in the `buildspec.yml` file in different stages and publish the artifacts and required reports. To start the build process for the `packt-aws-code-pipeline` CodeBuild project we created earlier in this chapter, perform the following steps:

1. Click on the **Build projects** in the left panel of CodeBuild and then select the `packt-aws-code-pipeline` project. Click on the **Start build** button and select **Start now**:

Figure 8.19 – packt-aws-code-pipeline start build process

2. Once you click on the **Start now** button, CodeBuild will start the build execution and take you to the build status page as shown in the following screenshot. You can see that a unique ID is generated for this build and the build status is marked as **In progress**:

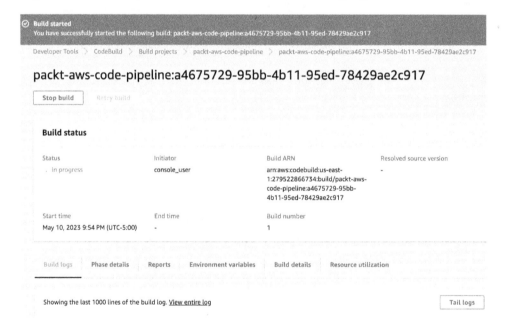

Figure 8.20 – packt-aws-code-pipeline in progress build

3. We can see the logs of the build on the **Build logs** tab. The build will take a few minutes to complete. Once the build is successful, the status will change to **Succeeded**. If for some reason a build fails, then the status will be marked as **Failed** and we can find the reason for the failure in the logs, shown as follows:

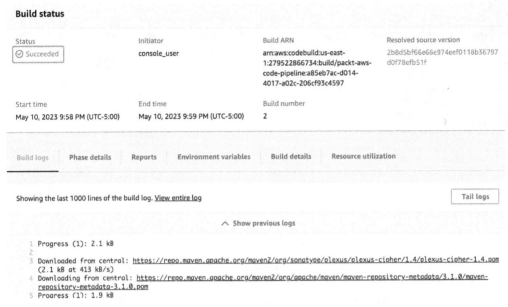

Figure 8.21 – packt-aws-code-pipeline success build and build logs

4. The **Phase details** tab for a given build displays the different phases and how much time each phase took. This tab helps us determine the progress of the build during execution:

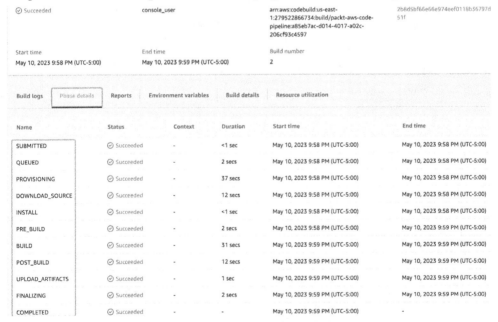

Figure 8.22 – packt-aws-code-pipeline different build phases

5.   The **Reports** section displays the test cases and code coverage reports generated during the build execution. Reports are generated and published under the group that we defined in the `buildspec.yml` file. Click on the report name link:

Figure 8.23 – packt-aws-code-pipeline test report status

6.   Once we click on the report name link, it takes us to the detailed report view, where we can see how many test cases were run and how many passed or failed in graphical and tabular form. This page provides us with the option to filter the results by the different test case statuses:

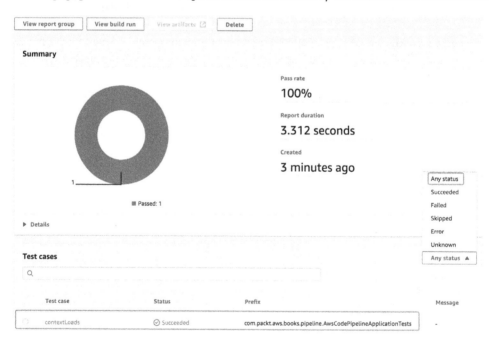

Figure 8.24 – packt-aws-code-pipeline test report details

7.   The **Environment variables** tab shows the different environment variables utilized in a particular build execution. In our case, we didn't define any variables, so this section is empty:

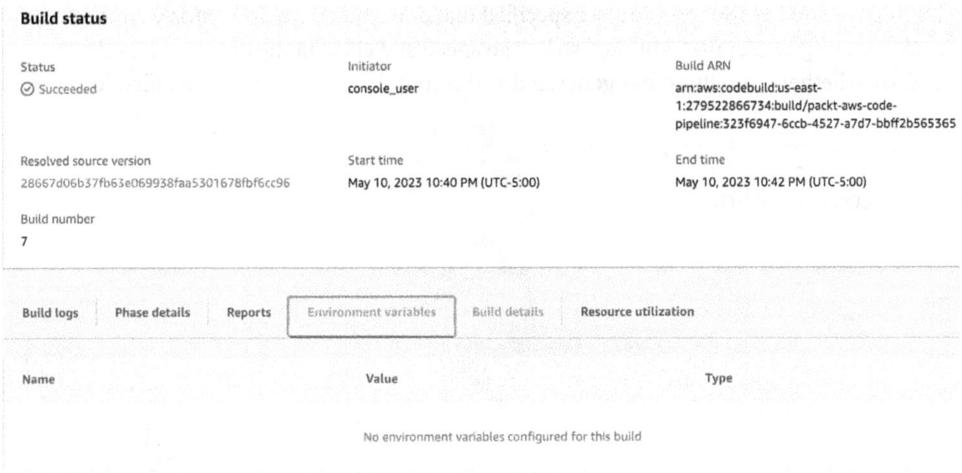

Build status

Status	Initiator	Build ARN
⊘ Succeeded	console_user	arn:aws:codebuild:us-east-1:279522866734:build/packt-aws-code-pipeline:323f6947-6ccb-4527-a7d7-bbff2b565365

Resolved source version	Start time	End time
28667d06b37fb63e069938faa5301678fbf6cc96	May 10, 2023 10:40 PM (UTC-5:00)	May 10, 2023 10:42 PM (UTC-5:00)

Build number
7

Build logs	Phase details	Reports	Environment variables	Build details	Resource utilization

Name	Value	Type

No environment variables configured for this build

Figure 8.25 – packt-aws-code-pipeline environment variables details tab

8. The **Build details** tab displays the build environment information based on the configuration, so if you were to change this information at a later time, you could still come back to a particular build's details and see what the configuration was when this particular build was executed:

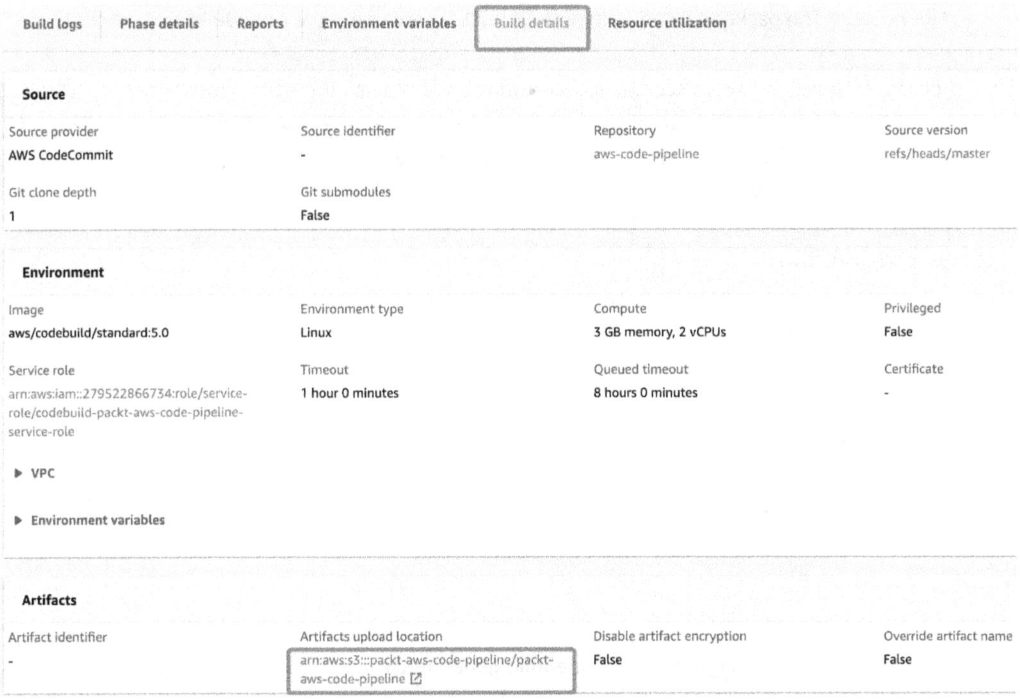

Build logs	Phase details	Reports	Environment variables	Build details	Resource utilization

Source

Source provider	Source identifier	Repository	Source version
AWS CodeCommit	-	aws-code-pipeline	refs/heads/master

Git clone depth	Git submodules		
1	False		

Environment

Image	Environment type	Compute	Privileged
aws/codebuild/standard:5.0	Linux	3 GB memory, 2 vCPUs	False

Service role	Timeout	Queued timeout	Certificate
arn:aws:iam::279522866734:role/service-role/codebuild-packt-aws-code-pipeline-service-role	1 hour 0 minutes	8 hours 0 minutes	-

▶ VPC

▶ Environment variables

Artifacts

Artifact identifier	Artifacts upload location	Disable artifact encryption	Override artifact name
-	arn:aws:s3:::packt-aws-code-pipeline/packt-aws-code-pipeline 🗗	False	False

Figure 8.26 – packt-aws-code-pipeline build details and artifact configuration

9.   During the build project creation, we specified that our artifact should be stored in the `packt-aws-code-pipeline` S3 bucket, so let's go ahead and click on the link mentioned previously and see whether CodeBuild has generated and stored the artifact in the specified location:

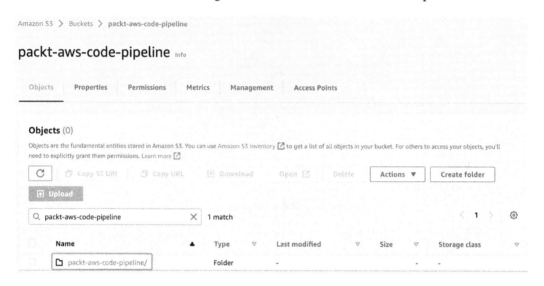

Figure 8.27 – The packt-aws-code-pipeline S3 bucket containing the CodeBuild artifact

10.   Click on the folder name to see the actual artifact. We can see the `aws-code-pipeline-xxx.jar` file is available as an outcome of the `mvn package` Maven command we defined in the `post_build` step of the `buildspec.yml` file:

Figure 8.28 – CodeBuild generated artifact

11. The **Resource utilization** tab of the build project allows us to visualize the memory and CPU usage, along with other metrics of the underlying infrastructure used to build the project:

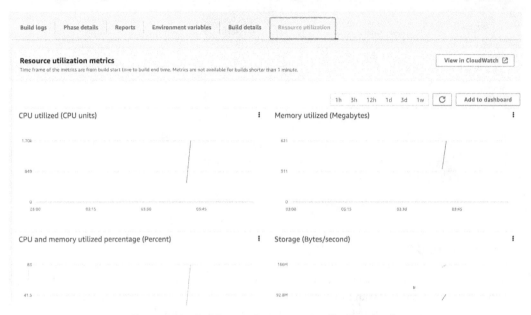

Figure 8.29 – Build execution resource utilization details

Each build attempt, be it successful or a failure, generates a record in the CodeBuild service that goes into the build history. CodeBuild allows you to see the build job history for up to 1 year. Let's now see how you can find the history of a build job.

Build history

The build history allows us to view data on historical job execution in the CodeBuild service. To see the history of executed jobs, you can click on the **Build history** link on the left panel of CodeBuild. This will show you the different build executions listed for different build projects. You can click on any of the links to go to the details of a particular build execution:

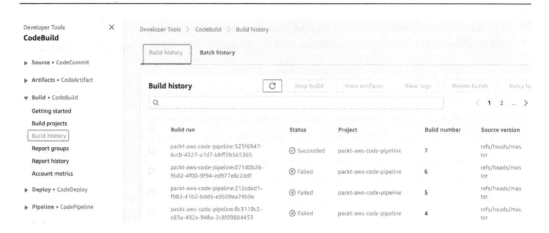

Figure 8.30 – packt-aws-code-pipeline build history

Report groups and history

The **Report groups** section allows us to visualize the reports generated as part of the build execution and published to the report group. In our case, we published the JUnit test case using the Maven surefire plugin. Click on the **Report group** tab in the left panel and select the aws-code-deploy-test-report report group we created earlier in this chapter:

Figure 8.31 – Report group details

AWS provides two types of reports, one for test case execution and another for code coverage. Once you click on the report group, you can see the consolidated view of the test cases or test coverage and test case trends over a given period of time:

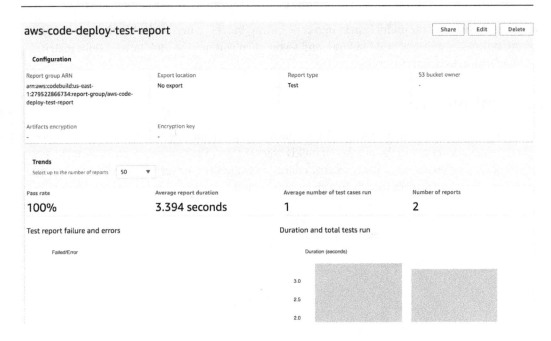

Figure 8.32 – aws-code-deploy-test-report group trend details

On the same page, you can see the history of successful and unsuccessful reports published under this group. Alternatively, you can click on the **Report history** link in the left panel to see the report history. If you click on a particular report link, you can see how many test cases were executed and those that passed and failed:

Figure 8.33 – Test group and code coverage report history

Test group reports provide useful insight into code quality, but if you need to look at the overall build execution times, number of build failures, and so on, then you need to look into account metrics. We'll look at this in the next section.

Account metrics

The **Account metrics** section provides graphical information about the usage of the CodeBuild service. It provides metrics related to the number of builds executed, build minutes used, and successful and failed build execution counts. Account metrics data is provided as time series, so you can adjust the time filter to examine metrics specific to a given time period:

Figure 8.34 – AWS CodeBuild usage metrics

Account metrics allow you to understand the overall build performance and see how much time on average your build executions are taking and how many builds are failing. Based on this data you can review the build steps to ensure that you are working efficiently and not using build minutes unnecessarily, as AWS charges you on the basis of build minutes used. AWS allows you to set up build notifications to get status updates for build executions. In the next section, we'll see how to set up build notifications.

Build notifications

CodeBuild allows you to configure notifications through SNS and Amazon Chatbot (Slack) to get status updates on build progress. You can define what status changes should trigger these notifications. Perform the following instructions to configure build notifications:

1. In the `packt-aws-code-pipeline` CodeBuild project, click on **Settings** under **Build projects**:

Figure 8.35 – AWS CodeBuild project settings

2. Click on the **Notifications** tab and then the **Create notifications rule** button to create a notification:

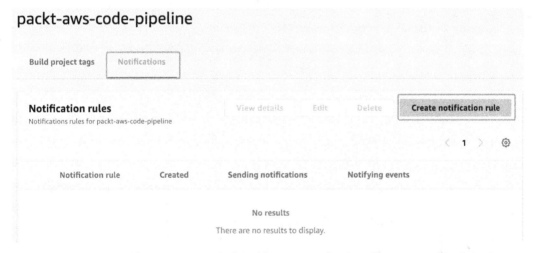

Figure 8.36 – CodeBuild project notifications tab

3.   Provide a notification name, specify the type of details you want to see in the notification, and select the events that should trigger the notification. In the following screenshot we have selected all events, so we will get notifications as the build moves through each status. Select and configure the notification target – you can select **SNS topic** or **Amazon Chatbot (Slack)**. We select **SNS topic** and our existing SNS topic from the **Choose target** dropdown. If you need help with creating an SNS topic, you can follow the instructions provided in *Appendix, Creating SNS notification* section. Click on the **Submit** button to create the notification rule:

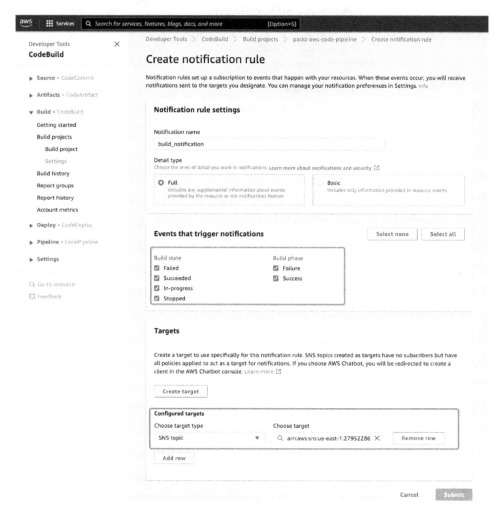

Figure 8.37 – CodeBuild create notifications rule

4.   The following screen shows confirmation that a notification rule has been created and the notification target is active:

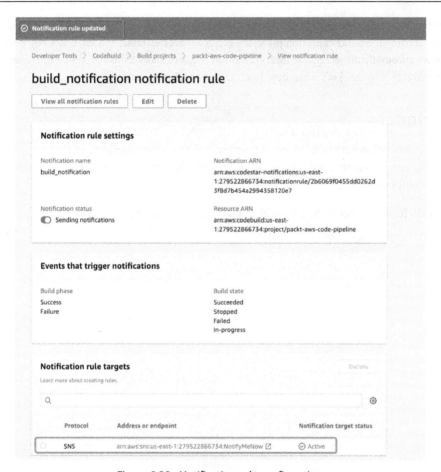

Figure 8.38– Notification rule confirmation

5.   Once you start the build project, the notifications will start for the different build status changes:

Figure 8.39– CodeBuild SNS notification sample event

The preceding screenshot shows a build notification triggered and sent to an email address via SNS notification. Starting the build execution manually is not always desirable, as sometimes you need your builds to start automatically at regular intervals. Build triggers allow us to schedule build executions for our CodeBuild projects. Let's learn how to create build triggers next.

Build triggers

Build triggers allow us to automate the build start process on a scheduled interval. You can configure the build to start on a fixed hourly, daily, weekly, or custom schedule. Perform the following instructions to enable the build triggers:

1. Click on **Build project** in the left panel and select the `packt-aws-code-pipeline` project:

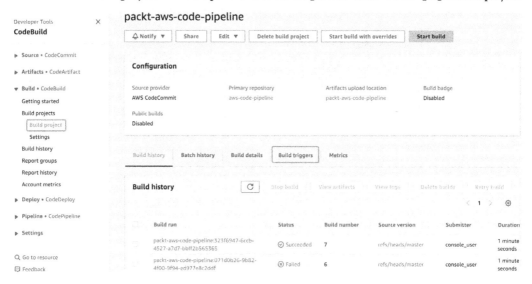

Figure 8.40 – packt-aws-code-pipeline build triggers

2.   Click on the **Build triggers** tab for the selected project and then click on the **Create trigger** button:

Figure 8.41 – AWS CodeBuild triggers

3.   Provide a trigger name and description. Set **Frequency** to **Daily** and select the hour and minutes at which you want to start the build. All times are in the UTC time zone for the build trigger. You can select an **Hourly**, **Daily**, or **Weekly** frequency or choose **Custom** to provide a cron job style expression. You can configure a build timeout if you want your build to be stopped if it is not completed within a certain time limit:

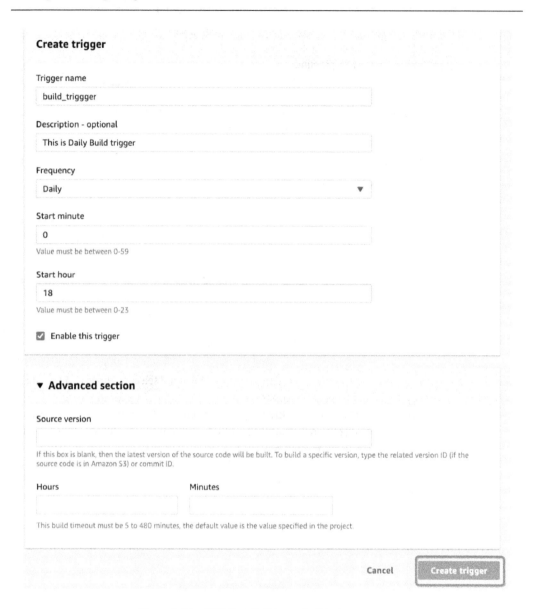

Figure 8.42 – AWS CodeBuild new build trigger setup

4. Once you click on the **Create trigger** button, it will take you to the configuration screen where you can edit the trigger by selecting it. This trigger will automatically start the build when the configured time is reached:

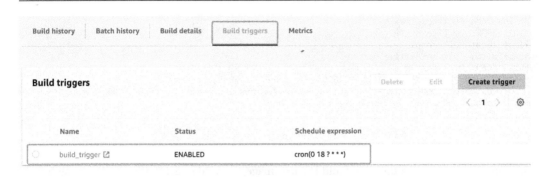

Figure 8.43 – AWS CodeBuild new build trigger confirmation

CodeBuild triggers execute builds periodically at the same configured interval as defined in the configuration. So, if you are using these triggers only for testing purposes then delete the build trigger after testing to avoid unnecessary costs.

Local build support

Whenever you make changes to your buildspec.yml file, you have to commit it to the repository and start the CodeBuild job. This creates a lot of overhead for developers and increases the build minute costs for projects. To overcome this, CodeBuild supports local builds through CodeBuild agents. To set up your local environment to test the buildspec.yml file, implement the following steps:

1. As a first step, you need to download the Docker image for the CodeBuild runtime selected for your build project in the previous section. You can pull these images from the AWS Docker registry as shown below or you can build these images yourself by cloning the source from the AWS CodeBuild Git repository at https://github.com/aws/aws-codebuild-docker-images.git. Here, we are focusing on using the existing Docker image. To install Docker locally, follow the instructions from *Appendix, Docker Desktop Installation*, where we have provided a step-by-step guide. Execute the following command:

```
docker pull public.ecr.aws/codebuild/amazonlinux2-x86_64-standard:3.0
```

2. Once you execute the preceding command, you should see a confirmation that the CodeBuild Docker image has been downloaded to your local system, as shown in the following screenshot.

```
0b8b37db8ddd: Pull complete
86f1b23c5392: Pull complete
73618c4c40ca: Pull complete
3e55ab15d4c6: Pull complete
f1951154f469: Pull complete
Digest: sha256:203f7b0db002081679e59460059f9078f49d9b6600d8a1fb961fe964bdac5a56
Status: Downloaded newer image for public.ecr.aws/codebuild/amazonlinux2-x86_64-standard:3.0
public.ecr.aws/codebuild/amazonlinux2-x86_64-standard:3.0
```

Figure 8.44 – AWS CodeBuild runtime Docker image

3.   Once we have pulled the CodeBuild Docker image, we need to get the latest local CodeBuild agent to execute builds locally. Execute the following command to download the local build agent:

```
docker pull public.ecr.aws/codebuild/local-builds:latest
```

4.   As the next step, we need to download the CodeBuild script to execute local builds. Execute the following command:

```
curl -O https://raw.githubusercontent.com/aws/aws-codebuild-docker-images/master/local_builds/codebuild_build.sh
```

5.   The following screenshot shows the script downloaded in our source code directory. Change the file permissions as required and execute the script with chmod +x codebuild_build.sh:

Figure 8.45 – Local build script file

6. Now we have everything in place to start building the CodeBuild project locally and make sure that the changes we have made to our `buildspec.yml` file will work. Execute the downloaded `codebuild_build.sh` script with the following command, where the first parameter is the CodeBuild runtime environment image and the second parameter is the CodeBuild artifact directory where we want CodeBuild to generate our output artifacts as part of the build process:

```
./codebuild_build.sh -i public.ecr.aws/codebuild/amazonlinux2-x86_64-standard:3.0 -a .
```

7. The following screenshot shows that CodeBuild local build has executed successfully and you can see the `artifacts.zip` generated artifact file in our source directory. If you make any syntax error (or other error) in the `buildspec.yml` file, the local build will fail and you will see an error output on the console:

Figure 8.46 – AWS CodeBuild local build execution

Testing CodeBuild projects locally provides developers great flexibility to test their changes before they push code to their repository and use build minutes on the server. The capability to run builds locally saves developers time and reduces project costs. Downloading the CodeBuild runtime image and CodeBuild local agent is a one-time job, and in future instances you can just execute the script straight away.

Summary

In this chapter, we learned about the AWS CodeBuild service and examined the benefits and limitations this service has. We also created a new CodeBuild project, set up notifications, and mastered build triggers to run the build on a regular, fixed schedule. In addition to this, we also learned about the `buildspec.yml` file syntax and created a `buildspec.yml` file for our `aws-code-pipeline` CodeCommit project. We started the build process and reviewed the generated artifact along with different build phases and metrics.

In the next chapter, we will learn about the AWS CodeDeploy service and how to deploy the artifact we generated in this chapter through CodeBuild project build.

Part 3: Deploying the Pipeline

This section focuses on the deployment of a sample microservice application to the AWS environment and provides a deep dive into the AWS CodeDeploy and CodePipeline services.

This part has the following chapters:

- *Chapter 9, Deploying to an EC2 Instance Using CodeDeploy*
- *Chapter 10, Deploying to ECS Clusters Using Code Deploy*
- *Chapter 11, Setting Up CodePipeline*
- *Chapter 12, Setting Up an Automated Serverless Deployment*
- *Chapter 13, Automated Deployment to an EKS Cluster*
- *Chapter 14, Extending CodePipeline Beyond AWS*

Deploying to an EC2 Instance Using CodeDeploy

In this chapter, you will learn about the AWS CodeDeploy service and how this service allows us to automate the software deployment process. We will learn about the service's benefits and limitations and learn the terminology related to the CodeDeploy service. In this chapter, we will be covering the following topics:

- What is AWS CodeDeploy?

- Understanding the **application specification file (AppSpec file)**

- Deployment lifecycle events

- Deployment groups

- Deployment strategies

- The CodeDeploy agent

- Sample app deployment to EC2 instances

In the previous chapters, we developed a microservice using Java, and we generated the artifact for that microservice using the CodeBuild service. In this chapter, we will use the same sample microservice, and try to deploy it to the AWS EC2 instance using the CodeDeploy service. Before discussing the deployment in more detail, we need to understand what the AWS CodeDeploy service is, so let's focus on that first.

What is CodeDeploy?

Deployment is the ultimate destination for any software so that it can start serving consumers. Previously, to deploy a software package, you would have to prepare a set of instructions that needed to be followed by the operations team and they performed the code deployment manually. Consistently following these manual instructions is not scalable and prone to error, and we may face several deployment failures followed by manual rollbacks.

The AWS CodeDeploy service automates the deployment part of the application. CodeDeploy is a managed deployment service that provides the capability to deploy software artifacts across different platforms.

AWS CodeDeploy automates the software deployment process to different types of infrastructures, including Amazon EC2, AWS Fargate, AWS Lambda, and on-prem servers using CodeDeploy agents. AWS CodeDeploy makes it easy to automatically push newer versions of an application to your servers without causing downtime to the existing version of the application.

The following diagram explains how the CodeDeploy service takes an artifact uploaded into an S3 bucket and deploys to the EC2 instances, AWS Lambda, or on-prem systems:

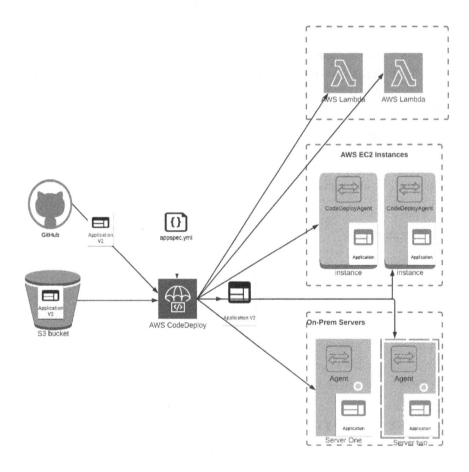

Figure 9.1 – The AWS CodeDeploy service deployment model

The CodeDeploy service works based on the client service architecture, so CodeDeploy services connect to the CodeDeploy agents installed on the instances, push the changes, and deploy the application based on the instructions provided in the `appspec.yml` file.

Now that we understand what the CodeDeploy service is, let's explore some of the benefits it provides.

The benefits of CodeDeploy

Some of the benefits of AWS CodeDeploy Service are listed as follows:

- CodeDeploy is a fully managed service, so you don't have to worry about adding storage or compute capacity or patching the servers to use this service.

- The CodeDeploy service is not limited to the AWS cloud, so you can deploy to on-prem servers as well.

- The CodeDeploy service helps automate the deployment process, so it can produce consistent results, and you don't have to follow a set of manual commands.

- The CodeDeploy service helps us to minimize downtime and maximize availability during deployment by introducing changes incrementally using rollover updates of the applications. You control how you want to deploy the application on your servers, and you can define the number of instances that should be updated at a time and other instances serving live traffic.

- The CodeDeploy service provides step-by-step updates on the deployment process, so in case of a deployment failure or issues, you don't have to troubleshoot everything; you can look at a particular step directly.

- Blue-green deployment is supported by the CodeDeploy service, so you can independently deploy a newer version of the software without affecting the currently running version, and once the newer version is validated, you can switch traffic to a newer version of the application.

- The CodeDeploy service is based on the existing deployment principles and provides easy integration with other AWS services, so it is easy to adopt. In addition, it is platform neutral and provides the same deployment experience whether you are deploying to on-prem servers or EC2 instances.

- There is no additional cost to use the AWS CodeDeploy service; you just need to pay for the infrastructure you are using to deploy the code. The cost of deploying to on-prem systems is minimal.

Now that we have a fair idea of the different benefits provided by the CodeDeploy service, let's look at some of the limitations of the service.

The limitations of CodeDeploy

While AWS CodeDeploy is a highly scalable and serverless service, it has the following limitations at the time of writing this book:

- The CodeDeploy service can only perform deployments with AWS resources located in the same Region. For deployment to other Regions, you need to redefine the application in another Region and copy the application package to S3 in the local Region you want to deploy.

- The CodeDeploy service allows a maximum of one concurrent deployment to a deployment group.

- The CodeDeploy service supports a maximum of 100 concurrent deployments within an AWS account.

- A maximum of 1,000 applications can be created within an AWS account. If you are a large organization with more than 1,000 applications, you must request a service quote increase.

- The CodeDeploy service is limited to deploying the archive files with the `.zip` or `.tar` extensions for EC2/on-prem instances.

- CodeDeploy service allows you to deploy to multiple autoscaling groups together, but you can specify a maximum of 10 autoscaling groups.

- CodeDeploy service allows a maximum of 500 instances in a single deployment; if you have a larger deployment, then you will face this limitation.

- Blue-green deployment allows a maximum of 120 hours between deploying an application revision and traffic switching. If you expect more than 120 hours to validate your application revision, you will face this limitation.

A full list of limitations can be found here: `https://docs.aws.amazon.com/codedeploy/latest/userguide/limits.html`.

To understand the CodeDeploy service better, we need to learn about a few concepts related to CodeDeploy. So, let's first focus on understanding what an application is in the context of the CodeDeploy service.

What is an application?

An application is a software package that needs to be deployed on a set of servers. A software package can be just a binary or a combination of configuration and binary, which needs to be deployed to EC2 instances, on-prem servers, **Elastic Container Service** (**ECS**), or Lambda function. To do anything using AWS CodeDeploy, as a first step, we need to create an application.

An application revision is a specific version of the software package that we deploy to the server.

Deployment strategies

Deployment strategies allow us to determine how we want to roll out the application revision to the existing workload. For EC2 instance-based deployment, CodeDeploy supports two types of deployments. So, let's look at them one by one.

In-place deployment

In-place deployment doesn't create a separate instance for the new application revision; it deploys the new version of the application on the same instances, and each instance is taken offline for the duration of the deployment to that particular instance. Once the deployment is completed to that instance, it is added back to handle the traffic.

In this strategy, you can cause an outage to your users if you haven't chosen a rolling update and try to install the new version simultaneously on all the instances in the deployment group. The deployment setting allows you to choose an existing deployment configuration or even create a new one so that you can choose how you want your deployment to take place. In addition, you can update each instance at a time, half at a time, or all at once.

Blue-green deployment

The blue-green deployment strategy doesn't interfere with the existing workload during deployment; it deploys the newer application version to newly created instances. Traffic is switched to the new instances only once when those instances are ready to take the traffic. This way, the existing application instances are not disturbed during the deployment and keep running at full capacity.

In blue-green deployment, you can choose at what point you want to switch the traffic from the blue version to the green version of the application and CodeDeploy can switch the traffic automatically or manually.

Deployment group

In the CodeDeploy service, a deployment group is a collection of the EC2 instances or AWS Lambda functions you want to target for deployment. You can associate the EC2 instances to a deployment group either using an autoscaling group using tags, or both.

A CodeDeploy application can have more than one deployment group, and usually, you create a separate deployment group for each environment, such as development, staging, and production.

Deployment configuration

Deployment configuration determines the deployment behavior within a deployment group. Deployment configuration specifies how the CodeDeploy service should handle the deployment and any failure during deployment. Using deployment configuration, you can control what percentage or how many instances you want to target for deployment at a time within your deployment group and how gradually you scale the deployment to the remaining workload.

For AWS Lambda and ECS deployment types, you can choose between canary or linear deployment types and increase the percentage of traffic to the new revision of the application at a configured time.

AWS CodeDeploy provides several deployment configurations by default and, using them, you can design a deployment strategy to perform a zero-downtime deployment model. By default, if you don't specify any deployment configuration, then CodeDeploy will deploy to one instance at a time for EC2 instance-based deployments. If the existing deployment configuration doesn't satisfy your needs, you can create your own deployment configurations. To create a new configuration, let's follow these steps:

1. Click on the CodeDeploy service and click **Deployment configurations** in the left panel. Click on the **Create deployment configuration** button:

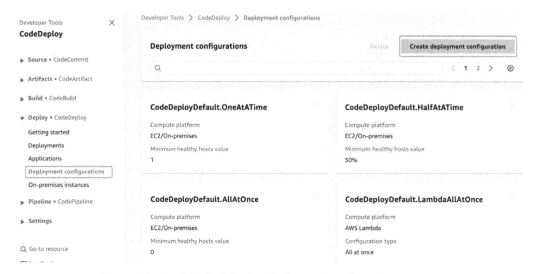

Figure 9.2 – The AWS CodeDeploy Deployment configurations screen

2. Provide the deployment configuration details such as the **Deployment configuration name** and the **Compute platform** options. If you select **Amazon ECS** or **AWS Lambda** as the **Compute platform** option, then you will be asked for additional details such as **Step** and **Interval** to gradually update traffic. If you choose **Amazon ECS** as the **Compute platform** option, you need to select either the percentage of instances or the number of instances to perform the deployment with at a time:

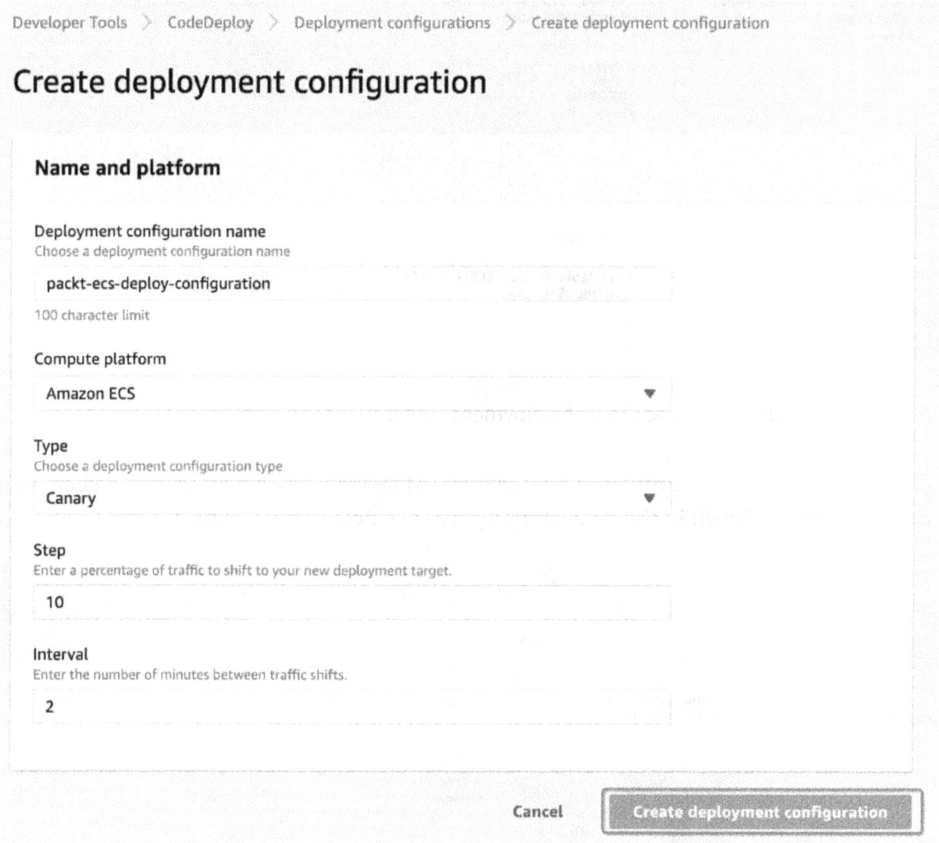

Developer Tools > CodeDeploy > Deployment configurations > Create deployment configuration

Create deployment configuration

Name and platform

Deployment configuration name
Choose a deployment configuration name

> packt-ecs-deploy-configuration

100 character limit

Compute platform

> Amazon ECS ▼

Type
Choose a deployment configuration type

> Canary ▼

Step
Enter a percentage of traffic to shift to your new deployment target.

> 10

Interval
Enter the number of minutes between traffic shifts.

> 2

Cancel   Create deployment configuration

Figure 9.3 – The Deployment configurations settings

3. Once you click on the **Create deployment configuration** button, it will create a deployment group configuration, and you can use this configuration while creating a deployment group:

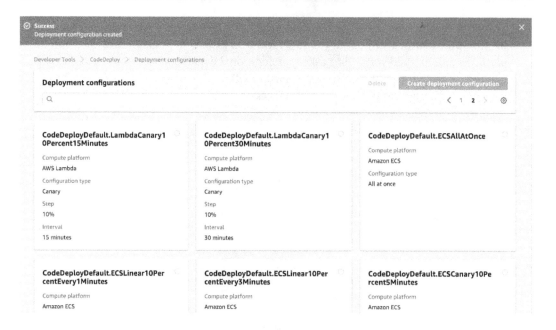

Figure 9.4 – The Create Deployment configuration confirmation

4. To delete a deployment configuration item created by you, you can select it and click on the delete button, as shown in the following screenshot. Click on the **Delete** button:

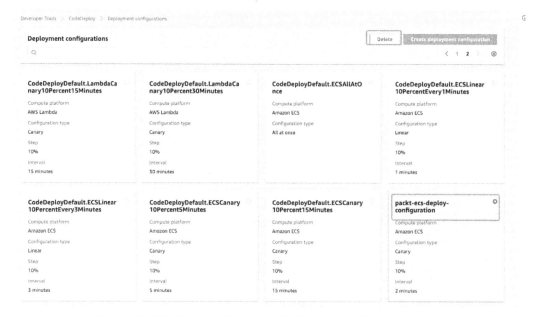

Figure 9.5 – The Delete option on the Deployment configuration screen

5. In order to confirm the deletion, type `delete` in the provided input box and click the **Delete** button:

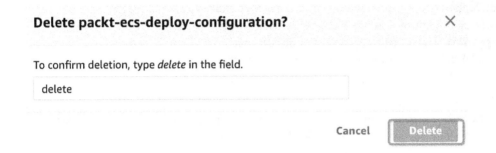

Delete packt-ecs-deploy-configuration?   ✕

To confirm deletion, type *delete* in the field.

```
delete
```

Cancel   **Delete**

Figure 9.6 – The Delete button on the Deployment configuration review

6. Once the configuration is deleted, it will show a confirmation similar to the following screenshot:

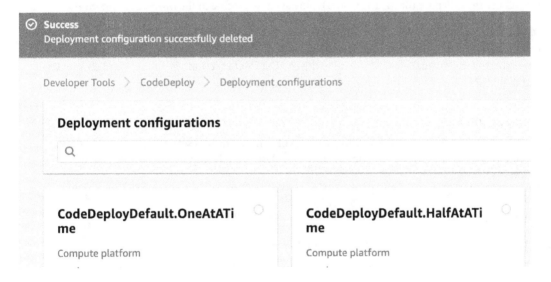

⊘ **Success**
Deployment configuration successfully deleted

Developer Tools > CodeDeploy > Deployment configurations

Deployment configurations

🔍

CodeDeployDefault.OneAtATi me

Compute platform

CodeDeployDefault.HalfAtATi me

Compute platform

Figure 9.7 – The deleted Deployment configurations confirmation

CodeDeploy service relies on the CodeDeploy agent to deploy to EC2 or on-prem instances. Let's take a quick look at the CodeDeploy agent and how to install it.

The CodeDeploy agent

The CodeDeploy agent is a software package installed on the EC2/on-prem instances and used by the CodeDeploy service to communicate to the instance and perform the deployment. By default, the CodeDeploy agent isn't installed on the EC2 instances until you have baked in an **Amazon Machine Image (AMI)** with the agent installed. For our example application, we will be installing and starting the CodeDeploy agent using the **User data** section of the EC2 template.

> **Important note**
>
> A CodeDeploy agent is only needed for EC2/on-prem instance deployments, for Amazon ECS or AWS Lambda deployments, you don't need to install the agent.

There are two ways to install a CodeDeploy agent on an EC2 instance – you can either install it using Amazon Systems Manager or directly using the command line/installer on the instance. It is easier to install and manage your instances using Amazon Systems Manager, but for the sake of simplicity, we will use the other approach. The CodeDeploy agent is specific to the **operating system (OS)** type.

In our case, we are going to deploy our application to the Amazon Linux-based system image so we can install and deploy the CodeDeploy agent using the following commands. We don't need to run these commands manually as we will be using the **User data** section within EC2 and automatically installing the CodeDeploy agent using the Terraform template when provisioning the infrastructure.

> **Note**
>
> The **User data** section allows you to pass a set of commands or a script to your EC2 instances that you want to execute at EC2 instance startup. **User data** helps to automate the startup steps on EC2 instances.

You can follow these steps to manually install the CodeDeploy agent on a Linux-based system:

1. Log in to the instance using the terminal window.

2. Install the `wget` utility to download the agent using the following command:

   ```
   sudo yum install wget -y
   ```

3. Once yum is installed, you can download the CodeDeploy agent from the S3 bucket of the Region where your instance is being provisioned. In our case, we are using the us-east-1 Region; you can find the exact S3 location of the agent by using this link: `https://docs.aws.amazon.com/codedeploy/latest/userguide/codedeploy-agent-operations-install-cli.html`. Run the following command to download the agent:

   ```
   wget https://aws-codedeploy-us-east-1.s3.amazonaws.com/latest/
   install
   ```

4. Change the file permissions to execute using this command:

    ```
    chmod +x ./install
    ```

5. Execute the CodeDeploy installer using the following command:

    ```
    sudo ./install auto
    ```

6. Once the CodeDeploy agent is installed on the instance, you can start it by running the following command:

    ```
    service codedeploy-agent start
    ```

For any Windows system installation, you can download the CodeDeploy agent msi installer using the following link: https://docs.aws.amazon.com/codedeploy/latest/userguide/codedeploy-agent-operations-install-windows.html.

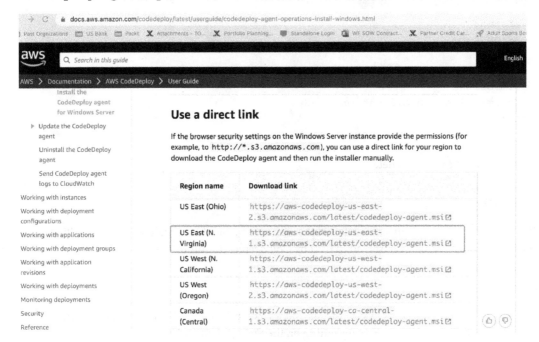

Figure 9.8 – The CodeDeploy agent Windows download page

The CodeDeploy agent uses a configuration file called appsepc.yml to execute scripts on the EC2 instances using lifecycle events to perform the deployment. So, let's explore the appspec.yml file in detail in the next section.

What is an AppSpec file?

The CodeDeploy service uses the YAML or JSON formatted AppSpec file to provide instructions for the deployment. This file is usually referred to as the `appspec.yml` file and is provided with the deployment package.

When we deploy a package, the CodeDeploy agent reads this file and executes the instructions provided in the `appsepc.yml` file. This file has two sections – one provides information regarding where the CodeDeploy agent needs to copy the artifact or package for installation, and another section contains the hooks to execute scripts for a specific deployment lifecycle.

The `appspec.yml` file syntax changes a bit based on the target compute platform. In this chapter, we are focusing on deployment to EC2 instances, so we will discuss the `appspec.yml` file syntax specific to the EC2/on-prem instance perspective, and in later chapters, we will cover the other two compute type-specific changes. The following is the structure of the `appspec.yml` file:

```
version: 0.0
os: os-name
files:
  source-destination-files-mappings
permissions:
  permissions-specifications
hooks:
  deployment-lifecycle-event-mappings
```

Let's learn about each tag in detail.

version

This section specifies the current version of the `appspec.yml` file specification. As of writing this book (December 2022), it is reserved to `0.0` by CodeDeploy for future use.

os

This section specifies the type of OS of the instances you are deploying. This is a required section and you can specify Linux or Windows depending on the OS type you are using.

files

This section provides information to the CodeDeploy service about the files that need to be installed on the EC2/on-prem instance from the application revision during the installation phase of deployment.

In the `files` section, you can provide multiple source and destination pairs. In the `files` section, `source` refers to a file or directory you want to copy from the application revision and `destination` refers to the location on the EC2/on-prem instance where you want to copy. The `source` path is related to the `appspec.yml` file, while for the `destination` path, you need to use a fully qualified location such as `/root/some/destination/somefile.sh` for a Linux-based system, or `C:/some/folder/` for a Windows-based system.

The `file_exists_behavior` section is optional and provides instructions to the CodeDeploy service about what needs to be done when a file already exists in the destination location but wasn't a part of the previous successful deployment. The following is a list of the different options `file_exists_behavior` can accept:

- `DISALLOW`: This is the default option if nothing else is specified. This option will cause the deployment to fail.

- `OVERWRITE`: This option will override the existing file version with the newest version from the current deployment.

- `RETAIN`: This option won't override the existing file and the deployment will skip the new version file and utilize the existing one for the current deployment.

Now that we understand the `files` section, let's take a look at how to provide appropriate permissions to these files once they are copied over to the EC2/on-prem instance.

permissions

This section is only applicable for Linux-based systems and it signifies how special permissions need to be applied to the files copied over to the EC2/on-prem instance using the `files` section. Following is an example of what the `permissions` section looks like:

```
permissions:
  - object: /tmp/my-app
    pattern: "**"
    owner: ec2-user
    group: ec2-user
    mode: 755
```

In this example, `object` key specifies the set of files or directories on which you want to apply a particular permission; you can define multiple sets of objects.

`pattern` identifies the files or directories to which `permission` is applicable within the `object` key; by default, `**` is the pattern if you don't specify explicitly. The `owner` section defines `owner` for the copied files or directories and `group` specifies to what Linux group this object will belong. You can specify the level of access to the `object` key using the numeric `permissions` value in the mode section.

hooks

The hooks section in the appspec.yml file allows us to define the script that needs to be executed at different lifecycle events of the deployment. The lifecycle events are specific to the compute type used for the deployment; here we are focusing on EC2/on-prem instance types, so the following are the different lifecycle events available for EC2 instance deployments:

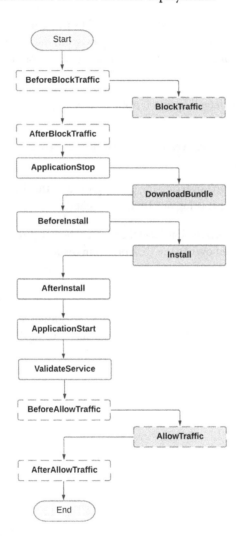

Figure 9.9 – The CodeDeploy lifecycle events

In the preceding diagram, deployment lifecycle events marked in the darker colored boxes are specific to the CodeDeploy agent, and you can't specify a script to execute. Events marked in dotted boxes are specific to the instances running behind a load balancer. In the `appspec.yml` file, you can specify the lifecycle events as a separate line and then specify the script you want to execute, as shown in the following example:

```
hooks:
  ApplicationStart:
    - location: start.sh
      timeout: 100
      runas: ec2-user
  ApplicationStop:
    - location: stop.sh
      timeout: 100
      runas: ec2-user
```

For each lifecycle event, you have a location section in which you can specify one or more scripts you want to execute. For each script, you have a timeout section to specify how long the CodeDeploy agent must wait for the script to be executed before its timeout. In the `runas` section, you specify the name of the user that the CodeDeploy agent will use to execute the script. Following this overview of the `hooks` section, let's look at each lifecycle event one by one so you understand what events you can utilize and are better suited for your application:

- `BeforeBlockTraffic`: This deployment lifecycle event runs any tasks before taking the instance out from the load balancer. You can specify a script to be executed before the instance is taken out of the load balancer pool. After this deployment event, the instance is deregistered from the load balancer.

- `BlockTraffic`: This deployment lifecycle event is reserved for the CodeDeploy agent, and you can't specify a script to run as a part of this event. During this event, traffic is blocked from accessing the instance.

- `AfterBlockTraffic`: This deployment lifecycle event allows you to specify and run a script once traffic is stopped from the instance and the instance is deregistered from the load balancer.

- `ApplicationStop`: This deployment lifecycle event occurs before the actual deployment. This event triggers scripts to gracefully stop an existing running application on the EC2/on-prem instance. If there is no previous revision available on the instance, this lifecycle event can be ignored, and you don't have to specify a script. This event happens even before the package is downloaded to the instance, so on the very first run of the application, this lifecycle won't execute the script, as it won't exist.

- `DownloadBundle`: This event is reserved for the CodeDeploy agent to download the application revision to a temporary location before copying it to the location you specified. This event is not allowed to be used in the `appspec.yml` file. If you encounter a deployment error in this phase and you are deploying the application from S3, ensure you have the right permissions in place, and the S3 bucket is in the same Region as the EC2 instance.

  CodeDeploy copies the package to the `C:\ProgramData\Amazon\CodeDeploy\ {deployment-group-id}\{deployment-id}\deployment-archive` location on a Windows-based system, while it copies the package to the `/opt/codedeploy- agent/deployment-root/{deployment-group-id}/{deployment-id}/ deployment-archive` temporary location on a Linux-based system.

- `BeforeInstall`: This deployment lifecycle event runs any pre-install script before deployment to the instance. This event can be used to download any dependency or decrypt files, get any password, or anything else required before deploying the application.

- `Install`: This deployment lifecycle event is reserved for the CodeDeploy agent to copy the application revision from the temporary location to the specified location.

- `AfterInstall`: This deployment lifecycle events allow you to specify and run a script after the deployment. You can use this event to configure your application, change file permissions, clean up any temporary files/downloads, and so on.

- `ApplicationStart`: This deployment lifecycle event restarts any service stopped during the `ApplicationStop` event. You can use this phase to run any script from your application bundle.

- `ValidateService`: This deployment lifecycle event runs any script to validate the application deployment.

- `BeforeAllowTraffic`: This deployment lifecycle event triggers any script after validating the service event and before the instance is registered with a load balancer to accept traffic.

- `AllowTraffic`: This deployment lifecycle event is used by the CodeDeploy agent to allow traffic to the instance after deployment. This event can't be used to execute a script.

- `AfterAllowTraffic`: This deployment lifecycle event executes any script once the instance is registered to the load balancer and starts accepting traffic.

Now that we understand the different deployment lifecycle events, it is time to create an `appspec. yml` file for the `aws-code-deploy` microservice application we created in the last few chapters.

Creating the appspec file

To create an `appspec.yml` file in our `aws-code-pipeline` microservice, perform the following steps:

1. Open the **Visual Studio (VS) Code integrated development environment (IDE)**, click **File | New File…**, and provide `appspec.yml` as the name:

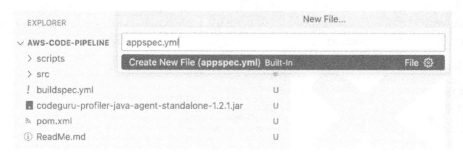

Figure 9.10 – VS Code creates a new appspec.yml file

2. Add the following code to the `appspec.yml` file and click **File | Save**. In this file, we are specifying the version of the CodeDeploy appspec file as `0.0` and `os` as Linux as we plan to deploy our application on a Linux system-based EC2 instances. In the `files` section, we wanted to copy everything from our deployment ZIP file to the home folder of `ec2-user`, and we have specified that `ec2-user` is the owner of all the files and has the required permissions to execute scripts. In the `hooks` section, we have specified two deployment lifecycle events that will be used by the CodeDeploy agent to execute the `start.sh` script on the `ApplicationStart` deployment event and the `shutdown.sh` script at the `ApplicationStop` deployment event to stop our microservice:

```
version: 0.0
os: linux
files:
  - source: /
    destination: /home/ec2-user/
permissions:
  - object: /
    pattern: "**"
    owner: ec2-user
    group: ec2-user
    mode: 755
hooks:
  ApplicationStart:
    - location: start.sh
      timeout: 100
```

```
      runas: ec2-user
ApplicationStop:
 - location: stop.sh
   timeout: 100
   runas: ec2-user
```

3. Now we have created the `appspec.yml` file, this file will be used by the CodeDeploy agent to deploy our application. Now, let's create the `start.sh` script file to start the microservices on the `ApplicationStart` lifecycle event. Create a folder named `scripts` and inside that, create a new file called `start.sh` and add a shell script like the one shown in the following code:

Figure 9.11 – The CodeDeploy start.sh script to start the application

This script starts the Spring Boot microservice Java process on port `80`.

4. We now need to create another script named `stop.sh` to stop the application and clean up the files from the system. Create a new file called `stop.sh` in the `scripts` folder and add the following code. In this file, we are killing the running Java process and removing the existing `aws-code-pipeline-x.x.jar` file, `appspec.yml`, and other scripts:

Figure 9.12 – The CodeDeploy stop.sh script to clean up the application

5. We now have all three `appspec.yml`, `start.sh`, and `stop.sh` files. Let's go ahead and check these files into our CodeCommit repository:

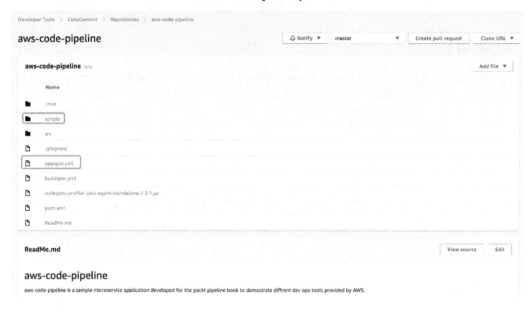

Figure 9.13 – The CodeCommit repository screen

6. Now we have our application code ready to deploy using the CodeDeploy service. In the next chapters, we will be setting up CodePipeline, which will take the artifacts produced from the CodeBuild service and use that for the deployment using the CodeDeploy service. In this chapter, we will take an alternate approach and deploy the application manually from the S3 bucket. As a next step, create a ZIP file containing the `appspec.yml`, `start.sh`, and `stop.sh` scripts and the `aws-code-pipeline-0.0.1-SNAPSHOT.jar` Spring Boot microservice JAR file:

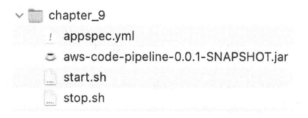

Figure 9.14 – The CodeDeploy ZIP file contents

7. Open the AWS console and upload the newly created `aws-code-pipeline.zip` zip file to the `packt-aws-code-pipeline` S3 bucket we created in previous chapters:

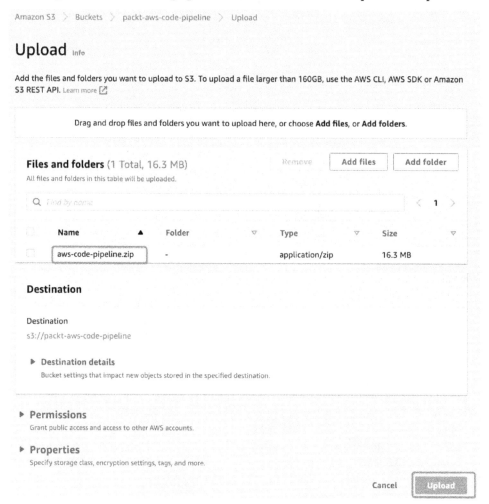

Figure 9.15 – The S3 ZIP file upload screen

8. Once the `aws-code-pipeline.zip` file is uploaded successfully, click on the filename, as shown as follows:

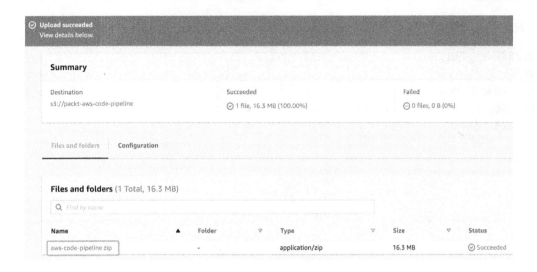

Figure 9.16 – The S3 ZIP file upload confirmation

9. Now click on the **Copy S3 URI** button as shown in the following screenshot and copy the **S3 URI** object; in our case, it is `s3://packt-aws-code-pipeline/aws-code-pipeline.zip`:

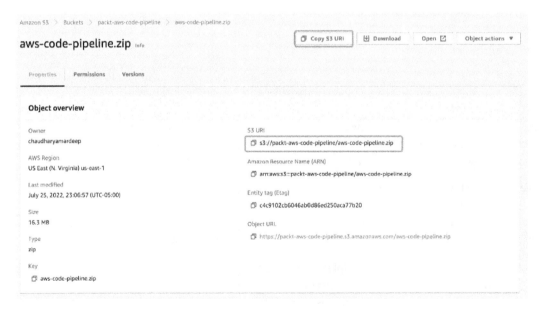

Figure 9.17 – The S3 ZIP file upload URL

For the CodeDeploy service to deploy the ZIP file, we first need to create the required infrastructure for deployment. Let's look into that next.

The deployment infrastructure

The following diagram shows the deployment architecture of the aws-code-pipeline microservice application. We will create a load balancer that will face any HTTP request on port 80, and then route requests to the healthy EC2 instance. We will create two EC2 instances and use the CodeDeploy service to deploy the application to these instances one at a time:

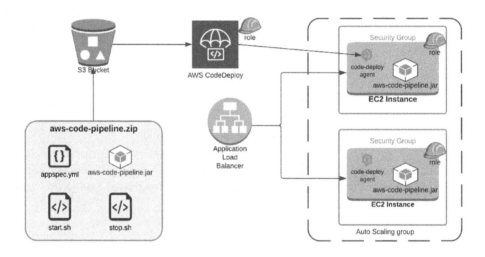

Figure 9.18 – A sample microservice deployment architecture

EC2 instances need to have the code deploy agent for the CodeDeploy service to communicate and deploy the application to the instance. EC2 instances also need to assume a role to connect to the S3 bucket to download the application ZIP file instructed by the CodeDeploy service.

We have developed a Terraform template to create the required infrastructure for this chapter. Complete the following steps to create the required infrastructure. We assume that you have the Terraform **command line interface** (CLI) setup on your system, which we set up in *Chapter 4*. If your environment is not set up, follow the instructions again and set up Terraform on your system:

1. Download the terraform template from the GitHub link: https://github.com/ PacktPublishing/Building-and-Delivering-Microservices-on-AWS/ tree/main/chapter_9/terraform. Once you have downloaded the Terraform template, make edits to the main.tf file and edit a few items, such as the **virtual private cloud** (VPC) ID, subnet IDs, image_ID, and key_name, to launch the EC2 instance. You can get these details from your AWS console.

2. Once you update the preceding settings, run the `terraform init` command. This will download the required providers and modules for AWS:

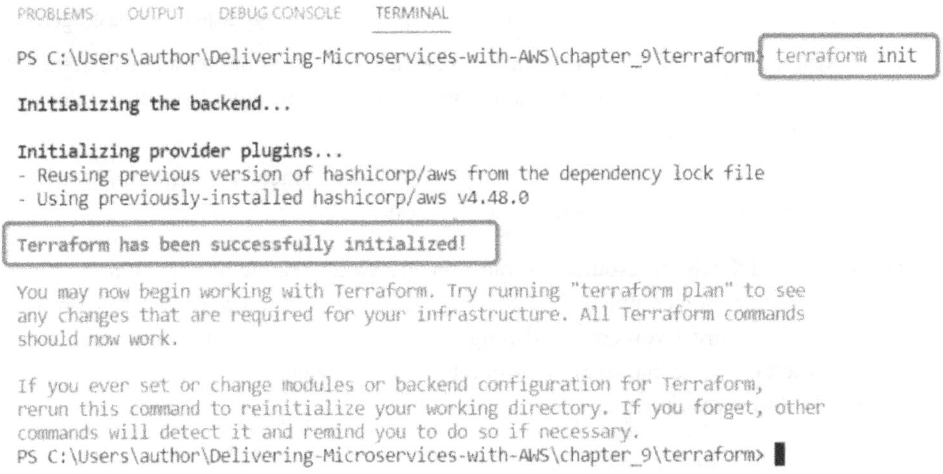

```
PROBLEMS    OUTPUT    DEBUG CONSOLE    TERMINAL

PS C:\Users\author\Delivering-Microservices-with-AWS\chapter_9\terraform terraform init

Initializing the backend...

Initializing provider plugins...
- Reusing previous version of hashicorp/aws from the dependency lock file
- Using previously-installed hashicorp/aws v4.48.0

Terraform has been successfully initialized!

You may now begin working with Terraform. Try running "terraform plan" to see
any changes that are required for your infrastructure. All Terraform commands
should now work.

If you ever set or change modules or backend configuration for Terraform,
rerun this command to reinitialize your working directory. If you forget, other
commands will detect it and remind you to do so if necessary.
PS C:\Users\author\Delivering-Microservices-with-AWS\chapter_9\terraform> 
```

Figure 9.19 – Terraform initialization

3. Now run the `terraform plan` command to get the information about all the resources this template will create in your AWS account. The following is a list of resources it is going to create:

   I.   `chapter_9_lb_sg`: This is the security group used by the load balancer.

   II.   `chapter_9_ins_sg`: This is the security group used by EC2 instances to allow traffic on ports 80 and 443, and **Secure Shell (SSH)** to the instance.

   III. `chapter_9_alb_tgt_group`: The load balancer target group to attach traffic.

   IV.   `chapter_9_deployer_role`: The AWS **identity and access management (IAM)** service role used by the CodeDeploy service to perform deployment; this role will be used later in this chapter while configuring the application.

   V.   `chapter_9_deploy_agent_role`: This is the IAM role assumed by the EC2 instance for the CodeDeploy agent to connect to the AWS CodeDeploy service and download the required files from Amazon S3.

   VI.   `chapter_9_deploy_policy`: The IAM policy to be attached to the `chapter_9_deploy_agent_role` role.

   VII. `chapter_9_alb`: This is the application load balancer that handles the traffic and distributes it to the `chapter_9_alb_listner:` and `Application load balancer listener` EC2 instances.

VIII. `chapter_9_instance_profile`: The instance profile to attach the IAM role to the instance's type.

IX. `chapter_9_deploy_policy_attach`: This attaches the `chapter_9_deploy_policy` policy with the CodeDeploy `chapter_9_deploy_agent_role` agent role.

X. `chapter_9_ec2_launch_template`: EC2 launches the template to create EC2 instances with the same settings and applies user data to download the CodeDeploy agent on instance launch.

XI. `chapter_9_asg`: The autoscaling group for EC2 instances so a minimum capacity of the instances can be made available all the time.

4. You have reviewed the list of resources Terraform will create. Creating these resources will incur a cost to you, so after you complete the exercise, run the `terraform destroy` commands to delete all the resources you created through Terraform. For now, let's create the resources by running the `terraform apply` command. This command will ask for a confirmation; just type `yes` and press *Enter*:

```
aws_alb.chapter_9_alb: Still creating... [1m10s elapsed]
aws_alb.chapter_9_alb: Still creating... [1m20s elapsed]
aws_alb.chapter_9_alb: Still creating... [1m30s elapsed]
aws_alb.chapter_9_alb: Still creating... [1m40s elapsed]
aws_alb.chapter_9_alb: Still creating... [1m50s elapsed]
aws_alb.chapter_9_alb: Still creating... [2m0s elapsed]
aws_alb.chapter_9_alb: Creation complete after 2m2s [id=arn:aws:elasticloadbalancing:us
aws_alb_listener.chapter_9_alb_listner: Creating...
aws_alb_listener.chapter_9_alb_listner: Creation complete after 0s [id=arn:aws:elasticl

Apply complete! Resources: 12 added, 0 changed, 0 destroyed.
PS C:\Users\author\Delivering-Microservices-with-AWS\chapter_9\terraform>
```

Figure 9.20 – Terraform apply confirmation

5. The preceding screenshot confirms that our infrastructure is created successfully. Now, let's log in to the AWS console and review that our EC2 instances have been created:

Figure 9.21 – Terraform provisioned EC2 instances

6. Now let's go to the **Load balancer** section. We should be able to see that a `chapter-9-alb` load balancer has been created. Let's copy the public DNS of this load balancer:

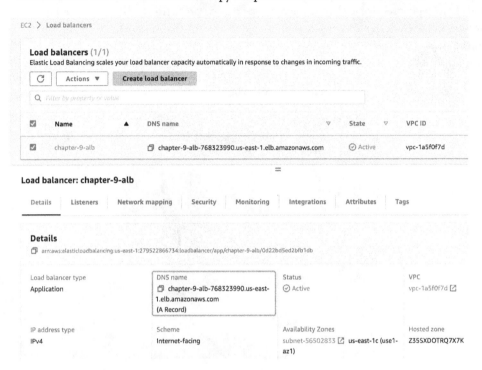

Figure 9.22 – The load balancer provisioned by Terraform

7. Copy and paste the load balancer DNS in the browser, and you should see a response similar to what is shown in the following screenshot. This is the correct output, as currently, we haven't deployed anything to our EC2 instances, and those are not returning the valid health status, so the load balancer is not able to reach the instance on port 80, so we are getting the 503 error:

503 Service Temporarily Unavailable

Figure 9.23 – The load balancer browser URL screen

In the next section, we will create the CodeDeploy application and the DeploymentGroup to deploy the `aws-code-pipeline` microservice to these instances. Once deployment is successful and EC2 instances start passing health checks on port 80, the load balancer will forward traffic to the EC2 instances, and we can view the welcome message on the screen.

App deployment to EC2 instances

For the CodeDeploy service to deploy the `aws-code-pipeline.zip` file, we need to create an application in the CodeDeploy console to initiate deployment. To create a CodeDeploy application, follow the steps shown as follows:

1. Log in to the AWS console and under the **CodeDeploy** service, click on the **Getting started** link in the left panel. Click on the **Create application** button to start creating a CodeDeploy application:

Figure 9.24 – CodeDeploy getting started

2. Provide an application name and select the **Compute platform** type. In this case, we are going to deploy this application to the EC2 instances, so select **EC2/On-premises** and click the **Create application** button:

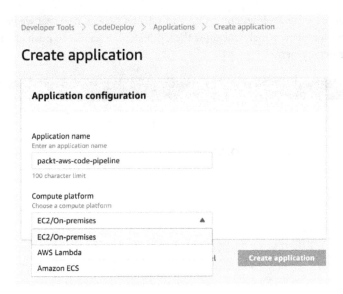

Figure 9.25 – CodeDeploy creates a new application

3. The CodeDeploy application is successfully created; now we need to create the deployment group to which we can target the deployment. As we discussed earlier in this chapter, a deployment group identifies what EC2 instances we are targeting for the deployment. We can have more than one deployment group in the CodeDeploy application. Click on the **Create deployment group** button:

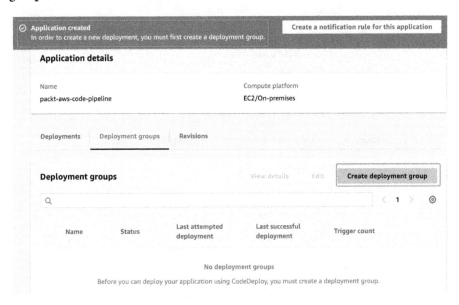

Figure 9.26 – The CodeDeploy new application confirmation

4.   Provide a name to the deployment group and select a service role that we created as part of our Terraform template, which will allow the CodeDeploy service to push changes to the EC2 instances for deployment:

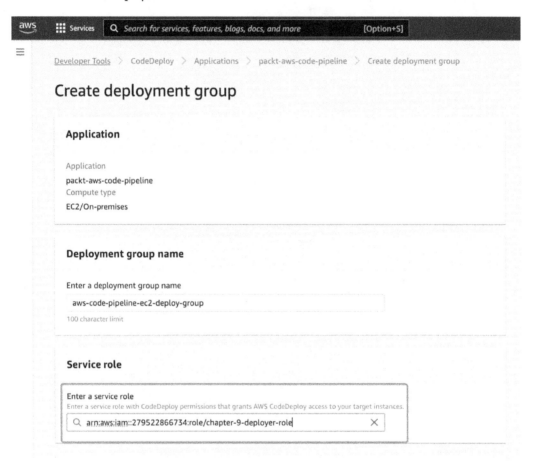

Figure 9.27 – CodeDeploy creates a new deployment group

5. In the next section, select the **Deployment type** option. Here, we are selecting the **In-place** deployment type. We will experiment with blue-green deployment in later chapters:

Deployment type

Figure 9.28 – Deployment type selection

6. When it comes to environment selection, we have multiple options to identify what EC2 instances we want to target for deployment in this deployment group. Here, we will select the autoscaling group for deployment and select `chapter9_asg` in the **Amazon EC2 Auto Scaling groups** drop-down list, which we created as part of the Terraform template. Once you select the autoscaling group, the CodeDeploy service will automatically identify the two running EC2 instances for this autoscaling group and display that information:

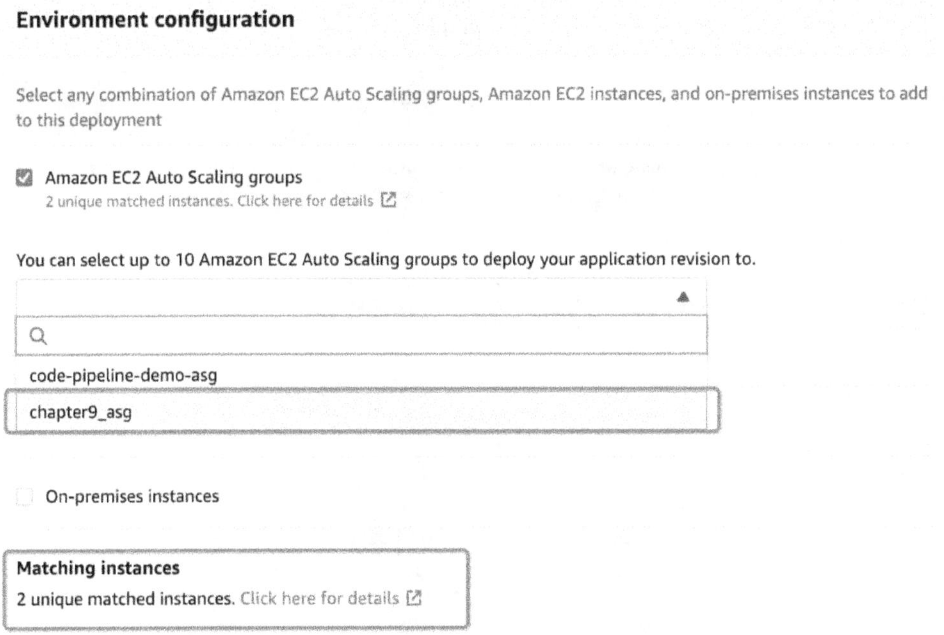

Environment configuration

Select any combination of Amazon EC2 Auto Scaling groups, Amazon EC2 instances, and on-premises instances to add to this deployment

☑ Amazon EC2 Auto Scaling groups
     2 unique matched instances. Click here for details ☑

You can select up to 10 Amazon EC2 Auto Scaling groups to deploy your application revision to.

    ▲

🔍

code-pipeline-demo-asg

chapter9_asg

☐ On-premises instances

Matching instances
2 unique matched instances. Click here for details ☑

Figure 9.29 – Deployment group Environment configuration

7. In the **Deployment settings** section, you can select an existing deployment setting to roll out application updates to the existing running instances. Here, we select the **CodeDeployDefault. HalfAtATime** strategy, which will take half of the total instances out of the load balancer and then perform the deployment to those instances; once the deployment is completed to those instances it will pick up the remaining ones for deployment. In our case, since there are only two instances, it will deploy to one instance at a time:

Figure 9.30 – Deployment settings

8. The **Load balancer** section allows us to select the load balancer target group we created using Terraform, and it manages the traffic to the instances. CodeDeploy first blocks the traffic to the instance before attempting deployment to the instance:

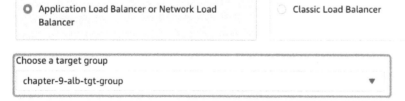

Figure 9.31 – Deployment group Load balancer selection

9. The **Advanced** section allows us to set up triggers and alarms for the deployment performed for this deployment group. The **Rollbacks** section allows us to choose the behavior for the deployment if any deployment step fails. You can create a trigger to send a notification to a **Simple Notification Service** (**SNS**) topic for any deployment event, such as start, stop, fail, ready, or succeed. You can also associate an existing alarm to stop the deployment if that alarm, such as cost, goes off. For the sake of simplicity here, we won't create a trigger or alarm and disable rollbacks, but for a production application, you should have these things configured. Now, click on the **Create deployment group** button:

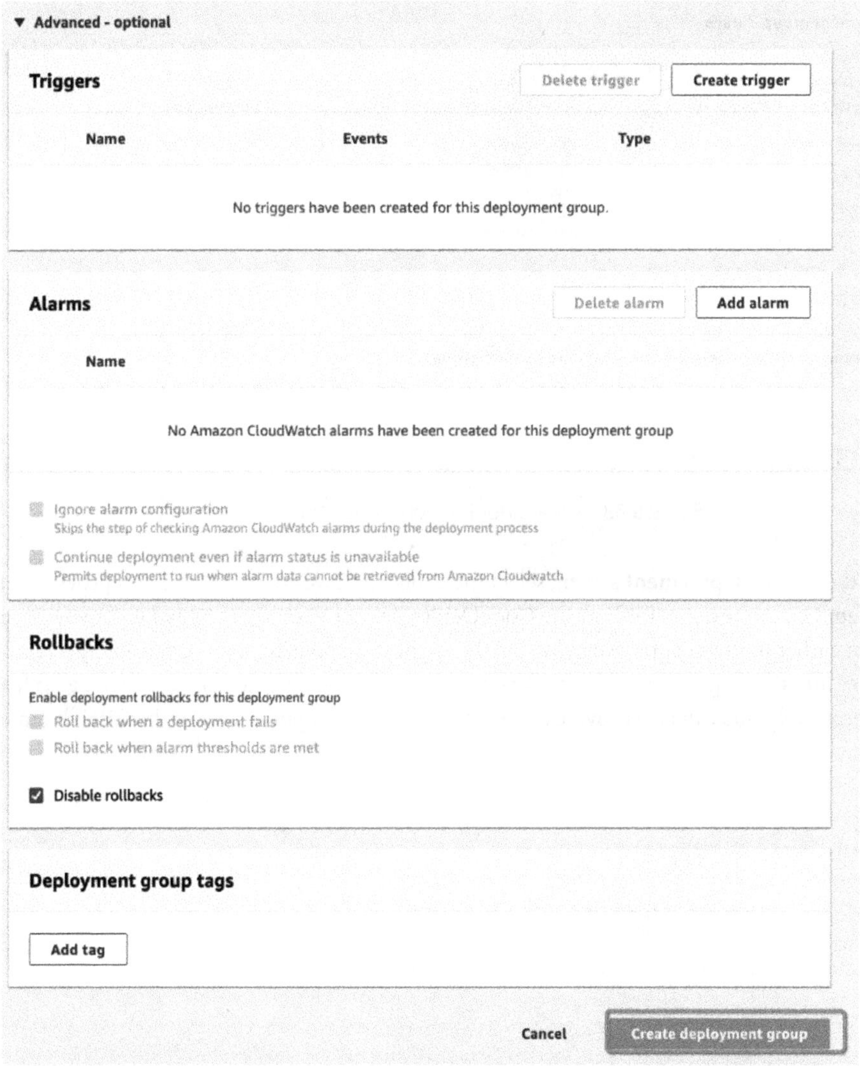

Figure 9.32 – Deployment group Triggers, Alarms, and Rollbacks configuration

10. The following screenshot shows a confirmation message stating that our deployment group was created successfully. Once the deployment group is ready, we need to create a CodeDeploy deployment to start the deployment to EC2 instances. Click on the **Create deployment** button:

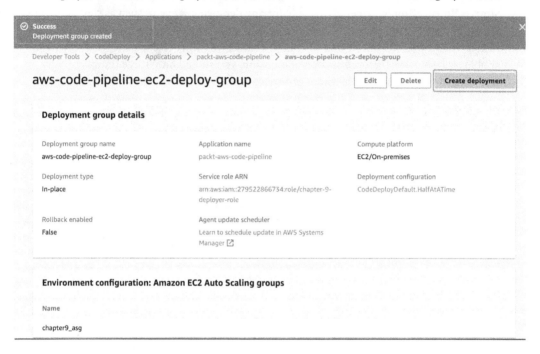

Figure 9.33 – Create deployment group confirmation

11. On the **Create deployment** screen, select the application we created earlier and provide details of the application revision for how the CodeDeploy service should download it. As you remember from earlier in this chapter, we uploaded the application revision `aws-code-pipeline.zip` file at the `s3://packt-aws-code-pipeline/aws-code-pipeline.zip` location. Alternatively, you can also provide application revision by uploading it to the GitHub repository:

Figure 9.34 – The CodeDeploy new deployment creation

12. In the next section, you can provide a description and override the deployment behavior for content and lifecycle events you defined in the `appspec.yml` file:

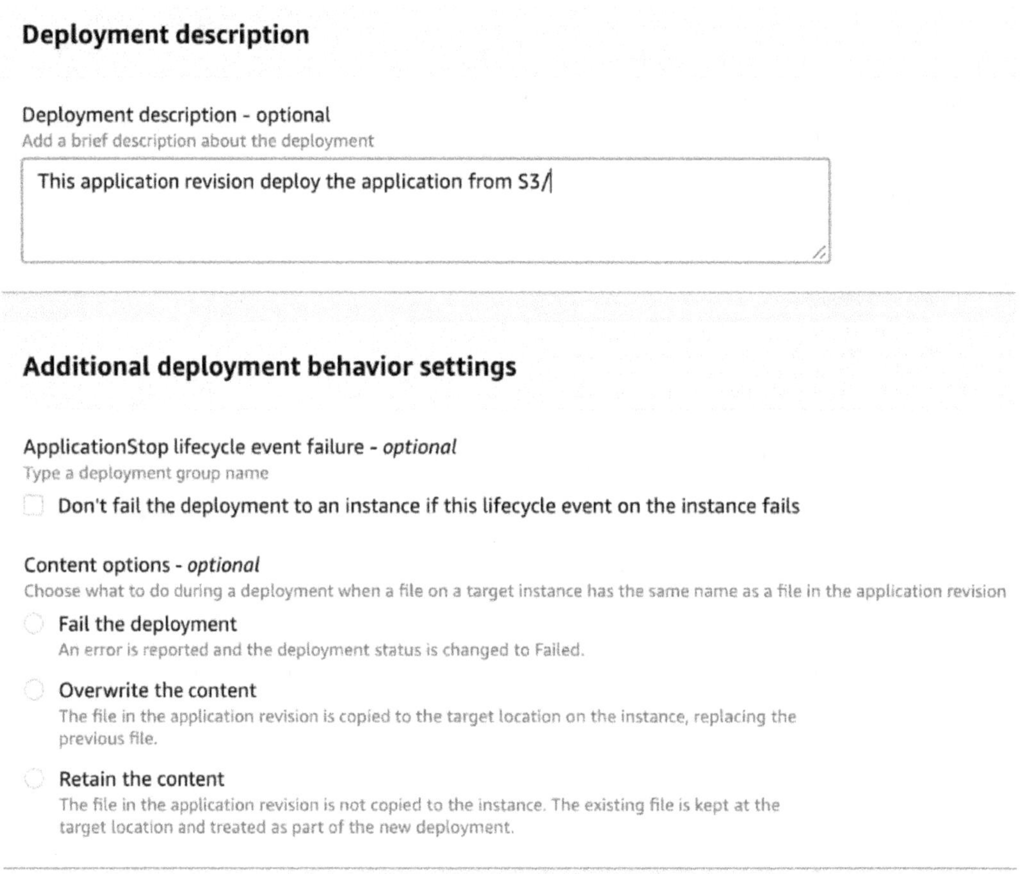

Deployment description

Deployment description - optional
Add a brief description about the deployment

```
This application revision deploy the application from S3/|
```

Additional deployment behavior settings

ApplicationStop lifecycle event failure - *optional*
Type a deployment group name

☐ Don't fail the deployment to an instance if this lifecycle event on the instance fails

Content options - *optional*
Choose what to do during a deployment when a file on a target instance has the same name as a file in the application revision

○ Fail the deployment
   An error is reported and the deployment status is changed to Failed.

○ Overwrite the content
   The file in the application revision is copied to the target location on the instance, replacing the
   previous file.

○ Retain the content
   The file in the application revision is not copied to the instance. The existing file is kept at the
   target location and treated as part of the new deployment.

Figure 9.35 – Deployment of existing contents overrides the behavior selection

13. The next section allows us to override the default **Deployment configuration** selection defined in the deployment group and choose how to deploy to the instances. You can also override the rollback behavior of this deployment. Click on the **Create deployment** button:

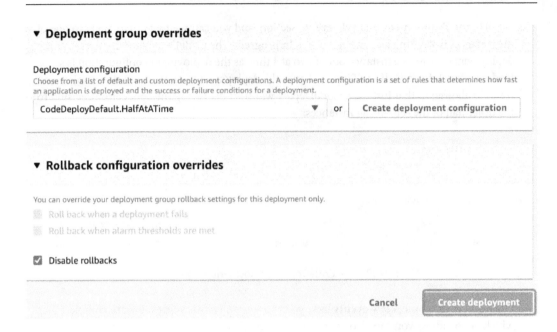

Figure 9.36 – New deployment rollback and deployment group override

14. The following screenshot shows that the deployment has been created successfully and is started to our EC2 instances:

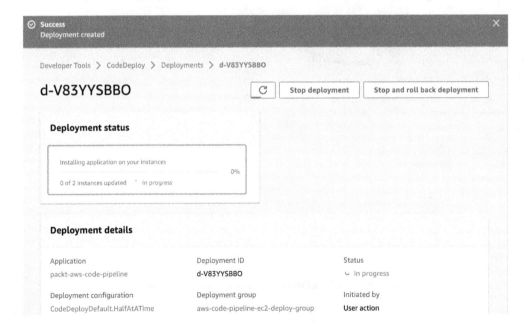

Figure 9.37 – New deployment confirmation

15. Scroll to the **Deployment lifecycle events** section, and you can see that deployment to one of the instances is **Pending** and another one is **In progress**. The CodeDeploy service performs the deployment only to one instance out of two at a time as the deployment configuration is set to **CodeDeployDefault.HalfAtATime** and once the deployment to one instance is complete and traffic is allowed to that instance, CodeDeploy will start deploying the application revision to the other half of the remaining instances:

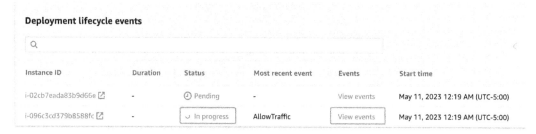

Deployment lifecycle events

Instance ID	Duration	Status	Most recent event	Events	Start time
i-02cb7eada83b9d66e ⧉	-	⏱ Pending	-	View events	May 11, 2023 12:19 AM (UTC-5:00)
i-096c3cd379b8588fc ⧉	-	↻ In progress	AllowTraffic	View events	May 11, 2023 12:19 AM (UTC-5:00)

Figure 9.38 – The CodeDeploy deployment progress

16. Once you click on the **View events** link, you can see which lifecycle events are complete and which are pending; you can also see any failed lifecycle events:

Revision details

Revision location	Revision created	Revision description
s3://packt-aws-code-pipeline/aws-code-pipeline.zip	4 minutes ago	Application revision registered by Deployment ID: d-V83YYSBBO

Event	Duration	Status	Error code	Start time	End time
BeforeBlockTraffic	less than one second	⊘ Succeeded	-	May 11, 2023 12:19 AM (UTC-5:00)	May 11, 2023 12:19 AM (UTC-5:00)
BlockTraffic	1 minute 31 seconds	⊘ Succeeded	-	May 11, 2023 12:19 AM (UTC-5:00)	May 11, 2023 12:20 AM (UTC-5:00)
AfterBlockTraffic	less than one second	⊘ Succeeded	-	May 11, 2023 12:20 AM (UTC-5:00)	May 11, 2023 12:20 AM (UTC-5:00)
ApplicationStop	less than one second	⊘ Succeeded	-	May 11, 2023 12:20 AM (UTC-5:00)	May 11, 2023 12:20 AM (UTC-5:00)
DownloadBundle	less than one second	⊘ Succeeded	-	May 11, 2023 12:20 AM (UTC-5:00)	May 11, 2023 12:20 AM (UTC-5:00)
BeforeInstall	less than one second	⊘ Succeeded	-	May 11, 2023 12:20 AM (UTC-5:00)	May 11, 2023 12:20 AM (UTC-5:00)
Install	less than one second	⊘ Succeeded	-	May 11, 2023 12:20 AM (UTC-5:00)	May 11, 2023 12:20 AM (UTC-5:00)
AfterInstall	less than one second	⊘ Succeeded	-	May 11, 2023 12:20 AM (UTC-5:00)	May 11, 2023 12:20 AM (UTC-5:00)
ApplicationStart	less than one second	⊘ Succeeded	-	May 11, 2023 12:20 AM (UTC-5:00)	May 11, 2023 12:20 AM (UTC-5:00)
ValidateService	less than one second	⊘ Succeeded	-	May 11, 2023 12:20 AM (UTC-5:00)	May 11, 2023 12:20 AM (UTC-5:00)
BeforeAllowTraffic	less than one second	⊘ Succeeded	-	May 11, 2023 12:20 AM (UTC-5:00)	May 11, 2023 12:20 AM (UTC-5:00)
AllowTraffic	1 minute 31 seconds	⊘ Succeeded	-	May 11, 2023 12:20 AM (UTC-5:00)	May 11, 2023 12:22 AM (UTC-5:00)
AfterAllowTraffic	less than one second	⊘ Succeeded	-	May 11, 2023 12:22 AM (UTC-5:00)	May 11, 2023 12:22 AM (UTC-5:00)

Figure 9.39 – The CodeDeploy deployment lifecycle events

17. Once the deployment is completed to one of the instances, the load balancer will start passing the health check as now it starts getting a response from that particular instance on port 80 as per our configuration in the Terraform template:

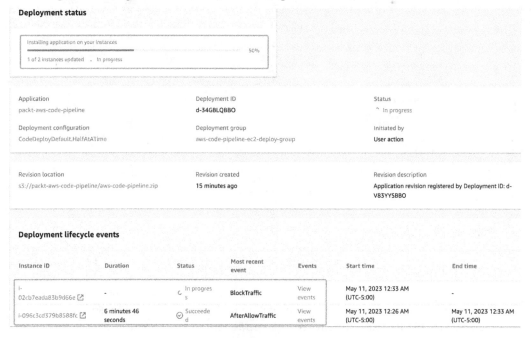

Figure 9.40 – CodeDeploy half-deployment progress

18. Now, let's go to the EC2 dashboard, click on the **Load balancers** link in the left panel, and copy the **DNS name** value:

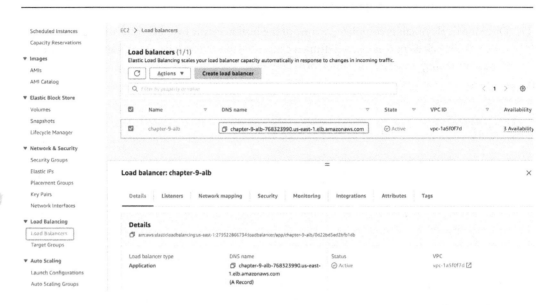

Figure 9.41 – The load balancer details screen

19. In our case, the load balancer DNS name is `http://chapter-9-alb-768323990.`
    `us-east-1.elb.amazonaws.com`; for you, this URL will be different. Let's paste this
    URL in the browser address bar and you will notice that we have a `200` response with the
    application name and version from our `aws-code-pipeline` microservice:

Figure 9.42 – The load balancer browser screen

20. Now, if you go back and look at the **Deployment status** value, it is **Succeeded**, and our
    `aws-code-pipeline` microservice is deployed to both the EC2 instances:

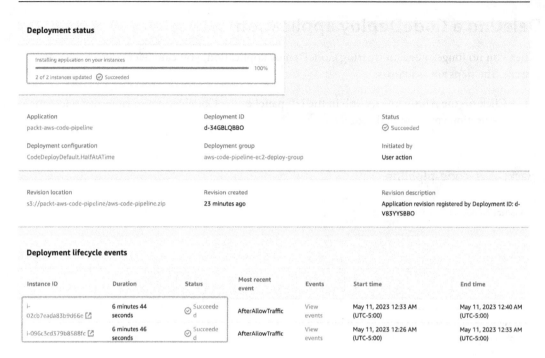

Figure 9.43 – The CodeDeploy deployment Succeeded status

21. You can click on the **Deployments** link in the CodeDeploy left panel to look at the deployment history for previous deployments. You can click on the deployment and view the details of the particular deployment:

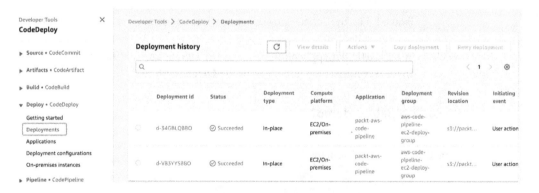

Figure 9.44 – The CodeDeploy Deployment history page

Now that we have created a CodeDeploy application and performed a successful deployment, let's learn how to delete an existing CodeDeploy application if it's no longer needed.

Deleting a CodeDeploy application

When you no longer need an existing CodeDeploy application, you can delete it using the AWS console. The steps are as follows:

1. Click on the **Applications** link in the left panel of the CodeDeploy console and select the application you want to delete:

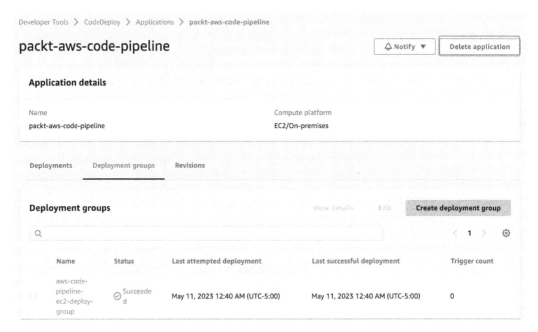

Figure 9.45 – The CodeDeploy Delete application screen

2. Once you click on the particular application, click on the **Delete application** button. On the popup screen, provide a confirmation and click on the **Delete** button:

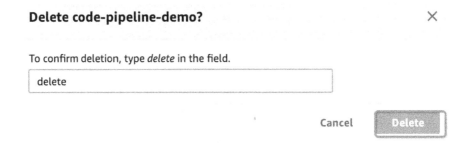

Figure 9.46 – The CodeDeploy Delete application review page

3. The following screenshot shows a confirmation message that states that the CodeDeploy application has been deleted successfully:

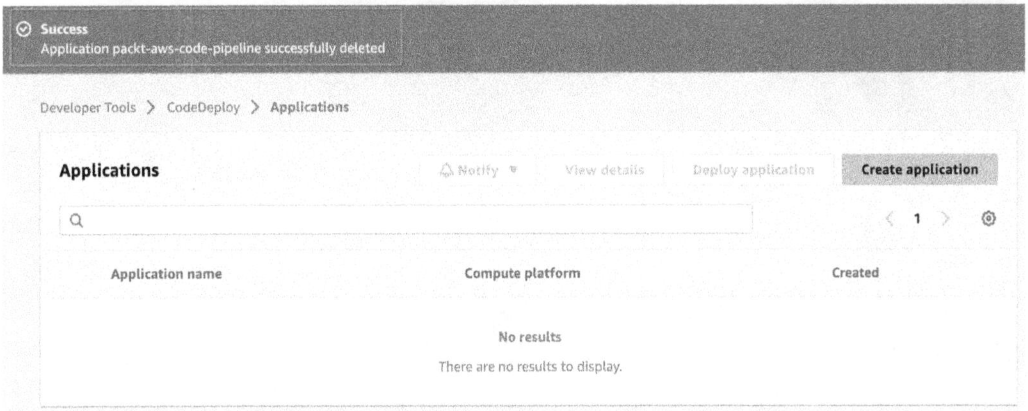

Figure 9.47 – The CodeDeploy Delete application confirmation

You don't need to delete the actual application we created in this chapter. We will use the same CodeDeploy service in the next chapter to create a CodePipeline. You can find all the example code and scripts for this chapter in our GitHub repository here: `https://github.com/PacktPublishing/Building-and-Delivering-Microservices-on-AWS/tree/main/chapter_9`.

Summary

In this chapter, we learned about the AWS CodeDeploy service, its benefits, and its limitations. We also created a new CodeDeploy application and deployment group and deployed our sample `aws-code-pipeline` microservices to the EC2 instances we created. In addition to this, we also learned about the `appspec.yml` file syntax and created an `appspec.yml` file in our `aws-code-pipeline` CodeCommit project. We created the infrastructure using the provided Terraform template and tested the sample application deployment on the EC2 instances. In the next chapter, we will extend our knowledge of CodeDeploy and will deploy the application to Amazon ECS.

Deploying to ECS Clusters Using CodeDeploy

In this chapter, you will learn about containers and the container services provided by AWS. You will get an overview of Docker and Elastic Container Registry (ECR) and learn about **Elastic Container Service (ECS)** as provided by AWS for container management. We will make changes to our example `aws-code-pipeline` application and deploy it to ECS. In this chapter, we will be covering the following topics:

- What are containers?
- Docker overview
- What is ECS?
- What is Amazon ECR?
- Manually deploying an application to ECS
- Configuring CodeDeploy to install apps to ECS

In the previous chapter, we deployed our example application to the EC2 instances using the CodeDeploy service. In this chapter, we will deploy the example `aws-code-pipeline` microservice to ECS using CodeDeploy. Before we deploy our containers to ECS, we need to understand more about containers themselves, so let's get started.

What are containers?

If you have worked in the software industry for some time, you might be aware of the rivalry between software developers and test engineers. One of the common terms used by developers in response to a defect is that *"it works on my machine."* Test engineers frequently find defects in a test environment that developers are not able to replicate in the development environment. These things happen due to environment disparities. A minor runtime version difference between environments can cause a big difference in behavior. Containers help to solve these kind of issues.

A container is an executable unit of software that packages together the application's code, libraries, and any dependencies so the application can be run anywhere reliably without needing to worry about specific dependencies.

A container engine allows us to run containers in a host system and uses the given operating system's virtualization features to isolate the process and control CPU, memory, and disk usage, along with managing inter-process security.

The following diagram explains the architecture of containers and virtual machines. On the left side, we see all the containers share the same operating system while the container platform (Docker in this instance) is responsible for controlling the resource allocation for each container application. On the right side of the diagram, we can see the virtual machine architecture. We have an optional **host operating system** and on top of that, a hypervisor is required to host and create the different virtual machines. Each virtual machine requires a guest operating system to run an application, which takes a lot more resources compared to the container architecture on the left:

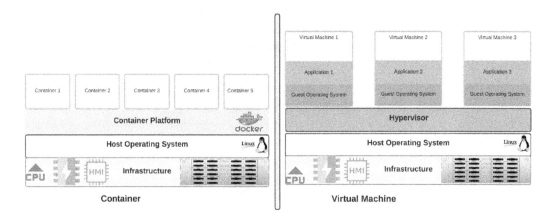

Figure 10.1 – A comparison between the container and virtual machine architectures

Containers are small, fast, and portable as they don't require the entire guest operating system to run the application. Unlike virtual machines, containers share the kernel of the host operating system and create the required layers needed by the specific container instance.

There are a variety of container technologies available today including Docker, Podman, LXD, BuildKit, and Containerd. Docker is one of the most popular container technologies so let's start by learning more about it.

An overview of Docker

Docker is an open source container platform that enables developers to build, package, ship, and run containers. Docker is written in the Go programming language and uses the Linux operating system's kernel namespace feature to provide isolation between different containers. Each container is a workspace in a different namespace as a means of ensuring isolation.

Docker allows us to package an application's source code and its dependencies together in the form of an image, which is then used to run an instance of the application on a container. You can run as many instances as you want using the same image file; all of them will be created identically using the same versions of the packaged dependencies and runtime. So, when your application moves from the development to the QA environment, an application instance is created and deployed the same way.

To work with Docker, you don't need a full-fledged operating system – Docker just needs the Docker runtime environment to be available to create and run Docker containers.

This book is not about Docker in particular, but we will still cover the Docker architecture from a high level so you understand it better when we later create a Docker image and run our aws-code-pipeline example application as a Docker container.

Docker architecture

Docker is a client-server architecture implementation where a Docker client connects to a local or remote dockerd daemon running on a Docker host using the REST API over Unix sockets or a network interface. The following diagram shows the architecture of Docker where a Docker client executes commands to build a Docker image and connects to the Docker daemon, which builds the image and pushes it to the Docker registry:

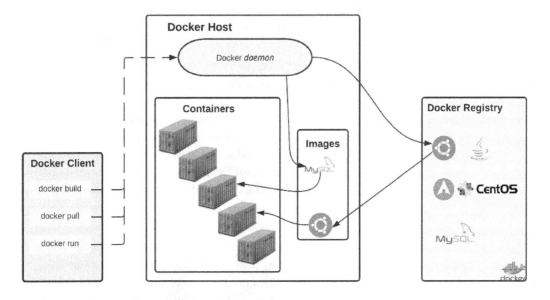

Figure 10.2 – Docker architecture

The docker pull command pulls the existing images from the registry, while the docker run command executed by the Docker client connects to the Docker daemon to create the containers and run on the Docker host. In order to create Docker containers, you need a Dockerfile, which is a text file that explains how your image is to be built. You create your Dockerfile from a base image and on top of that, you add your application and its dependencies:

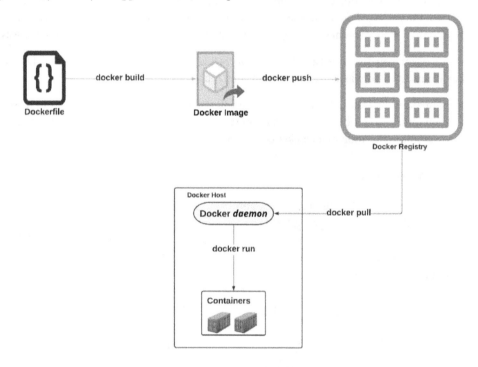

Figure 10.3 – Docker image build process

Once your Dockerfile is ready, you execute the docker build command to create a Docker image and then push it to a central repository such as Docker Hub or AWS's Elastic Container Registry. Whenever you need to create a container for your application, you pull images from the repository using the docker pull command and with the help of Docker daemon create the containers. Docker daemon is responsible for creating and managing the containers and assigning resources to these containers. Let's learn a bit more about the different Docker architecture components.

Dockerfile

As stated, a Dockerfile is a text document containing instructions about how your Docker image should be built. This text file contains the command that needs to be executed to assemble an image for your application. A Docker image is made up of layers where each layer represents a command in the Dockerfile. These layers are stacked on top of each other, and the next layer represents the delta from the previous layer.

The following is an example Dockerfile to build a Docker image for an Apache server. In this example, we are starting from the base Ubuntu image and then we execute a few shell commands to update Ubuntu and install the Apache server. The RUN instruction executes commands in a new layer on top of the current one and commits the results. The EXPOSE instruction tells Docker that the container should be exposed on port 80 at runtime. The ENTRYPOINT and CMD instructions define the commands to be executed when the container is run. Here we use the CMD command to start the Apache server in the background:

```
#Example Docker image to create an Apache server
FROM ubuntu
RUN apt update
RUN apt install -y apache2
RUN apt install -y apache2-utils
EXPOSE 80
CMD ["apache2ctl", "-D", "FOREGROUND"]
```

To understand the Dockerfile in depth, examine the Dockerfile reference guide at https://docs. docker.com/engine/reference/builder/. Once a Dockerfile is built, it gets converted into a Docker image, so let's examine what an image is in the context of Docker.

Image

A Docker image is a template from which containers are created. An image is made of a set of instructions used to create the container at runtime and is often based on another image with some more instructions added to customize its behavior. The following diagram shows a layered architecture where a Spring Boot application aws-code-pipeline.jar image is created from a base Ubuntu image, and on top of that, Java is installed as a layer, after which we have to package our JAR file:

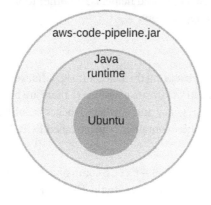

Figure 10.4 – Docker image

Whenever you change the Dockerfile and rebuild an image, only those layers that have been changed will be rebuilt. Containers are created using these images, so next let's understand what a container is.

Container

At runtime, when you need a runnable process, a container represents that process. An image is a template to create the containers and each container is an instance of that image. A container can be created, started, stopped, or deleted, and you can connect a container to one or more storage systems or networks. A container is created from the image specified by the Docker deamon process when you execute the `docker run` command.

Docker daemon

The Docker daemon (`dockerd`) is a Docker server-side component that accepts Docker API requests, and in response to those requests, it manages the Docker objects and their life cycles.

Docker client

The Docker client is what you interact with when you execute a `docker` command, such as `run`, `build`, or `stop`. When you run a command using the Docker CLI, the Docker client sends those commands to the Docker daemon (`dockerd`) process running on a local or remote server. The Docker client sends API requests to the Docker daemon and can interact with more than one Docker daemon.

Docker Desktop

Docker Desktop is an executable binary for different operating systems that enables you to easily build and test Docker containers. Docker Desktop includes the Docker daemon (`dockerd`), Docker client (`docker`), Docker Content Trust, credential helpers, and other tools so you can easily get started with developing native container applications.

The Docker registry

The Docker registry is a repository for storing the Docker images. Images can be pushed and versioned, and then later pulled from the registry to create containers. There are both public and private Docker registries. Docker Hub is a public registry and is where Docker is configured to look for images by default. AWS provides another popular container registry called **ECR**. We will learn about ECR later in this chapter.

When we execute a `docker pull` or `docker run` command, the Docker daemon looks for images locally, and if no image is found, then it connects to the Docker registry and downloads the image.

Docker commands

Now that you have understood the different concepts related to Docker, let's learn some of the basic commands that we will use later in this chapter to successfully create a Docker image for our example `aws-code-pipeline` application:

- `docker version`

  This command renders all version information in an easy-to-read format. This command will display version information for all installed Docker components, such as the client, Docker Desktop, and so on. You can specify a format parameter to print only specified information from the version information:

```
docker version [OPTIONS]
```

- `docker build`

  This command is used to build a Docker image from a Dockerfile and the set of files at the specified context path. The URL parameter can refer to a Git repository, a tarball archive, or a plain text file:

```
docker build [OPTIONS] PATH | URL | -
```

- `docker pull`

  This command is used to pull an existing image from the Docker registry. You can use the parameters to download a specific tag/version of an image, or just get the latest version by default:

```
docker pull [OPTIONS] NAME[:TAG|@DIGEST]
```

- `docker push`

  This command is used to push an image to the Docker registry so that others can also use it:

```
docker push [OPTIONS] NAME[:TAG]
```

- `docker start`

  This command is used to start a stopped container:

```
docker start [OPTIONS] CONTAINER [CONTAINER...]
```

- `docker stop`

  This command is used to stop one or more running containers. The main process inside the container will receive `SIGTERM`, and after a grace period has passed, a `SIGKILL` will be issued. You can define the grace period in the command for which to wait before the container is killed:

```
docker stop [OPTIONS] CONTAINER [CONTAINER...]
```

- `docker run`

    This command is used to run a container from an image. This is a two-step process; first, it creates a writable container layer from the specified image and then it starts that container:

```
docker run [OPTIONS] IMAGE [COMMAND] [ARG...]
```

We covered some of the basic commands here, but Docker has many more. In order to get a complete list of commands, you can try the `docker help` command, which will provide you a full list with descriptions of each one.

You've now had a fair overview of Docker containers and you can easily run containers using the Docker commands. Running a few containers across multiple servers is simple enough, but things get complicated when you have thousands of containers running across hundreds of servers and you're responsible for managing their life cycles and resource requirements. To manage this complexity, you need an *orchestrator*, which allows you to run containers across multiple servers in multiple data centers with ease. Docker Swarm, Kubernetes, and AWS's ECS are some of the most popular products for managing containers at scale. In the next section, we will learn about Amazon ECS.

What is ECS?

We discussed in the previous section that running and managing smaller workloads is easy, but when workload sizes increase, maintenance becomes a challenge. Keeping track of all the available servers in a cluster and ensuring the available capacity, server health, resource utilization, and container placement is likely to become a headache.

To solve the challenges mentioned here, you need an orchestrator, which takes care of these issues for you. Amazon ECS is a fully managed container orchestration service that allows you to quickly deploy, manage, and scale containers in the cloud, providing the necessary tooling to manage both the containers and infrastructure. In order to understand ECS better, we need to first discuss a few related concepts of clusters, services, tasks, and task definitions.

Task definitions

In ECS, a **task definition** describes your application containers and their configuration. The task definition is a file in JSON format that works like a template to create one or more containers to define your application. A task definition can have a maximum of 10 containers, but you can define your application across multiple task definitions if required.

In a task definition, you specify parameters for things such as which operating system to use, what container to use, memory requirements, the number of CPUs or storage volumes, and port mappings.

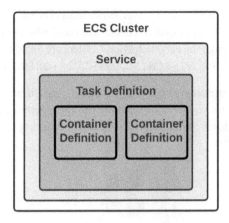

Figure 10.5 – ECS cluster components

The launch type is an important configuration item within the task definition, as it defines what type of infrastructure your workload should be deployed on. A task definition is a required component for ECS, and they are executed as tasks or services.

Tasks

A **task** is an execution unit of the task definition in an ECS cluster. A task represents the runtime execution of the task definition for your application. When ECS deploys a task, it is an instantiation of the task definition based on the configuration. You can specify how many tasks you want to deploy, and a task can be executed as a single unit or as part of a service.

A task is standalone in nature, performing its assigned work and then terminating. Tasks are designed with batch processing in mind, for which long-running process services are used.

Services

In an ECS cluster, a **service** allows you to run and maintain a configured number of task definition instances simultaneously. A service allows you to run and maintain long-running tasks in a stateless fashion and maintains the desired state. If for some reason any task stops or fails, the service will launch a new one to maintain the desired capacity.

ECS clusters

An ECS cluster is a logical grouping of tasks and service. A cluster is used to logically isolate applications from one another so that your underlying infrastructure is not the same.

A cluster's infrastructure can be created in a serverless fashion (Fargate), or on AWS EC2 instances, on-prem servers, or virtual machines. A default cluster is created when you use an ECS instance and infrastructure is provisioned through the Fargate service. Clusters are Region-specific and can go through different states such as ACTIVE, PROVISIONING, DEPROVISIONING, FAILED, and INACTIVE.

ECS provides flexibility to its consumers on top of the underlying infrastructure and as a consumer you can decide how much control you need by selecting the appropiate launch type. As shown in the following diagram, ECS provides three launch types based on differing computing needs:

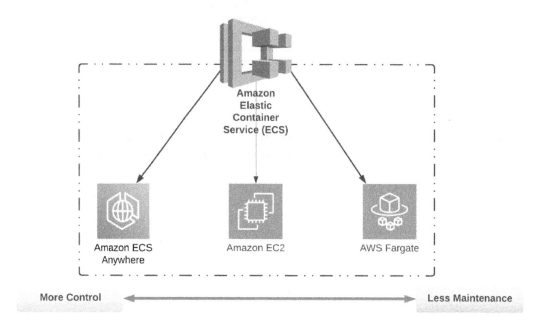

Figure 10.6 – ECS launch types

Let's dive into these different launch types in more detail.

Amazon ECS Anywhere

This launch type configuration provides the greatest flexibility in terms of control. In this setup, you own the complete infrastructure and are responsible for the physical security of the ECS cluster, running ECS from your private on-prem cloud. You need to install the ECS agent on your servers or virtual machines so that they can register with ECS as the capacity provider and ECS can start placing tasks on these machines as part of your cluster. This use case is more suitable when you have your existing infrastructure and you want to utilize that to start with ECS, and when you need more resources, you can expand to the cloud. ECS does the heavy lifting by managing the orchestration of containers for you, making it easier to manage.

Amazon ECS on Amazon EC2

In this launch type configuration, you deploy the ECS cluster on Amazon EC2 instances you own. ECS handles the deployment of the clusters and manages the containers, but you are responsible for managing the EC2 instances. When creating an ECS cluster, you will need to specify an auto-scaling group through which ECS will create the assigned machines as a capacity provider.

This use case is suitable when you have existing EC2 instances and your workload calls for sustained and dedicated utilization. So, you can run a dedicated number of tasks/services all the time.

AWS Fargate

This is the serverless offering for ECS where you do not directly own any private server or EC2 instance. With this setup, you don't have to worry about hardware procurement or software patching; AWS will manage everything for you. You just need to configure your cluster, choose AWS Fargate as your capacity provider, and configure the number of tasks you need to run in parallel, and ECS will execute those tasks on the Fargate infrastructure.

This launch type provides you great flexibility in terms of scaling – you can scale up and down on demand and you pay only for what you use. Fargate is the default capacity provider for ECS and you can extend your capacity provider to other launch types in addition to Fargate. In this launch type, no setup or maintenance of infrastructure is required and it is suitable for workloads of variable capacity.

Now we have understood the different components related to ECS and its different launch type configurations, let's look at how all these components come together and function.

ECS architecture

ECS is an agent-based service where the ECS agent is the driving force, controlling the cluster and running containers on the host machines. All machines in the cluster need to run the ECS agent so it can execute the tasks scheduled by ECS.

The following diagram describes the architecture of an ECS instance. As a developer, you create container images and push them to a container registry such as Docker Hub, **ECR**, or a private registry. To get started, you need to create an ECS cluster and define the capacity provider for your cluster.

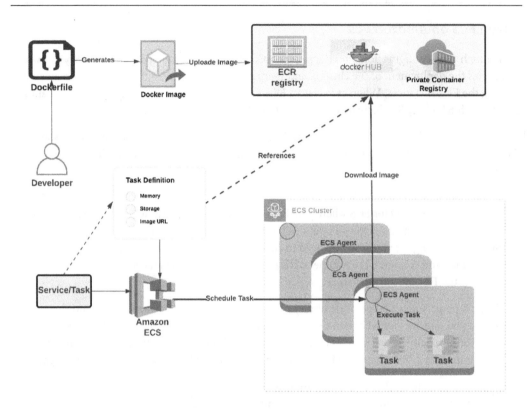

Figure 10.7 – ECS architecture

You also define a **task definition**, consisting of the details of memory, storage, the container image, and the number of tasks required. Based on this task definition, ECS will schedule and place containers on the appropriate instance. The ECS agent is responsible for downloading the appropriate image from the repository and executing the tasks. ECS monitors the tasks through ECS agents and ensures that you have the number of tasks requested. All container-optimized **Amazon Machine Images** (**AMIs**) are pre-baked with ECS agents and if you use a different AMI, you can easily install the ECS agent on it.

ECS uses the task placement strategy and constraints to determine where to place a new task based on the requirements specified in the task definition. Similarly, when you need to scale down the number of tasks, ECS needs to determine which tasks specifically to take down. ECS uses a spread strategy for placement by default, but you can customize the placement strategy and constraints. However, the ECS Fargate launch type spreads the tasks evenly across Availability Zones and doesn't support customization of the placement strategy.

> **Note**
>
> Task placement strategy is a mechanism by which ECS determines the instance to run a new container on (when scaling up) or the container to be terminated (when scaling down). There are three placement strategies supported by ECS – **binpack, random**, and **spread**. In the binpack strategy, containers are placed on a given instance until memory and CPU reach the maximum utilization as specified in the container's task definition. In the spread strategy, tasks are distributed evenly based on fields such as instances or Availability Zones. In the random strategy, tasks are placed and terminated in random order.

A set of related tasks can be configured as a task group, which can be used as a placement constraint while ECS tries to place your containers. For example, say you are running a multi-tier application within one cluster, comprising a web tier and database tier. You can define your web and database tiers as two separate task groups and apply the spread placement strategy, so ECS will ensure that your web-tier and database-tier containers are placed across different Availability Zones such that even if one Availability Zone goes down, your application will still work.

ECS can be configured to use Amazon **Elastic Load Balancer** (**ELB**) to distribute traffic evenly to all running tasks. ECS can also auto-scale tasks to meet traffic demands and utilize different CloudWatch metrics to determine when to scale up or down.

Now that we have a fair understanding of the ECS architecture and the different concepts related to this service, let's examine the AWS ECR image repository and how to publish a Docker image to it in the next section, and then later we will deploy this image to the ECS cluster.

What is Amazon Elastic Container Registry?

Amazon **ECR** is a fully managed container registry for storing and sharing container images for reliable deployment. As a developer, you create your container images and push them to ECR, and later you configure ECS task definitions to pull your images from the ECR repository for deployment.

ECR is a serverless container registry, so you don't have to provision or manage any servers to store your images. You only pay for the storage of your images, which makes it very affordable. ECR scales automatically to meet demand and provides end-to-end encryption so you can safely store and share images outside AWS as well. Your repositories are private, and you control access to images through the AWS IAM service.

In order to use ECS, we need to create an ECR repository, and later we will modify and publish our example `aws-code-pipeline` microservice as a Docker image to this ECR repository. Work through the following instructions to create the ECR repository:

1. Log in to the AWS console with the user we created in our earlier chapters and search for `Elastic Container Repository`. Click on the **Get Started** button on the ECR home page, which will take you to the repository's **General settings** page. A repository can be configured as private or public, but note that once you create a repository, this setting can't

be changed. Public repositories are publicly accessible, while access to private repositories is configured through IAM and repository policies:

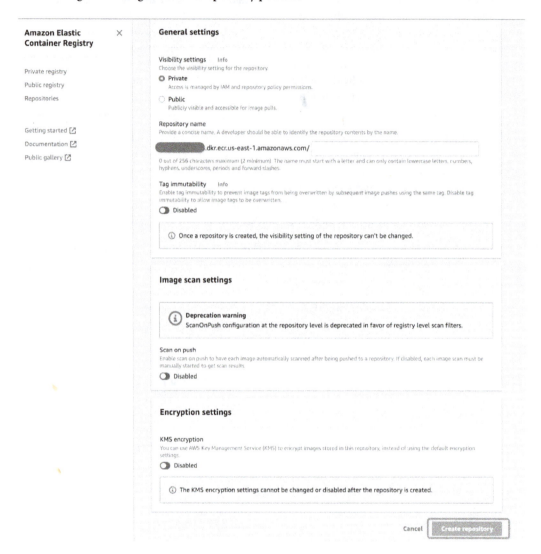

Figure 10.8 – ECR repository creation

Provide a name to the repository, and you can enable **Tag immutability**, **Scan on push**, and **KMS encryption** for the images pushed to this repository, but we are keeping everything default here, selecting **Private** for our repository, and providing `packt-ecr-repo` as the name.

2. Click on the **Create repository** button and it will take you to a list of your repositories. You should see that your repository has been created successfully. By default, AWS enables AES-256 encryption. Now we have our ECR repository ready, in order to push a container image to it, we need to follow a set of commands. You can find these commands by clicking on the **View push commands** button onscreen, but before that, we need to install Docker on our machine and modify our `aws-code-pipeline` application to build it as a Docker image.

3. Follow the instructions from *Appendix* to install Docker Desktop. Once Docker Desktop is installed and ready, we can use Docker to run containers or push a Docker image. Before we can push a Docker image to `packt-ecr-repo`, we need to first create an image from our `aws-code-pipeline` service. Follow the instructions in the next step to create the Docker image.

4. Open the VS Code IDE and create a Dockerfile in the root directory of the source code. Then, add the following lines of code:

```
FROM openjdk:11
ARG JAR_FILE=target/aws-code-pipeline*.jar
COPY ${JAR_FILE} app.jar
ENTRYPOINT ["java","-jar","/app.jar"]
EXPOSE 80
CMD ["echo" "AWS code pipeline application executed
successfully!"]
```

In this Dockerfile, we start with the Java 11 base image and then rename our `aws-code-pipeline*.jar` Spring Boot executable file `app.jar`. The application is executed using the ENTRYPONT keyword and will run on port 80 so we have to make sure to expose to port 80 from the Docker container.

5. Now our Dockerfile is ready, let's go ahead and build it using the `docker build -t aws-code-pipeline` command.

6. Once the Docker image is built successfully, you can use the `docker images` command to see your available images. Now it is time to run the aws-code-pipeline application using the `docker run -p 80:80 aws-code-pipeline` command.

7. The preceding command will start the `aws-code-pipeline` application. You can validate it by typing `http://localhost:80/` in your browser's address bar.

8. At this point, our `aws-code-pipeline` Docker image has been validated locally and is ready to be pushed to the `packt-ecr-repo` ECR repository we created earlier. Our local Docker client needs to be authenticated to ECR before we can upload any images to the `packt-ecr-repo` repository. Execute the following command, making sure to replace xxxxxxxxxxxx with your AWS account ID:

```
aws ecr get-login-password --region us-east-1 | docker login
--username AWS --password-stdin xxxxxxxxxxxx.dkr.ecr.us-east-1.
amazonaws.com
```

9. Now the Docker client is logged in and connected to ECR. In the previous steps, we built the Docker image, so next, let's tag this image to the ECR repository. AWS ECR repositories are designed such that you need a separate repository for each image, so here we are going to use the tags to differentiate between images and use the same repository for storing different type of images. Execute the following command to tag your aws-code-pipeline image to the ECR repository. Please replace xxxxxxxxxxxx with your own account ID:

```
docker tag aws-code-pipeline:latest xxxxxxxxxxxx.dkr.ecr.us-east-1.
amazonaws.com/packt-ecr-repo:aws-code-pipeline
```

10. Now we need to run the final command to push the image to the ECR repository:

```
docker push xxxxxxxxxxxx.dkr.ecr.us-east-1.amazonaws.com/packt-ecr-
repo:aws-code-pipeline
```

11. Now, let's go back to our AWS console and validate that our newly pushed Docker image is in the **packt-ecr-repo** repository. Click on the image tag link and copy the image URI (in our case, xxxxxxxxxxxx.dkr.ecr.us-east-1.amazonaws.com/packt-ecr-repo:aws-code-pipeline), which we will use later in this chapter for our task definition:

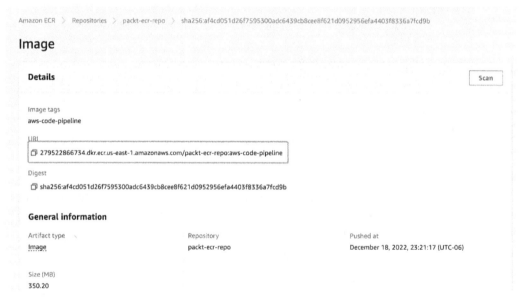

Figure 10.9 – Docker image details for packt-ecr-repo

At this point, we have our Docker image ready and available in our repository for deployment. In the next phase, we will deploy this image to ECS.

Manually deploying an application to ECS

To deploy an application to ECS, first we need to create an ECS cluster, then define task definitions and execute the task.

Creating an ECS cluster

To create an ECS cluster, we need to first log in to the AWS console, search for ECS, and then work through the following steps:

1. Click on the **Clusters** link in the left panel of the ECS home screen and click on the **Create cluster** button. Fill in the cluster name as you wish; we went for `packt-ecs-cluster`. Then, select your VPC and subnets – we chose our default VPC and all the subnets for the execution of tasks and services:

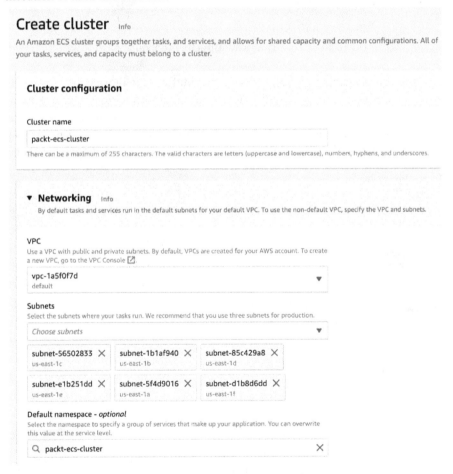

Figure 10.10 – ECS cluster name and networking

2. In the **Infrastructure** section, by default, the ECS cluster launches with the AWS Fargate serverless offering, but you can select additional compute capacity through AWS EC2 instances or on-prem servers. All your EC2 and on-prem instances need to run the ECS agent in order to be detected by ECS as part of your cluster. In our case, we choose **AWS Fargate (serverless)**:

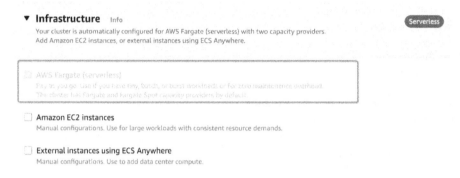

Figure 10.11 – ECS cluster infrastructure configuration

3. The **Monitoring** section allows you to configure additional monitoring for your containers. By default, CloudWatch automatically collects a selection of different metrics for resource usage. Container Insights offers enhanced collection of container-specific metrics for easy troubleshooting. You can also add different tags to the cluster for cost management and policy enforcement. Click on the **Create** button to create the cluster for our ECS deployment:

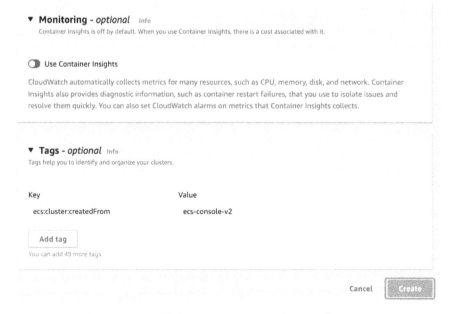

Figure 10.12 – ECS cluster monitoring and tag configuration

4. Our cluster creation is now in progress; it will take approximately 1-2 minutes and once ready, the cluster will be visible in the cluster list and a confirmation message will be shown:

Figure 10.13 – ECS cluster creation confirmation

5. In order to see the details of the cluster, click on the cluster name. This will take you to the cluster details page, where you can see the different services or tasks running in this cluster, along with the containers and the infrastructure they are running on:

Figure 10.14 – ECS cluster creation confirmation

Now we have our ECS cluster ready to run containers on. For the next step, we need to create a task definition based on which our tasks and services can be created.

Creating task definitions

As we learned earlier in this chapter, task definitions work like templates to execute and run tasks on ECS clusters. Implement the following steps to create a task definition for our `aws-code-pipeline` container application:

1. Click on the **Task definitions** link in the left panel and click on the **Create new task definition** button to start creating a task definition. The button displays a menu with two options; you can create the task definition with JSON or walk through the wizard to create one. Let's choose the latter:

Figure 10.15 – ECS task definition creation

2. Provide `packt-task-definition` as the name for this task definition. In the container definition section, enter the details of the container image and port mappings. We used the name `aws-code-pipeline` and pasted the image URI we copied back in *Figure 10.9* (`xxxxxxxxxxxx.dkr.ecr.us-east-1.amazonaws.com/packt-ecr-repo:aws-code-pipeline`). You can define more than one container definition in this section if your application needs to run more than one container as part of the same application:

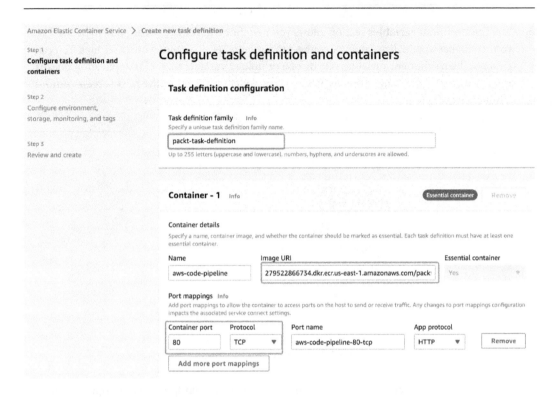

Figure 10.16 – ECS cluster creation confirmation

Container definition

A container definition allows you to define certain environment settings and resource requirements, which are passed to the Docker daemon when your container is launched. You can have more than one container definition as part of the same task definition. You should define more than one container as part of the same task definition if your containers share the same life cycle, refer to each other via the `localhost` port, or share resources and data volumes.

3. The **Environment variables** section allows you to define a list of key-value pairs to be used by your container to initialize values or connect to a remote service, among other settings. You can either add environment variables one by one or upload variable files. Our application doesn't require any environment variables at this point so let's proceed to the next screen without adding anything here:

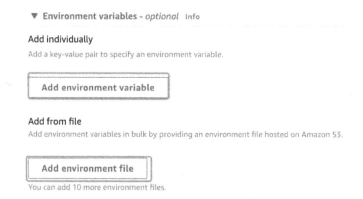

Figure 10.17 – Environment variables

4. The **HealthCheck** section allows us to configure the health endpoint within our container, which is used by ECS to determine whether a container is healthy or a new instance needs to be launched. This section allows us to customize the command to execute, the frequency and time of healthchecks, the grace period, and the number of retries when determining the health of the container. We haven't created a health endpoint in our application, so for now, let's leave this empty and click the **Next** button:

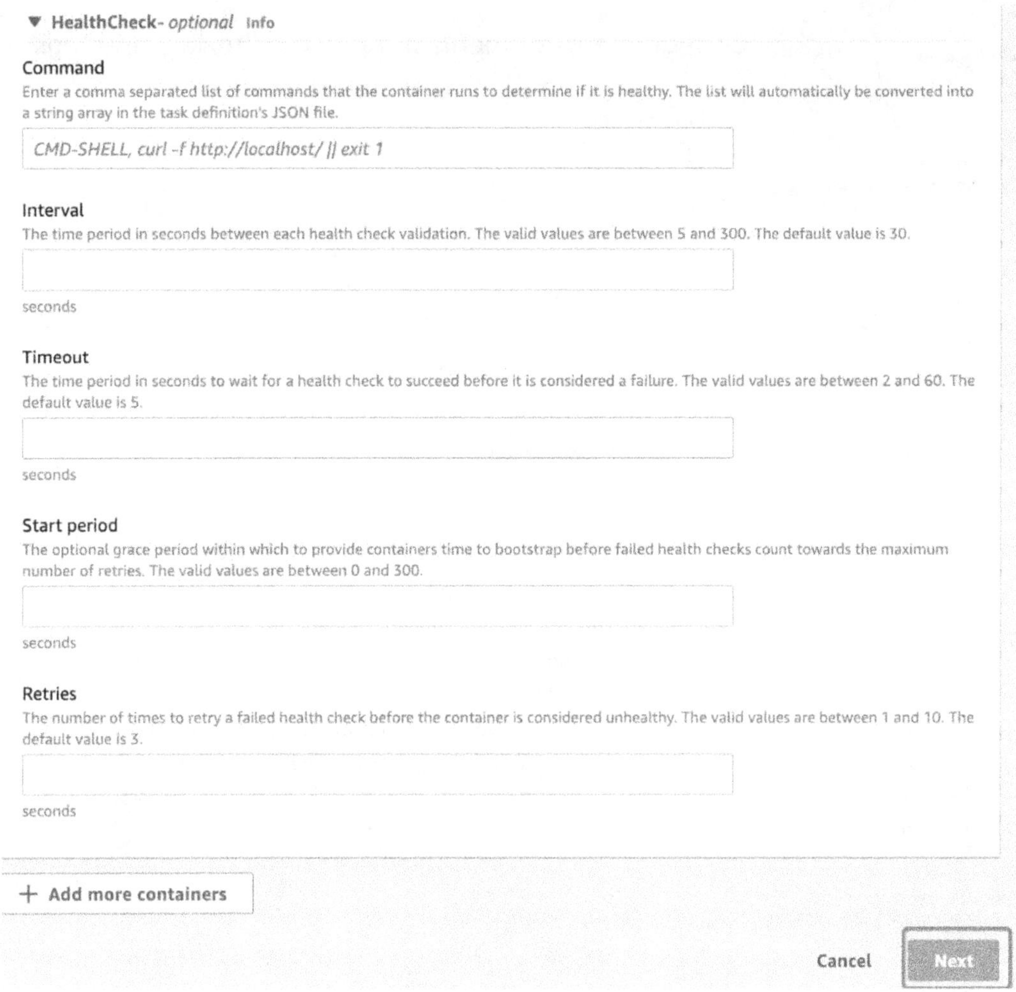

▼ **HealthCheck-** *optional* Info

Command
Enter a comma separated list of commands that the container runs to determine if it is healthy. The list will automatically be converted into a string array in the task definition's JSON file.

CMD-SHELL, curl -f http://localhost/ || exit 1

Interval
The time period in seconds between each health check validation. The valid values are between 5 and 300. The default value is 30.

seconds

Timeout
The time period in seconds to wait for a health check to succeed before it is considered a failure. The valid values are between 2 and 60. The default value is 5.

seconds

Start period
The optional grace period within which to provide containers time to bootstrap before failed health checks count towards the maximum number of retries. The valid values are between 0 and 300.

seconds

Retries
The number of times to retry a failed health check before the container is considered unhealthy. The valid values are between 1 and 10. The default value is 3.

seconds

+ **Add more containers**

Cancel   Next

Figure 10.18 – Health endpoint configuration

5.   The next page allows us to configure the environment, storage, and other settings. The following screenshot shows the details related to the environment in which our containers will be launched. We can select the infrastructure, operating system, and CPU and memory resource allocation for the task. We will launch our application in a Linux environment with 1 vCPU and 3 GB of memory:

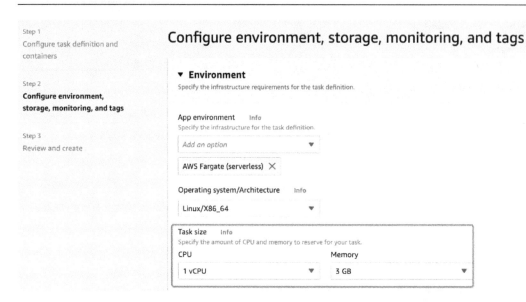

Figure 10.19 – Environment, storage, and monitoring configuration

6.   If you click on the arrow next to **Container size**, you can select the different containers defined in the previous section and assign resources at each container level. This configuration is important when you have more than one container within a single task definition, where you want to limit which container gets how many vCPUs and how much memory out of the totals defined at the task-size level. The vCPU and memory values should not exceed the totals defined for both at the task level:

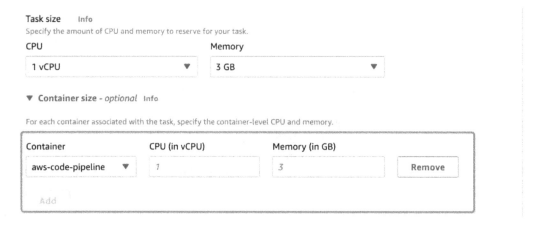

Figure 10.20 – Container level resource limit configuration

7. **Task role** and **Network mode** allow us to configure what roles we assign to our task to communicate with AWS services and APIs, and the role to be executed by the ECS container agent. **Network mode** defaults to `awsvpc` in the case of the AWS Fargate environment, but you can choose between `bridge network mode` or `awsvpc` for EC2 instances. We are keeping the default values here and moving to the next section:

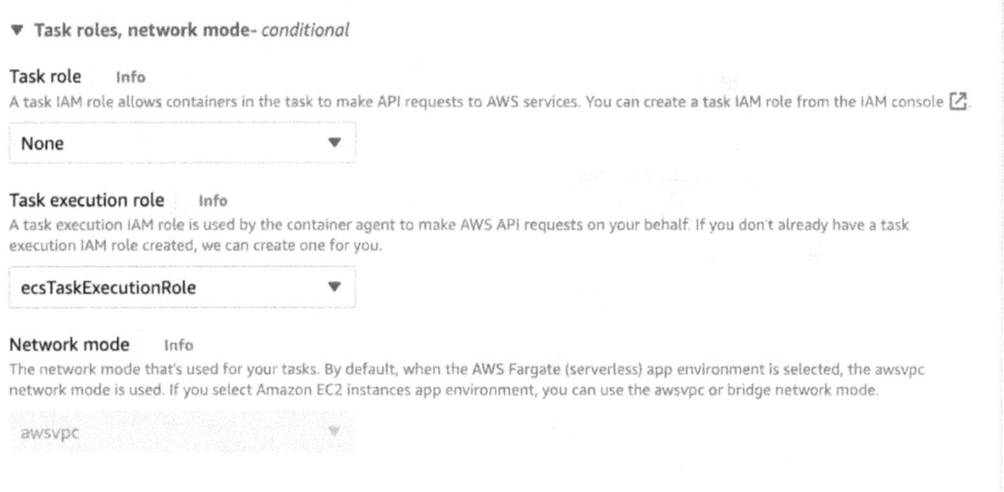

▼ **Task roles, network mode-** *conditional*

Task role   Info

A task IAM role allows containers in the task to make API requests to AWS services. You can create a task IAM role from the IAM console ☑.

None ▼

Task execution role   Info

A task execution IAM role is used by the container agent to make AWS API requests on your behalf. If you don't already have a task execution IAM role created, we can create one for you.

ecsTaskExecutionRole ▼

Network mode   Info

The network mode that's used for your tasks. By default, when the AWS Fargate (serverless) app environment is selected, the awsvpc network mode is used. If you select Amazon EC2 instances app environment, you can use the awsvpc or bridge network mode.

awsvpc ▼

Figure 10.21 – Task execution roles and network configuration

8. The **Storage** section allows us to add a storage volume other than the ephemeral storage attached to the machine used to launch our task. If your workload has a requirement for dedicated storage, you can add a volume here, which can be mounted to your containers on a specified patch so your container can access it easily. In our case, we don't need any additional storage:

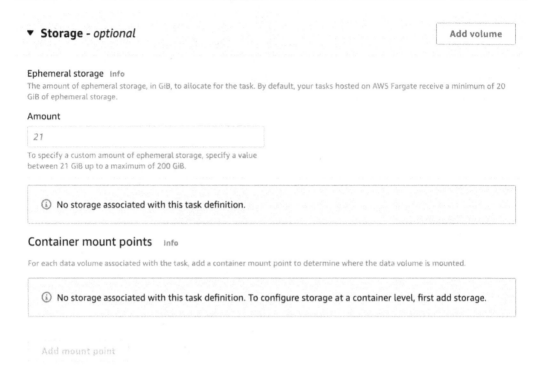

Figure 10.22 – Storage and mount configuration

9. The **Monitoring and logging** section allows us to customize how we collect logs from our containers. By default, CloudWatch is configured to handle logs for us. Alternatively, we have the option to move our logs to Firehose, Kinesis, OpenSearch, or S3. Let's keep the defaults here and move on to the next section:

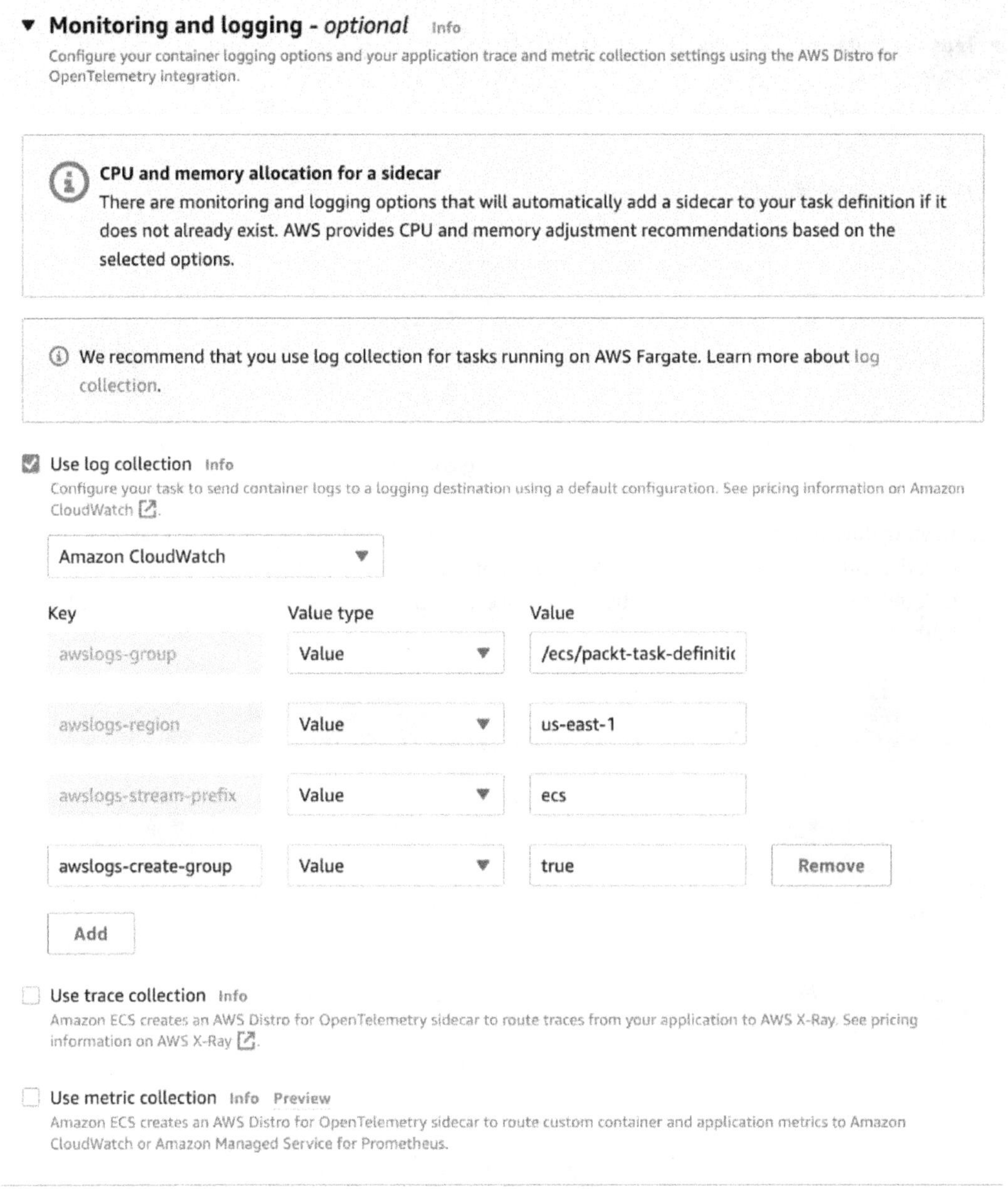

▼ **Monitoring and logging** - *optional*   Info

Configure your container logging options and your application trace and metric collection settings using the AWS Distro for OpenTelemetry integration.

ⓘ **CPU and memory allocation for a sidecar**

There are monitoring and logging options that will automatically add a sidecar to your task definition if it does not already exist. AWS provides CPU and memory adjustment recommendations based on the selected options.

ⓘ We recommend that you use log collection for tasks running on AWS Fargate. Learn more about log collection.

☑ Use log collection Info

Configure your task to send container logs to a logging destination using a default configuration. See pricing information on Amazon CloudWatch 🔗.

Amazon CloudWatch	▼

Key	Value type	Value	
awslogs-group	Value ▼	/ecs/packt-task-definitic	
awslogs-region	Value ▼	us-east-1	
awslogs-stream-prefix	Value ▼	ecs	
awslogs-create-group	Value ▼	true	Remove

Add

☐ Use trace collection Info

Amazon ECS creates an AWS Distro for OpenTelemetry sidecar to route traces from your application to AWS X-Ray. See pricing information on AWS X-Ray 🔗.

☐ Use metric collection Info Preview

Amazon ECS creates an AWS Distro for OpenTelemetry sidecar to route custom container and application metrics to Amazon CloudWatch or Amazon Managed Service for Prometheus.

Figure 10.23 – Monitoring and logging configuration

10. The **Tags** section allows us to add tags to the task definition and those tags can be added automatically to the tasks or services created from this task definition. Add a few tags to identify the task definition and click **Next**:

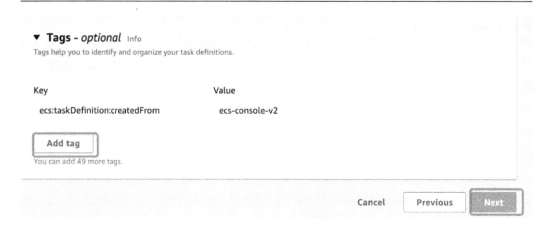

Figure 10.24 – Resource tag configuration

11. Review the information on the next screen and click on the **Create** button to create our first task definition. Once you click on the **Create** button, it will take about 60 seconds to create the task definition, and once it is ready, you will see a confirmation message like the one displayed in the following screenshot:

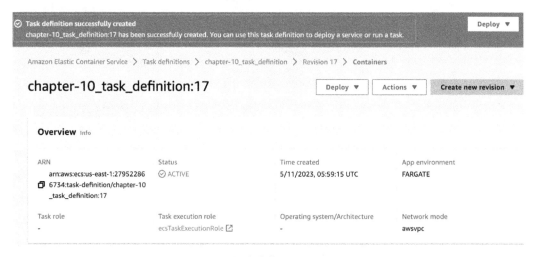

Figure 10.25 – Task definition review screen

The task definition is now ready for the execution of a service or a task. Follow the instructions in the next section to create and run a task.

Running a task

Now that our task definition is ready, the next step is to create a task or service to run the container workload. In our case, we just want to demonstrate the capability, so we don't need to create a whole service – we are going to just create a task here, but note that the process to create a service is similar. Implement the following steps to create a task:

1. Under **Task definitions**, select the `packt-task-definition` task definition for which we want to create a task, click **Deploy**, and select **Run task**:

Figure 10.26 – Task definition Deploy task screen

2. On the **Run task** screen, we have to select from the **Compute options** presented. Here we are selecting **Capacity provider strategy**, so our tasks can scale based on the strategy we have defined for our cluster. If you select **Launch type**, then your tasks will be launched directly on the type of infrastructure you select:

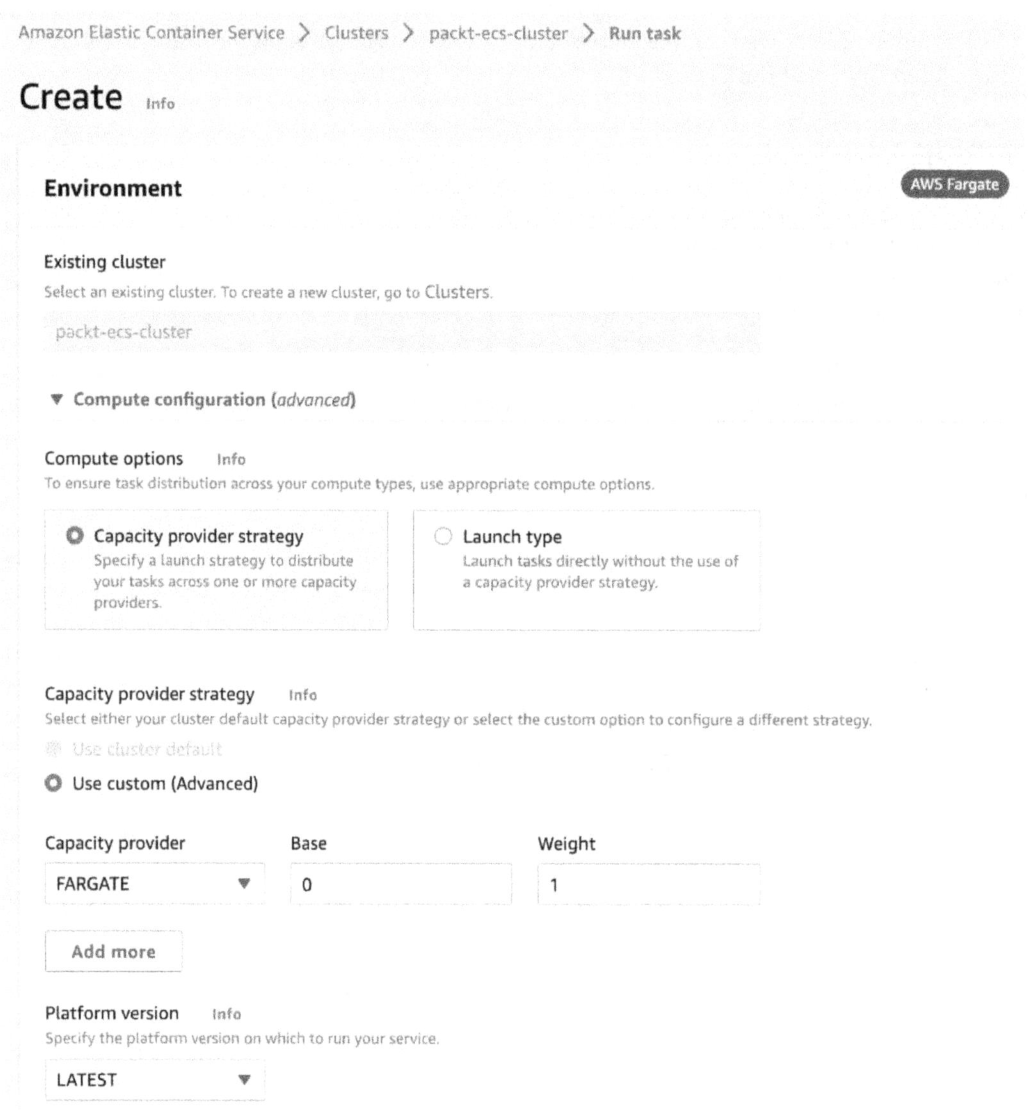

Amazon Elastic Container Service > Clusters > packt-ecs-cluster > **Run task**

Create Info

Environment

AWS Fargate

Existing cluster

Select an existing cluster. To create a new cluster, go to Clusters.

packt-ecs-cluster

▼ **Compute configuration** (*advanced*)

Compute options   Info

To ensure task distribution across your compute types, use appropriate compute options.

◉ **Capacity provider strategy**
Specify a launch strategy to distribute your tasks across one or more capacity providers.

○ **Launch type**
Launch tasks directly without the use of a capacity provider strategy.

Capacity provider strategy   Info

Select either your cluster default capacity provider strategy or select the custom option to configure a different strategy.

◉ Use cluster default

◉ **Use custom (Advanced)**

Capacity provider	Base	Weight
FARGATE ▼	0	1

Add more

Platform version   Info

Specify the platform version on which to run your service.

LATEST ▼

Figure 10.27 – Run task compute configuration setting

Capacity providers

A capacity provider defines the capability of the cluster infrastructure to scale in or out when needed. Each cluster is created with a default strategy, but you can override that strategy at the task/service level.

A capacity provider strategy manages the scaling of the infrastructure when launching tasks for you. Each cluster can have one or more capacity providers and a provider strategy to scale the infrastructure when needed. In your capacity provider strategy, you have base and weight options, where base defines how many tasks will be run at a minimum on the provider type, while weight identifies the ratio of tasks if there is more than one capacity provider. So, suppose you have 2 capacity providers, A with a weight of 1, and B with a weight of 4, then when ECS needs to scale up tasks, it will launch 4 tasks on capacity provider B for each on provider A.

3. The **Deployment configuration** section allows us to select the type of application we want to launch. **Task** is more of a standalone application that runs and then gets terminated, while services are groups of long-running tasks that can be stopped or restarted as needed with the required compute capacity maintained at all times. For our application, the **Task** type is sufficient as we are only using it for demo purposes:

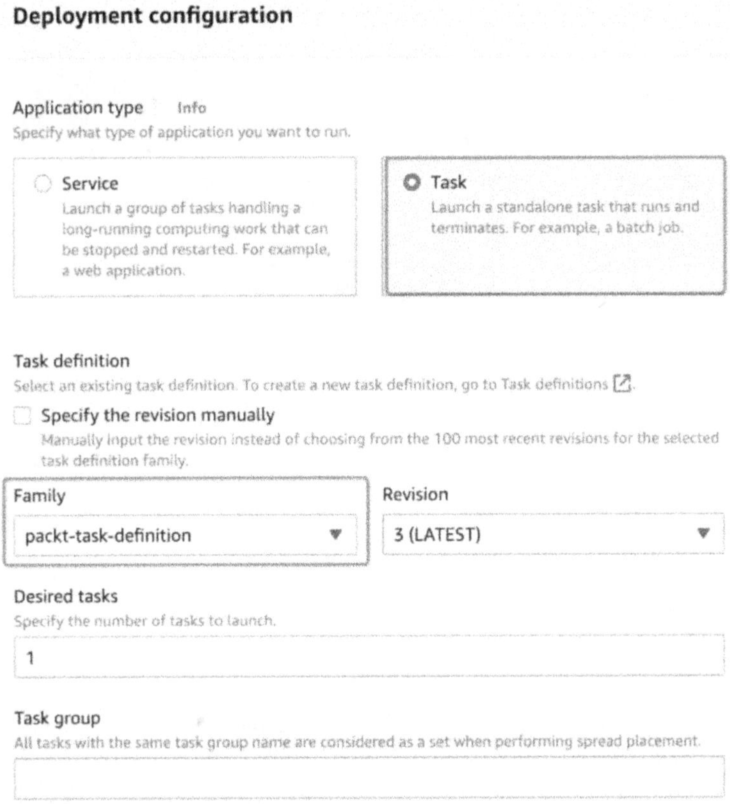

Deployment configuration

Application type   Info
Specify what type of application you want to run.

Service	Task
Launch a group of tasks handling a long-running computing work that can be stopped and restarted. For example, a web application.	Launch a standalone task that runs and terminates. For example, a batch job.

Task definition
Select an existing task definition. To create a new task definition, go to Task definitions.

☐ Specify the revision manually
    Manually input the revision instead of choosing from the 100 most recent revisions for the selected task definition family.

Family	Revision
packt-task-definition ▼	3 (LATEST) ▼

Desired tasks
Specify the number of tasks to launch.

1

Task group
All tasks with the same task group name are considered as a set when performing spread placement.

Figure 10.28 – Deployment configuration setting

Your task definition can have more than one version, so you can also define what revision of your task definition you want to launch. You can also specify the group if your task is part of a task group. The **Desired tasks** field specifies how many tasks you want to be launched. Suppose your task definition has two containers defined and you set **Desired tasks** to 3 – there will be a total of 6 containers launched within ECS. In our case, we have only one container defined in the task definition and we want only one task, so there will only be one container launched.

4. The **Networking** section allows us to define which VPC and subnets we want to launch the desired tasks and the security group to be used. Toggling on the **Public IP** option will assign a public IP address to our task with which we can access it:

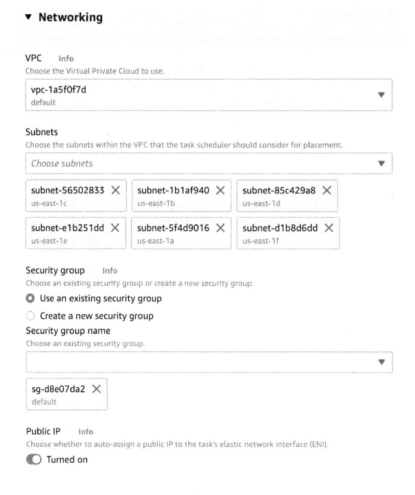

Figure 10.29 – Run task networking configuration

5.   The **Task overrides** section allows us to define a different role for the task and the ECS agent that we defined at the task definition level:

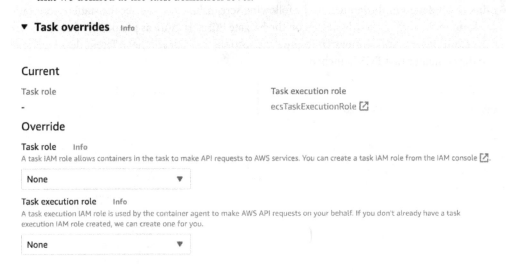

Figure 10.30 – Task role override configuration

6.   The **Containers overrides** section allows us to change the commands and environment variables defined at the task definition level. If you have more than one container, then it will provide you with the option to change the behavior of each one separately. Leave the default values in this section. The **Tags** section allows us to propagate the tags defined at the service or task definition levels for easy management of tasks. Click on **Create** to start launching the task:

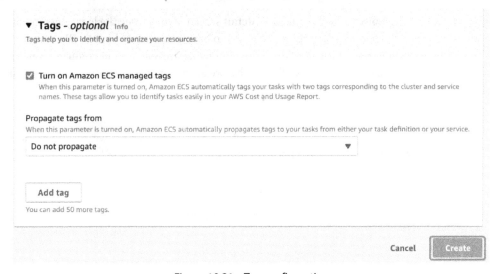

Figure 10.31 – Tag configuration

7.   Once you click on the **Create** button, it will launch the configured number of tasks. You can see the status on the cluster details screen under the **Tasks** tab. Click on the respective task link to get details of the launch. In the following screenshot, you can see the status of our task is **Running** and that it was launched on the Fargate infrastructure as per our configuration. In the **Container details for aws-code-pipeline** section, you can see the different details related to the container that ECS launched:

Figure 10.32 – Container details screen

8.   If you click on the **Logs** tab under the **Task detail** screen, you can see the logs generated by our application container. We configured this task to have an associated public IP address, so let's click on the link next to the IP address. The following browser output confirms that our container has been launched successfully and that we are able to get a response from the task on port 80:

aws-code-pipeline-1.0 ip-172-31-53-181.ec2.internal/172.31.53.181

Figure 10.33 – Container browser response

We now have our task running and working fine. For the next step, let's learn how to deploy these tasks using the CodeDeploy service so we don't have to do all of this manually every time.

Configuring CodeDeploy to install apps to ECS

In the previous section, we covered how to configure and install an application revision to ECS as a task. We configured the ECS cluster and a task definition, executed a task, and then accessed it through the browser. In this section, we are going to expand our use case and configure the AWS CodeDeploy service to perform the deployment of our aws-code-pipeline application container to ECS. We learned about the blue-green deployment pattern in *Chapter 2*, now we will implement that knowledge here. We will start our ECS application with a base version of the application, and then deploy a newer version to a separate task set and slowly migrate traffic to that newer task set until we have completely switched the traffic over. In this deployment model, you will see a flavor of canary deployment where we migrate the traffic slowly to the new version of the application over a period of time.

> **Canary deployment**
>
> A canary deployment is a mechanism for releasing new features to production. In a canary deployment, you install the new version of the application in the production environment without affecting the existing version. The newer version of the application takes a certain percentage of traffic based on different factors such as a feature flag, user subscribed preferences, or other conditions. Canary deployments help in testing an application or a feature with a smaller subset of users in live production, without impacting the wider user base if anything goes wrong. Once you have confidence in the newer version of the application, you scale up the percentage of traffic going to this new version. Canary deployments thus help reduce the risks that come with new releases.

We will use Terraform to provision the basic infrastructure, then we will configure CodeDeploy to install the newer application revision. As shown in the following diagram, we start with a base application setup where we have an **Application Load Balancer** (**ALB**) with a listener to listen on port 80, which has two target groups – one with a blue label and another one with a green label, both listening on port 80. Initially, we have the original task set running version V1 of the application on an ECS instance:

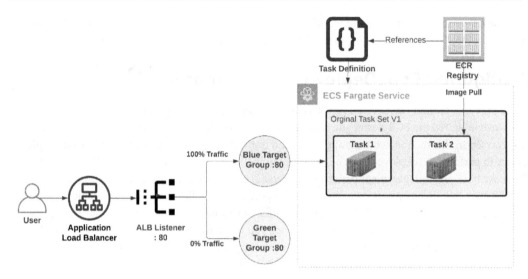

Figure 10.34 – ECS deployment with blue-green setup

Then, we will make changes to our sample application to print version V2 when you make a request and publish that to the ECR repository, and update the task definition in ECS.

> **Target groups**
>
> In AWS , a target group is a logical entity responsible for routing the traffic from a load balancer to a set of registered targets such as EC2 instances. A target can be part of one or two target groups and you configure the port number and protocol to route the traffic to a particular target group.

We will create a new application and deployment group to deploy this application version to ECS and slowly migrate traffic to the newer version. The CodeDeploy service will perform the deployment in blue-green fashion. CodeDeploy will launch a replacement task set with the newer version of the application without modifying the original task set, and our ALB listener will slowly start switching traffic to the replacement task set. Note that in CodeDeploy, you do also have the option to completely switch 100% of the traffic in one go, but we will switch a certain percentage at given intervals of time instead:

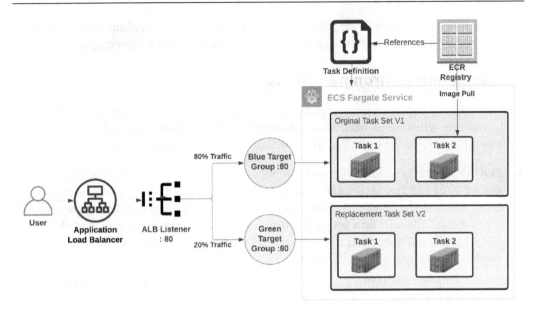

Figure 10.35 – CodeDeploy deployment to the replacement group with traffic switching in progress

Once the new task set is ready, all traffic will be switched to the green stack slowly and the listener will stop sending any traffic to the original blue task set. The following diagram demonstrates how, once the deployment is completed and 100% of the traffic has been switched to the green task set, CodeDeploy will terminate the original blue stack:

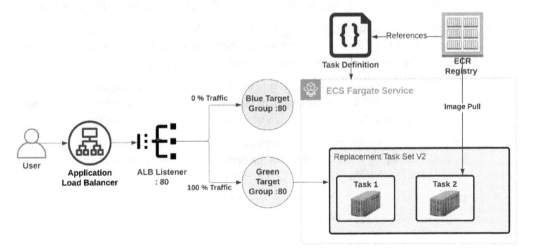

Figure 10.36 – CodeDeploy traffic fully switched to replacement group

Now that we have covered what we are going to do, let's get started with the implementation part. We have developed a Terraform template to create the required infrastructure for this chapter.

ECS cluster and environment setup

Perform the following steps to create the required infrastructure. We assume that you have the Terraform CLI set up on your system, as we walked through that in *Chapter 4*. You can refer back to that chapter and follow the instructions to set up Terraform on your system if required:

1. Download the Terraform template from GitHub at `https://github.com/PacktPublishing/Building-and-Delivering-Microservices-on-AWS/tree/main/chapter_10`. Once you download the Terraform template, make edits to the `main.tf` file and edit a few items where your personal values are required, such as the VPC ID, subnet IDs, `container_image_uri`, and `availability_zones`. You can get these details from your AWS console.

2. Once you update these settings, run the `terraform init` command. This will download the required providers and modules for AWS.

3. Next, run the `terraform plan` command to get information on all the resources this template is going to create in your AWS account. The following is a summary of these resources:

    I. `chapter-10_lb_sg`: This is the security group used by the load balancer and allows traffic on ports `80` and `8080`.

    II. `chapter-10_alb`: The ALB to handle traffic and distribute it to `chapter_10`.

    III. `chapter-10_alb_listner`: The ALB listener to transfer traffic to the target groups.

    IV. `chapter-10_blue_alb_tgt_group`: This is load balancer target group to listen to traffic on port `80` for the existing task set. Traffic will be switched from this blue target group to the green group slowly when we deploy the newer version of the application.

    V. `chapter-10_green_alb_tgt_group`: This is load balancer target group to listen to traffic on port `80` for the replacement task set. Traffic will be switched to this green target group from the blue group when we deploy the newer version of the application.

    VI. `Chapter-10_code_deploy_service_role`: The AWS IAM service role used by the CodeDeploy service to perform the deployment to the ECS cluster; this role will be used later in this chapter when configuring the application.

    VII. `Chapter-10_ecs_cluster`: A new ECS cluster with the **FARGATE** capacity provider to deploy applications.

VIII. `Chapter-10_ecs_capacity_providers`: The ECS cluster capacity provider list. This includes the **FARGATE** and **FARGATE SPOT** instances and defaults to **FARGATE**, but you can add additional providers such as an EC2 Auto Scaling group.

IX. `chapter-10_task_definition`: The task definition with the container settings to deploy the aws-code-pipeline application from the ECR repository.

X. `chapter-10_ecs_service`: An ECS instance to run the desired capacity of the tasks for aws-code-pipeline. This service starts with the original task set, which will be replaced by another deployment later in this chapter.

XI. `chapter-10_code_deploy_app`: The AWS CodeDeploy application to create deployment groups under it later in the chapter for deployment to the ECS cluster.

4. Having reviewed the list of resources Terraform is going to create, note that creating these resources will incur a cost to you, so after you complete the exercise, make sure to run the `terraform destroy` command to delete all the resources created through Terraform. For now, let's create the resources by running the `terraform apply –auto-approve` command.

5. Next, let's log in to the AWS console and review our ECS cluster and other resources. The following screenshot confirms that our `chapter-10_ecs_cluster` cluster is ready and one service underneath this is running. Click on the cluster name:

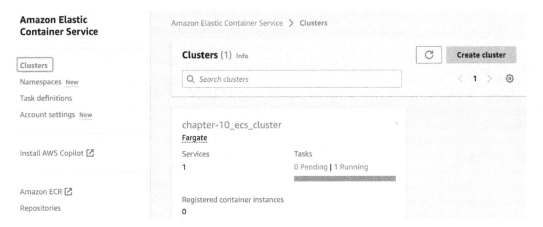

Figure 10.37 – ECS cluster

6. Click on the **Services** tab and confirm that our `chapter-10_ecs_service` service is running one task with revision 1 of the `chapter-10_task_definition` task definition. Later in this chapter, we will deploy revision 2 of this application through the CodeDeploy service:

Figure 10.38 – ECS with task definition revision 1

7. Click on the **Networking** tab of the chapter-10_ecs_service service and note that we have a target group, chapter-10-blue-alb-tgt-group, associated with this service, which is forwarding traffic on port 80 from the chapter-10-alb ALB. Click on the chapter-10-alb load balancer link and go to the **Listeners** tab:

Figure 10.39 – ECS networking details

8. chapter-10_alb_listner is running on port 80 and is associated with the chapter-10-blue-alb-tgt-group target group, forwarding 100% of the traffic in line with the current rule. Once you configure your blue-green deployment, you will see another target group get associated with this listener, and during deployment, traffic will start switching from the blue to the green target group:

Figure 10.40 – ALB listener and target group configuration

9. Now copy the load balancer DNS link and paste it into your browser. You should see the `aws-code-pipeline` application is running version 1.0:

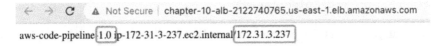

Figure 10.41 – aws-code-pipeline version 1.0

We have everything up and running in terms of the ECS setup. In the next section, let's go to our aws-code-pipeline service and change the version information.

Application version update

To create a new revision of our application and push that to our **ECR** instance, we need to change the application version, build a Docker image, push it to ECR, and then update the reference in the task definition file. Follow these instructions for a step-by-step guide:

1. Open up the VS Code IDE and modify the `app.version` property in the `application.properties` file under the `resources` section:

Figure 10.42 – application.properties version update

2. Run the mvn clean install command to build and generate the required artifact.

3. Use the instructions we went through earlier in this chapter in the *Amazon Elastic Container Registry* section, *steps 7-10*, or enter the following commands to create a Docker image for version 2 of the application and push it to the ECR repository. Remember to replace xxxxxxxxxxxxx with your AWS account ID:

```
docker build -t aws-code-pipeline .
aws ecr get-login-password --region us-east-1 | docker login
--username AWS --password-stdin xxxxxxxxxxx.dkr.ecr.us-east-1.
amazonaws.com
docker tag aws-code-pipeline:latest xxxxxxxxxxxx.dkr.ecr.
us-east-1.amazonaws.com/packt-ecr-repo:aws-code-pipeline-v2
docker push xxxxxxxxxxxx.dkr.ecr.us-east-1.amazonaws.com/packt-
ecr-repo:aws-code-pipeline-v2
```

4. Now, you can go to the ECR repository and validate the v2 revision of the application image:

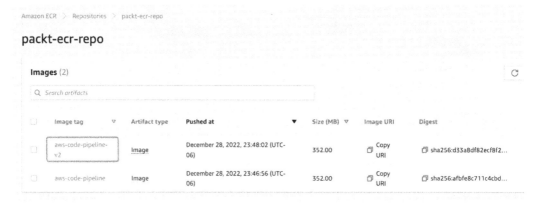

Figure 10.43 – aws-code-pipeline v2 Docker tag

Copy the URL `xxxxxxxxxxxxx.dkr.ecr.us-east-1.amazonaws.com/packt-ecr-repo:aws-code-pipeline-v2` for `aws-code-pipeline-v2` image tag, then let's go ahead and update the task definition.

5. Go to **Task definitions**, select `chapter-10_task_definition`, and click on the **Create new revision** button:

Figure 10.44 – Task definitions list

6. On the **Create new task definition revision** page, we need to update the Image URL `xxxxxxxxxxxxx.dkr.ecr.us-east-1.amazonaws.com/packt-ecr-repo:aws-code-pipeline-v2` for `aws-code-pipeline-v2` to point to the new Docker tagged image of our application:

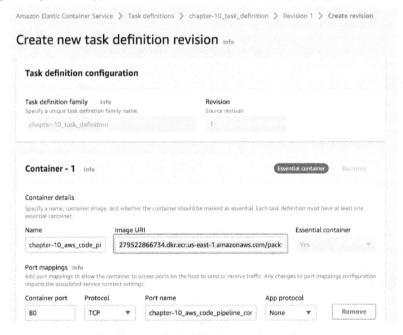

Figure 10.45 – Task definition revision update

7.   Click on the **Create** button and you should see confirmation once the application revision is ready. Copy the task definition revision ARN, as shown in the following screenshot, as you'll need this for the CodeDeploy service when you deploy this new revision. In our case, the ARN is `arn:aws:ecs:us-east-1:xxxxxxxxxxxx:task-definition/chapter-10_task_definition:2`:

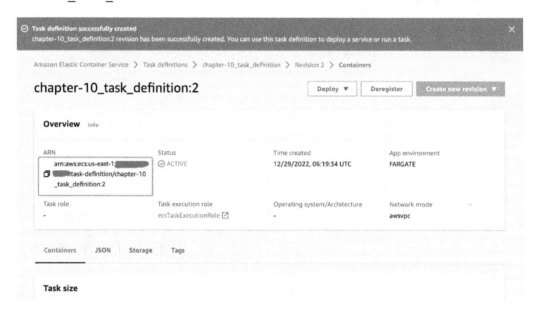

Figure 10.46 – Task definition revision update confirmation

We have updated our ECS task definition, so let's now focus on the CodeDeploy service. We will deploy the newer version of the application to the ECS cluster using CodeDeploy.

CodeDeploy setup

In order to use CodeDeploy to start deploying to the ECS cluster, we need to create an application and deployment group, and then deploy. Implement the following steps to perform the installation:

1.   Go to the CodeDeploy service, click on **Applications** in the left panel, and navigate to the `chapter-10_code_deploy_app` application. This application was created as part of the Terraform template. One thing to note is that this application uses **Amazon ECS** as its compute platform, as we will deploy it to the ECS cluster we created earlier in this chapter. Click on **Create deployment group**:

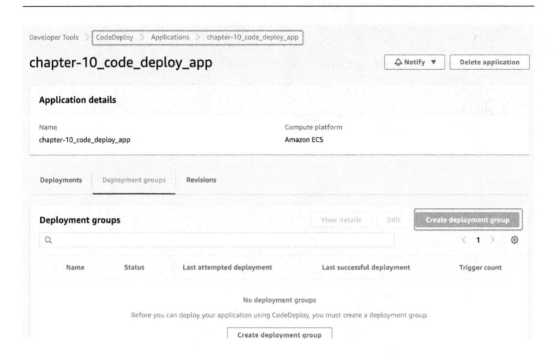

Figure 10.47 – CodeDeploy application details

2. On this screen, we need to provide a name for this new deployment group and select `chapter-10_code_deploy_service_role` for the service role. This role allows the CodeDeploy service to connect to ECS for deployment, and to EC2 to modify the load balancer settings to switch the traffic from the blue to the green environment:

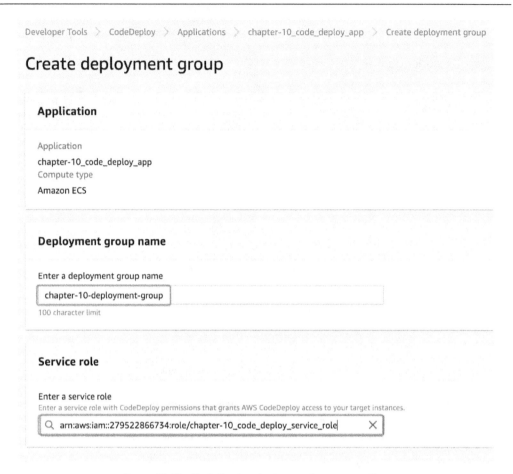

Figure 10.48 – CodeDeploy deployment group creation

3.   In the **Environment configuration** section, select the ECS cluster name and the service that we want to target for the deployment. We already created these using Terraform:

Figure 10.49 – ECS environment details for the deployment group

4. In the **Load balancers** section, select the load balancer we created, `-chapter-10-alb`-, and set the live production traffic to use port `80`. We have an optional **Test listener port** field, which can be used to test the application before traffic is switched to the live environment. To keep things simple here, we haven't configured this here:

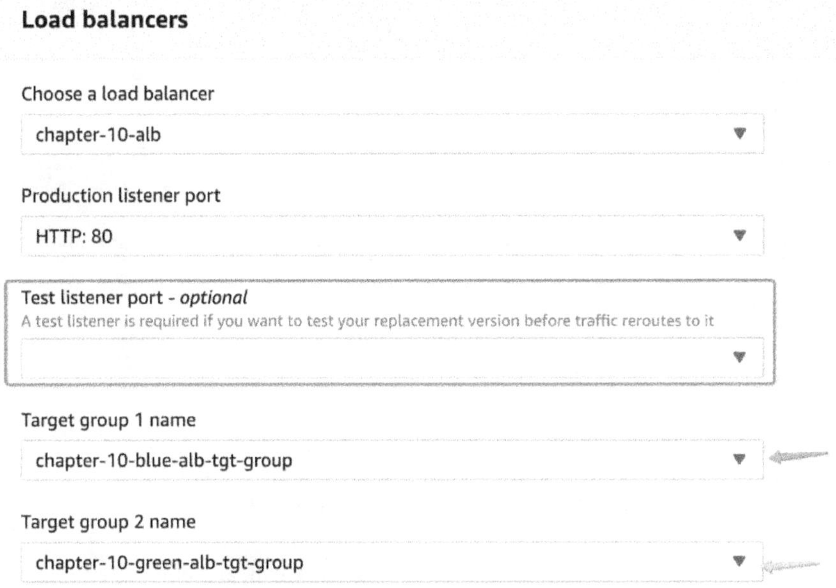

Load balancers

Choose a load balancer

chapter-10-alb	▼

Production listener port

HTTP: 80	▼

Test listener port - *optional*
A test listener is required if you want to test your replacement version before traffic reroutes to it

	▼

Target group 1 name

chapter-10-blue-alb-tgt-group	▼

Target group 2 name

chapter-10-green-alb-tgt-group	▼

Figure 10.50 – CodeDeploy load balancer settings

In the preceding screenshot, we have configured the CodeDeploy service with two target groups, one blue and the other green. Target group 1, `chapter-10-blue-alb-tgt-group`, is already associated with the `chapter-10-alb` load balancer and is taking 100% of the traffic, but the green target group is not associated with the load balancer listener. As soon as we configure this deployment group, CodeDeploy will go ahead and automatically modify the load balancer with 2 target groups and configure `chapter-10-blue-alb-tgt-group` with weight 1 and `chapter-10-green-alb-tgt-group` with weight 0. So technically, `chapter-10-green-alb-tgt-group` won't get any traffic. These traffic settings will be flipped later when we start the deployment.

5. The **Deployment settings** section allows us to configure how we want to route the traffic once our new application revision is deployed to the replacement task set. We have configured our traffic to be routed through `chapter-10-green-alb-tgt-group` immediately after deployment, but we will be re-routing 10% increments of the traffic each minute. Once 100% of the traffic is migrated, we will wait for 5 minutes before having ECS terminate our original task set. We can change this period to be as long as 2 days as of writing this book, giving us

the chance to fully validate the newer version, and perform a rollback if required, before terminating the original task set:

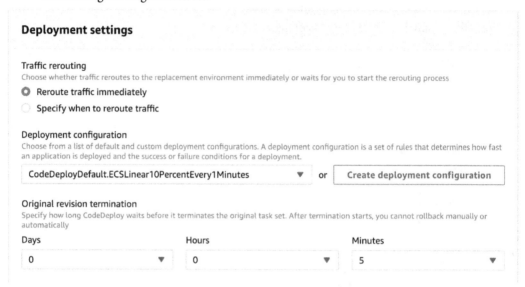

Figure 10.51 – CodeDeploy traffic switch settings

6. The **advanced** section allows us to configure the trigger, alarm, and automated rollback of the application in the case of a failure in deployment or an alarm being triggered. Let's go ahead and click on **Create deployment group**. The following screenshot shows our deployment group has been created:

Figure 10.52– CodeDeploy deployment group confirmation

7. Next, let's go ahead and click on the **Create deployment** button to start the deployment to the ECS cluster. Select the **Use AppSpec editor** option and **YAML** format to provide details of the `appsec.yml` file for the deployment. You can find the contents at `https://github.com/PacktPublishing/Building-and-Delivering-Microservices-on-AWS/blob/main/chapter_10/aws-code-pipeline/ecs-appspec.yml`. Change the task definition ARN to point to the task definition revision you want to deploy:

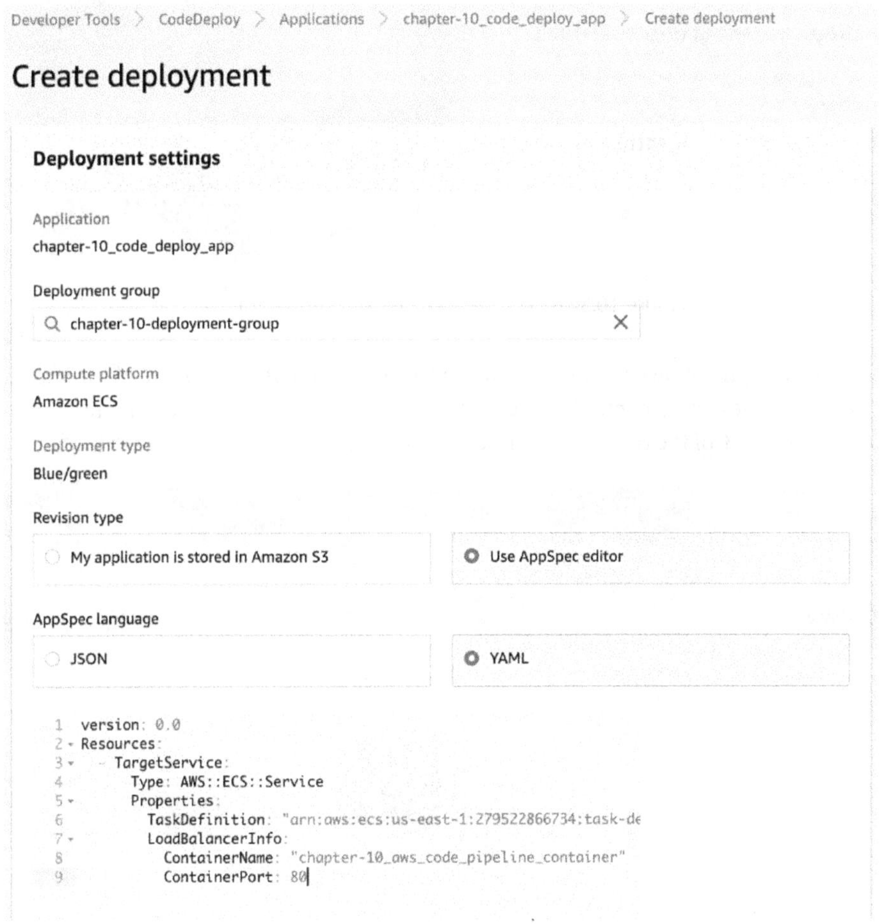

Figure 10.53 – CodeDeploy deployment AppSpec details

8. Optionally, we can provide a description for this deployment. We could also override some of the settings defined at the deployment group level here and determine how we want to handle rollbacks, but for now, let's just click on the **Create deployment** button:

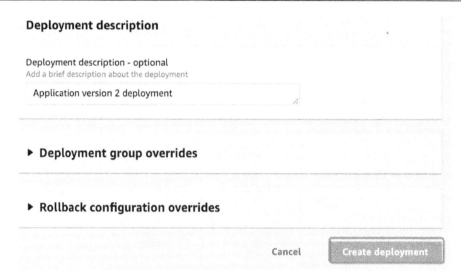

Figure 10.54 – CodeDeploy create deployment

9. Once the deployment task is ready, it will start the deployment of version 2 of the application. At the start of the deployment, all traffic is handled by the original task set, with the replacement task set getting 0% of the traffic. You can see the deployment status on the left side of the screen:

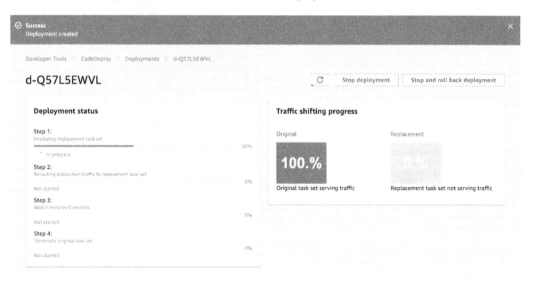

Figure 10.55 – CodeDeploy's ECS deployment in progress

10. You can see the different deployment life cycle events toward the bottom of the screen and watch the progress through the different deployment steps.

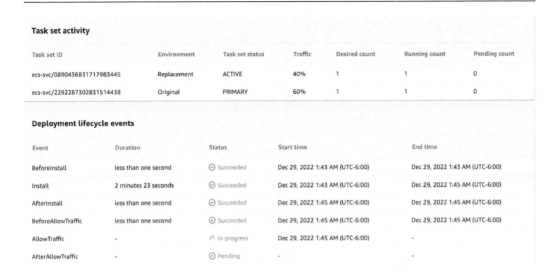

Task set activity

Task set ID	Environment	Task set status	Traffic	Desired count	Running count	Pending count
ecs-svc/0890436831717983445	Replacement	ACTIVE	40%	1	1	0
ecs-svc/2292287302831514438	Original	PRIMARY	60%	1	1	0

Deployment lifecycle events

Event	Duration	Status	Start time	End time
BeforeInstall	less than one second	⊘ Succeeded	Dec 29, 2022 1:43 AM (UTC-6:00)	Dec 29, 2022 1:43 AM (UTC-6:00)
Install	2 minutes 23 seconds	⊘ Succeeded	Dec 29, 2022 1:43 AM (UTC-6:00)	Dec 29, 2022 1:45 AM (UTC-6:00)
AfterInstall	less than one second	⊘ Succeeded	Dec 29, 2022 1:45 AM (UTC-6:00)	Dec 29, 2022 1:45 AM (UTC-6:00)
BeforeAllowTraffic	less than one second	⊘ Succeeded	Dec 29, 2022 1:45 AM (UTC-6:00)	Dec 29, 2022 1:45 AM (UTC-6:00)
AllowTraffic	-	◯ In progress	Dec 29, 2022 1:45 AM (UTC-6:00)	-
AfterAllowTraffic	-	⊙ Pending	-	-

Figure 10.56 – CodeDeploy deployment progress

11. After approximately 2 minutes, the new replacement task set will be ready and traffic will start being shifted to the new task set. If you scroll up at this point, you will see that 20% of the traffic has been moved to the new task, with 80% still being handled by the original task set. At this point, you have the option to stop the deployment and roll back 100% of the traffic to the original task set:

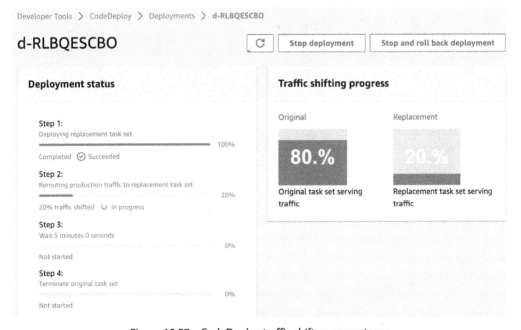

Developer Tools > CodeDeploy > Deployments > d-RLBQESCBO

d-RLBQESCBO   ↻ | Stop deployment | Stop and roll back deployment

Deployment status

Step 1:
Deploying replacement task set
━━━━━━━━━━━━━━━━━━━━━━━━━━ 100%
Completed ⊘ Succeeded

Step 2:
Rerouting production traffic to replacement task set
━━━━ 20%
20% traffic shifted ↻ In progress

Step 3:
Wait 5 minutes 0 seconds
0%
Not started

Step 4:
Terminate original task set
0%
Not started

Traffic shifting progress

Original

80.%

Original task set serving traffic

Replacement

20.%

Replacement task set serving traffic

Figure 10.57 – CodeDeploy traffic shift as percentages

12. Go to the **Listeners** section on the load balancer screen and you can observe that now we have two target groups attached to our listener, with the blue task group taking 80% of the traffic and the green one taking 20% of the traffic:

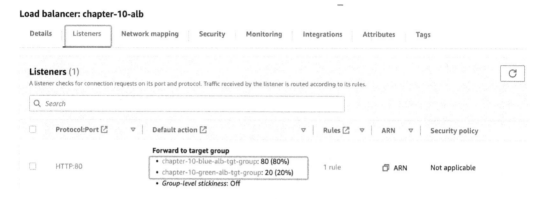

Figure 10.58 – Load balancer traffic distribution

13. If at this point, you copy the load balancer DNS address, paste it into your browser, and refresh the page a few times, you will notice a few calls going to version 2 of your application while the remaining ones will continue to go to the previous version as we have both task sets running at the moment. After waiting for approximately 10 minutes, all the traffic will have been moved to the new task set and the original task set will be terminated:

Figure 10.59 – CodeDeploy deployment complete

14. The load balancer settings will also update and all 100% of the traffic will now be switched to the green target group. From this point onward, all your traffic will be handled by version 2.0 of your application:

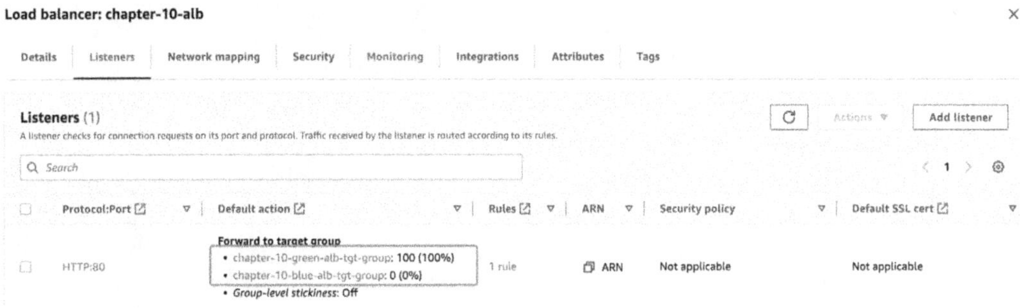

Figure 10.60 – CodeDeploy with 100% of the traffic switched to the green task set

At this point, we have successfully completed the blue-green deployment to ECS using the CodeDeploy blue-green deployment model. Make sure to go ahead and run the `terraform destroy` command to avoid additional charges to your account.

Summary

In this chapter, we learned about Docker containers and the ECS offering from AWS. We created a Docker image of our example `aws-code-pipeline` microservice and deployed it to the ECS infrastructure. In addition to this, we also expanded our knowledge of the CodeDeploy service and learned how to deploy an application to the ECS infrastructure using CodeDeploy. In the next chapter, we will learn about the AWS CodePipeline service and how to apply our knowledge to automatically deploy code to AWS from a developer check-in.

Setting Up CodePipeline Code

In this chapter, you will learn about the AWS CodePipeline service, its benefits, and its limitations. We will combine all our knowledge from previous chapters to create a pipeline for our sample `aws-code-pipeline` project. You will understand how CodeCommit, CodeBuild, and CodeDeploy services integrate using CodePipeline.

In this chapter, we will be covering the following topics:

- What is CodePipeline?
- The benefits of CodePipeline
- The performance and limitations of CodePipeline
- CodePipeline action types
- Creating a pipeline for a sample application
- Executing a pipeline

In previous chapters, we learned about the CodeDeploy service and how we can deploy applications to the EC2 instances and the **Elastic Container Service** (**ECS**). We also manually compiled our application using Maven, created a Docker image, pushed it to the **Elastic Container Repository** (**ECR**), and then deployed it. All of those steps were very much manual, and we needed orchestration to execute everything step by step. In this chapter, we will learn how we can orchestrate the different AWS services to perform release deployment, from a developer check-in code to a production deployment using AWS CodePipeline. Let's start by learning more about the AWS CodePipeline service.

What is AWS CodePipeline?

AWS CodePipeline is a continuous delivery service that orchestrates a code deployment process in an automated fashion. AWS CodePipeline is a managed service to automate application and infrastructure delivery. AWS CodePipeline provides integration with different AWS services, such as CodeCommit, CodeBuild, CodeDeploy, **Simple Notification Service** (**SNS**), and services outside AWS, such as Git Hub, Jenkins, Blazemeter, and Snyk. AWS manages all the resiliency and scaling of the CodePipeline service infrastructure, so you don't have to worry about maintaining it.

In *Chapter 3*, we learned about **Continuous Integration** and **Continuous Delivery (CI/CD)**. In this chapter, we will implement a pipeline for our sample application. The following diagram explains the typical flow of any pipeline. A pipeline involves application source code, which can be in any language. In the build step, you compile or package that source code to create an artifact that can be deployed or executed.

The test step involves testing the code to ensure that the application is able to perform the functionality it is intended to perform. In the deployment step, your application is deployed to your given environment and becomes available for use. Based on your organization's needs, there might be several environments that your application goes through before being deployed to production, such as a development environment, a test environment, a **User Acceptance Testing (UAT)** environment, and then a production environment.

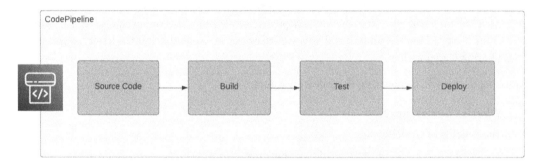

Figure 11.1 – A pipeline structure

You can add or remove steps based on your need, but the intention of a pipeline is to provide you a way to deliver your source code, from the exact moment it is checked into the source repository to deploying it to a production environment using automation.

As a developer, you need to model your release process, and CodePipeline will help you to automate this. As a developer, you define the workflow and describe how your code progresses through different stages. A pipeline is built using several stages, and each stage can have more than one action. AWS CodePipeline provides you with a graphical representation of these stages and actions, and you can visualize at what stage your pipeline is at during its execution. Let us look at what a stage and an action are in a pipeline:

- **Stage**: A stage in CodePipeline is a logical separation of the actions you want to perform in your workflow on application artifacts. Your pipeline is divided into several logically separated stages, such as code build, test, security scan, and deployment. You can further subdivide a stage into different actions, each action running in parallel or sequentially. Each stage is a logically separate environment to limit the concurrent changes within it.

In a pipeline, you define different stages, and your application artifact travels through these stages to complete the release process. Each stage can be fully automated or have manual steps to go from one stage to another. You can introduce approvals within stages, and the pipeline will wait for your approval before it can go to the next stage or action.

- **Action**: A pipeline stage can have more than one action, and each action is a set of instructions that you want to perform on application code at a particular point in the pipeline journey. An action can include things such as invoking a Lambda function or executing a code build action to perform a deployment. For example, a pipeline stage can have a set of actions to deploy application code to EC2 instances or an ECS cluster.

Now we understand what CodePipeline is, let's learn about some of the benefits it provides.

The benefits of using CodePipeline

AWS CodePipeline provides a lot more benefits if you compare it with other competing products. Jenkins, CircleCI, GitHub Actions, GitLab CI, Azure Pipelines, and CloudBees CI are some other popular CI/CD tools that help us to set up pipelines and deliver software. AWS CodePipeline has an advantage because it provides the following benefits over products from other competitors:

- AWS CodePipeline is a fully managed service, so you don't have to worry about purchasing hardware, software, installation, updates, and patching. AWS manages everything for you, and you can scale as much as you need and start quickly.

- AWS CodePipeline integrates well with your existing source code repositories, such as GitHub, GitHub Enterprise, CodeCommit, Bitbucket, and S3, so you don't have to learn any new integration tools. You can integrate your existing source code tool with CodePipeline.

- AWS CodePipeline provides you the flexibility to customize your pipeline workflow, and you can add or remove stages as needed.

- AWS CodePipeline gives you multiple choices to do a certain type of work; you are not bound to use a certain source code repository or use a certain build service to perform a build. AWS CodePipeline gives you the flexibility to use your existing tools – for example, if you have an existing Jenkins setup in your organization to build the source code, you can still use it and utilize the other AWS CodePipeline features.

- AWS CodePipeline provides prebuilt integration with existing tools, so you don't need to write custom logic to do that. For example, if you need to perform source code scanning, you can simply connect with the Snyk service to perform integration for you, or if you want to perform a load test for your application as part of a pipeline, you can do that in just a few clicks.

- CodePipeline is very cost-effective, and there is no commitment or infrastructure cost associated with it; you only pay for what you use. At the time of writing, you pay only $1 for each active pipeline per month.

- CodePipeline allows you to add manual approval steps within a pipeline, so you can do a final check before rolling out a feature to your production if you prefer.

- AWS CodePipeline integrates well with AWS events and SNS, so you can trigger notifications at any stage of a pipeline to notify interested parties.

- AWS CodePipeline is used in conjunction with the AWS IAM service, so you can manage and grant access to users who you want to manage the release process.

- AWS CodePipeline is a declarative pipeline, so you can define your pipeline in the form of a JSON document, which you can modify as needed and define as templates to create new pipelines.

- AWS CodePipeline allows you to extend a pipeline beyond the AWS infrastructure by using the CodePipeline open source agent with your servers.

Now that we have covered the benefits of CodePipeline, let's also look into its limitations.

The limitations of CodePipeline

While the AWS CodePipeline service is a highly available and scalable service, it has the following limitations at the time of writing. Some of these limitations can be increased through an increase in service quota:

- AWS CodePipeline history can be maintained for a maximum of 12 months.

- AWS CodePipeline needs a minimum of 2 stages and a maximum of 50 stages in a pipeline, so if your release cycle needs more than 50 stages, then you might have to combine a few together.

- AWS CodePipeline can have only a maximum of 1,000 pipelines in each region for a single account.

- Only a maximum of 300 pipelines can be configured for source repository polling within an account in a particular region. If you have a large organization and plan to use a pooling mechanism for change detection in your source code, then you need to plan for the use of multiple AWS accounts.

- The size of all output variables for a particular action is a maximum of 122,880 bytes combined.

- You can have a maximum of 500 actions in a pipeline and a maximum of 50 in a single stage.

- AWS CodePipeline supports manual approval action, but an approval action can wait a maximum of 7 days before it gets a timeout.

You can find a complete list of CodePipeline limitations here: `https://docs.aws.amazon.com/codepipeline/latest/userguide/limits.html`. Now that we have a good understanding of the CodePipeline service and its limitations, let's dive a little deeper into the different integrations it provides.

CodePipeline action types

CodePipeline is an orchestrator service that brings other services together to deliver your software end to end. CodePipeline provides integration with different tools to deliver the software, and these tools are organized by action types to cover the different aspects of the software delivery. Let's learn about each of these action types in more detail and understand all the capabilities that CodePipeline provides.

The source action type

For any CI/CD pipeline, you need source code to build an artifact and deploy that to your environment. In AWS CodePipeline, the source action type lets you define integration with a version control system to download the source code to build an artifact. Let's look at the different source code providers that CodePipeline provides out-of-the-box integration with:

- **Amazon S3**: An Amazon S3 bucket can be used as a source code provider for CodePipeline. CodePipeline expects your source to be a single file, so if your source code has more than one file, you need to upload that as a single zip file.

  You need to provide the name of the S3 bucket and object that contains the source for your application at the time of configuration. CodePipeline provides a configuration option to detect a change in an S3 bucket, through CloudWatch events or periodical pooling. The CloudWatch events approach is recommended by AWS to automatically start a pipeline as soon as a new object is uploaded to your S3 bucket.

- **Amazon Elastic Container Registry (ECR)**: Amazon ECR is another source code provider that can trigger a pipeline as soon as a new container image is pushed to the ECR repository. This action is often used with other source actions when source code is built and a new image is available for deployment in ECR. When you configure ECR as a source code provider, you need to define an ECR repository and a tag for the image you want to monitor; by default, it uses the latest tag for the image.

  This provider makes the `RegistryId`, `RepositoryName`, `ImageTag`, `ImageDigest`, and `ImageURI` environment variables available for downstream actions in the pipeline, which will be referred to through a namespace.

- **CodeCommit**: This provider action type makes source code available from your CodeCommit repositories and allows you to select the branch that you want CodePipeline to use. You can use the CloudWatch events to auto-detect the changes in the CodeCommit repository, or you can use AWS CodePipeline to periodically look for changes in your CodeCommit repository.

  The CodeCommit source action type provides the following output variables for your downstream actions: `CommitId`, `CommitMessage`, `RepositoryName`, `BranchName`, `AuthorDate`, and `CommitterDate`. By default, CodePipeline outputs the repository as a single zip file. Alternatively, you can choose a full clone option, which will pass the repository metadata information to a downstream action to clone the repository.

- **CodeStarSourceConnection**: This provider action type is used to trigger a pipeline when a commit is made to third-party Git repositories, such as Bitbucket, GitHub, or GitHub Enterprise Server. With this source code provider, you first need to establish a connection between the AWS CodeStar service and the Git provider using OAuth. Once a connection is established, you will be able to specify the repository and branch name you have chosen to build the source code.

  You also need to choose an output artifact type as a single zip file or a full repository clone option, for downstream actions. This source action provides the following output variables available for your downstream actions: `AuthorDate`, `BranchName`, `CommitId`, `CommitMessage`, `ConnectionArn`, and `FullRepositoryName`. In upcoming chapters, you will see a practical example of connecting to a Bitbucket repository.

The build action type

Once you have obtained your source code, you need to compile the code or package the source code, run test cases, or create container images for the release process. The build action type supports integration of CodeBuild and Jenkins action providers out of the box, but you can create custom actions to support providers beyond the default ones. You can do a lot more things with this action type – for example, perform a source code scan, run your integration test cases, download any application dependency, and package it with your code. Let's look at some of the major integrations supported by the CodePipeline build action type:

- **CodeBuild integration**: With the build action type, you can select CodeBuild as your provider for the build. As part of the configuration, you need to create a new CodeBuild project or select an existing one. The **Input artifacts** field allows you to select the name of the input files that will work as a source for this action. You can choose any input artifact from the previous actions or stages as your input:

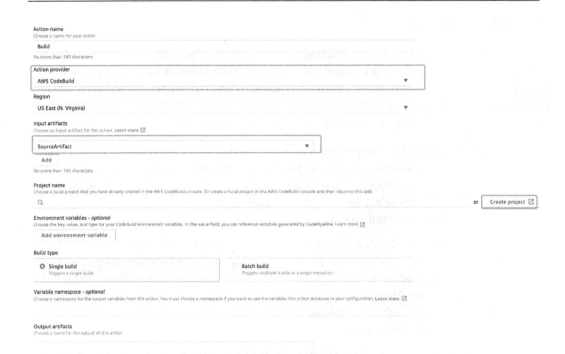

Figure 11.2 – AWS CodeBuild integration

The **Variable namespace** section allows you to provide a name, through which you can access any variable exported during the build phase in subsequent actions. You can access variables using the following format: #{namespace_name.env_var_name}.

- **Jenkins build integration**: In the Jenkins build action configuration, you specify details, similar to the CodeBuild action type, but here, you need to specify details about your Jenkins server. In order for CodePipeline to work with your Jenkins server, you need to install the AWS CodePipeline plugin on your Jenkins server and configure an appropriate role or IAM user.

  You need to use the same project name that you created in the Jenkins server and also use the same provider you configured in your Jenkins server CodePipeline plugin configuration. Output artifacts will allow you to specify a name for the artifact you produce with this action.

  In later chapters, we will be walking you through this process step by step and explaining how to integrate with a Jenkins server. If you need to use a build action other than Jenkins and CodeBuild, then you need to create a custom action.

We have learned about build action types; in the next section, let's focus on the different test integration tools supported by CodePipeline.

The test action type

When your pipeline transitions from a build stage to a test stage, there are different ways and different tools to test an application. There are several types of testing, such as unit testing, functional testing, and performance testing, and different tools used to perform these types of tests. AWS CodePipeline provides integration with several tools, including Jenkins, CodeBuild, AWS Device Farm, BlazeMeter, and Ghost Inspector UI testing. We will cover some of them in this section:

- **AWS Device Farm testing integration**: AWS CodePipeline provides integration to AWS Device Farm to test an application. AWS Device Farm is an application testing service that lets you test your mobile or web application on a wide variety of desktop browsers and real physical mobile devices. You can select the AWS Device Farm as a provider and specify the Device Farm project details to use as a test provider.

- **BlazeMeter testing integration**: BlazeMeter is a company that provides load and performance tests as a service. BlazeMeter allows you to test your application at scale to benchmark performance and simulate user load tests. BlazeMeter is compatible with open source Apache JMeter, so you can import your JMeter scripts. AWS CodePipeline provides very easy integration and a connector to connect with your BlazeMeter account project, and you can start a performance test as part of your build process.

- **Ghost Inspector UI testing integration**: Ghost Inspector is a testing service that creates automated browser tests for your websites and web applications. Ghost Inspector doesn't require you to code anything; it helps you to test your web application as an end user. You record your test script using the browser extension, and later, these actions are simulated by the pipeline, and your assertions will be performed. AWS CodePipeline provides easy integration with Ghost Inspector, and you can perform UI testing of your application with just a few clicks.

- **Micro Focus StormRunner/LoadRunner integration**: AWS CodePipeline supports integration with Micro Focus's LoadRunner, formerly known as StormRunner. LoadRunner is used to perform load and stress tests on your application to prepare for real production. LoadRunner creates realistic scenarios by simulating real-world network conditions and eliminating bottlenecks and dependencies. AWS CodePipeline provides an out-of-the-box connector to integrate LoadRunner with your pipeline. You can select **Micro Focus StormRunner Load** in the **Action provider** field, then you need to click on the **Connect to Micro Focus StormRunner Load** button to establish a connection between AWS and the LoadRunner website and select your project to perform testing.

- **Runscope API monitoring integration**: Runscope is an API monitoring and verification product. It helps to validate the structure of APIs and ensures that they follow the API contract. Runscope allows us to add simple assertions to JSON and XML data and validates the response from API calls. AWS CodePipeline provides a connector to integrate Runscope as part of a pipeline to validate your APIs. Just select Runscope as your action provider and connect to your Runscope account for testing.

We have covered all the test action types. Now, we will look at the different deploy action types supported by AWS CodePipeline.

The deploy action type

Deployment is an essential part of any pipeline, as deployment delivers your software to its final destination where your end users can access it. CodePipeline provides integration with several deployment types. Let's look at some of the major ones so that you can decide which one is more suitable for your pipeline:

- **Amazon S3 deploy action type**: CodePipeline supports deployment to Amazon S3 buckets as a deployment target. You can choose S3 as a deploy action to deploy a static website or just copy files to an S3 bucket:

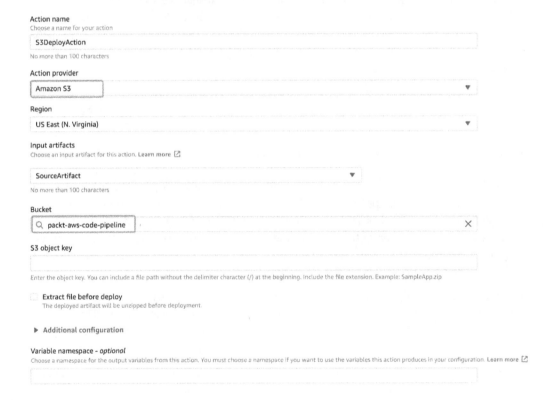

Figure 11.3 – Amazon S3 deploy action integration

You need to provide a bucket name and the object key for the deployment location.

- **AWS CloudFormation deploy action type**: AWS CodePipeline supports integration with the CloudFormation service as a deploy action. You can configure a pipeline to use CloudFormation as a deployment provider, and you can take action on AWS CloudFormation stacks and create, update, replace, or delete stacks and change sets as part of a pipeline. As a result of CloudFormation, the execution of a pipeline can create resources for use in subsequent stages.

- **AWS CodeDeploy action type**: AWS CodePipeline supports integration to the CodeDeploy service for the deployment of code to your target application. You need to choose AWS CodeDeploy as an action provider and provide the name of the CodeDeploy application and deployment group that you want to perform the deployment.

    We will use the AWS CodeDeploy action provider in our pipeline for deployment to EC2 instances later in the chapter.

- **Amazon Elastic Beanstalk deploy action type**: Elastic Beanstalk is a service provided by AWS to automate capacity provisioning, load balancing, auto-scaling, and application deployment for web applications and services developed in popular programming languages, such as Java, .NET, and Node.js, on familiar servers, such as Apache, **Internet Information Services** (**IIS**), and NGINX.

    AWS CodePipeline supports deployment to Elastic Beanstalk as part of a pipeline. You need to provide details of your existing Elastic Beanstalk application name and the environment to which you want to deploy, and then you can automate the deployment end to end.

- **AWS AppConfig deploy action type**: AWS AppConfig is the capability of AWS Systems Manager to manage application configuration. AppConfig allows you to deploy application configuration changes in a controlled environment to any application, with built-in validation checks and monitoring. AppConfig can deploy configuration changes to applications deployed on EC2 instances, containers, IoT mobile devices, or AWS Lambda. AWS CodePipeline supports integration with AppConfig to deliver configuration changes to your applications. You can provide details related to your AppConfig application name, environment, and configuration profile, and a pipeline will deploy these changes based on the strategy you select.

- **Amazon Alexa Skills Kit deploy action type**: Alexa skills are applications for Alexa-enabled devices and provide a new way to engage your users, through voice and text-based interaction. As an Alexa skill builder, you can build and add new skills and publish them to the Alexa Skills store, and your customers across the globe can use these skills.

    CodePipeline provides integration for Alexa skills and lets you deploy updates to them in an automated way. You can choose this deployment action type and provide details of the skill, and CodePipeline will deploy your Alexa skills to the Alexa Skills store.

- **AWS OpsWorks Stacks deploy action type**: AWS OpsWorks uses Chef for configuration management of applications. OpsWorks lets you define application architecture, components installation, configuration, and resources. AWS CodePipeline provides integration to deploy Chef cookbooks using OpsWorks. As part of the configuration, you provide details of the app and stacks, and CodePipeline will perform the deployment action to OpsWorks.

- **AWS Service Catalog deploy action type**: The AWS Service Catalog service allows organizations to centrally manage deployed services, applications, and resources. AWS Service Catalog allows us to create, manage and govern product catalogs, which can be deployed by customers as a self-service.

  CodePipeline allows you to deploy updates and versions of your product template version to AWS Service Catalog. You can select Service Catalog as a deploy action and provide details of the template and product; CodePipeline will then deploy these changes to AWS Catalog Service.

- **Amazon ECS and ECS (blue/green) deploy action type**: AWS CodePipeline provides integration with ECS as a deploy action. You can configure your pipeline to deploy your application changes to an existing ECS cluster, and CodePipeline will deploy updates to your ECS service. AWS CodePipeline provides two types of integration with ECS; one is specialized for blue/green deployment, which we will cover as part of our pipeline configuration later in this chapter.

We have covered all major deploy action providers. Next, we will cover **approval** action providers.

The approval action type

Automating an end-to-end deployment process is an ideal state for any CI/CD pipeline, but to achieve that, you need to ensure people have confidence in your automated test cases, buildquality gates to ensure that the release is of. Sometimes, even if you have everything, it is necessary to obtain approvals from stakeholders or business leaders before you deploy your release to a higher environment, such as production. AWS supports that approval flow as part of AWS CodePipeline. You can configure a manual approval action at any stage of your pipeline to obtain approval before subsequent actions or stages are executed.

An approval action takes an SNS topic and a custom URL as a configuration to deliver an approval notification. You can click on the **Approval** button to approve. We will demonstrate this feature in our pipeline later in the chapter.

The invoke action type

AWS CodePipeline supports the invocation of different action types as part of a pipeline to perform specialized tasks if needed. AWS CodePipeline supports the invocation of a Lambda function, AWS Step Functions, and Snyk invocation.

> **Note**
> Sync is a third-party code scanning service that finds and automatically fixes vulnerabilities in your code, open source dependencies, containers, and infrastructure as code through its security intelligence.

Now that we have covered the majority of integrations that AWS CodePipeline provides, it is time to use that knowledge to create an end-to-end delivery pipeline for our sample application, aws-code-pipeline.

Creating a pipeline

In this section, we are going to create an end-to-end pipeline for our aws-code-pipeline microservice, which we developed in previous chapters. We will create a multistage pipeline. The following diagram shows the different stages of the pipeline. Let's understand in detail what we will be doing at each stage:

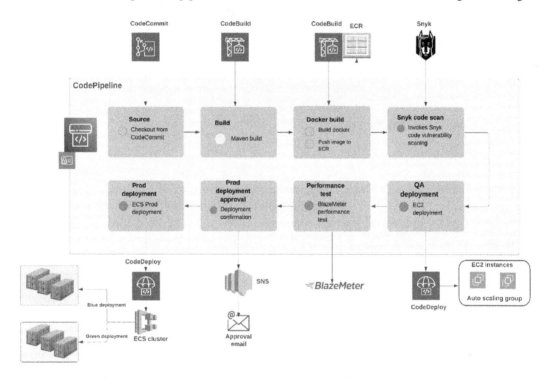

Figure 11.4 – The aws-code-pipeline microservice pipeline stages

The source stage

At the source stage, we integrate AWS CodePipeline to our CodeCommit repository, aws-code-pipeline, which we created in *Chapter 5*. If you don't have the CodeCommit repository, please follow the instructions and push the source code to the repository for further steps. You can download the source code from the GitHub repository here: https://github.com/PacktPublishing/Building-and-Delivering-Microservices-on-AWS/tree/main/chapter_11.

This stage checks out the source code from the CodeCommit repository and makes it available at subsequent stages as a zip file, named `SourceArtifact`.

The build stage

At the build stage, we are going to use the AWS CodeBuild service and use Maven commands to generate the Java artifact. The following `buildspec.yml` file describes how CodeBuild will run the `maven package` command to get the artifacts and include the Java JAR file, Dockerfile, `appspec.yml`, and other files in the output. We will use each one of the files in subsequent stages:

```yaml
! buildspec.yml
1    version: 0.2
2    phases:
3      install:
4        runtime-versions:
5          java: corretto11
6      pre_build:
7        commands:
8          - echo Starting pre-build phase
9          - java -version
10     build:
11       commands:
12         - echo Build started on `date`
13         - mvn test
14     post_build:
15       commands:
16         - echo Build completed on `date`
17         - mvn package
18         - ls
19         - ls target/surefire-reports
20   artifacts:
21     files:
22       - target/aws-code-pipeline*.jar
23       - appspec.yml
24       - ecs-appspec.yml
25       - taskdef.json
26       - docker_buildspec.yml
27       - Dockerfile
28       - scripts/*
29     discard-paths: yes
```

Figure 11.5 – buildspec.yml to create build artifacts

The Docker build stage

At the Docker build stage, we are going to build a Docker image out of our Java application `aws-code-pipeline*.jar` file. Later on, an ECS task definition will create containers using this Docker image for the ECS service. At this stage, we will push this Docker image to the ECR repository, `packt-ecr-repo`, which we created in previous chapters. This stage uses CodeBuild to perform these actions and the `docker_buildspec.yml` file for the CodeBuild project:

```
! docker_buildspec.yml U ✕

aws-code-pipeline > ! docker_buildspec.yml > { } artifacts > ⊡ discard-paths
  1    version: 0.2
  2    phases:
  3      pre_build:
  4        commands:
  5          - echo Building docker image for the application
  6          - aws ecr get-login-password --region us-east-1 | docker login --username AWS --password-stdin ██████████.dkr
                 ecr.us-east-1.amazonaws.com
  7      build:
  8        commands:
  9          - echo Build started on `date`
 10          - echo Building the docker image in build phase
 11          - ls
 12          - docker build -t aws-code-pipeline:latest .
 13          - docker tag aws-code-pipeline:latest ████████.dkr.ecr.us-east-1.amazonaws.com/
                 packt-ecr-repo:aws-code-pipeline
 14      post_build:
 15        commands:
 16          - echo Pushing docker image to the ECR recpository completed on `date`
 17          - docker push ████████.dkr.ecr.us-east-1.amazonaws.com/packt-ecr-repo:aws-code-pipeline
 18          - ls
 19    artifacts:
 20      files:
 21        - appspec.yml
 22        - ecs-appspec.yml
 23        - taskdef.json
 24        - scripts/*
 25      discard-paths: yes
```

Figure 11.6 – docker_buildspec.yml to create a Docker image

There are a couple of points to note; we are automatically building and pushing the Docker image to the ECR repository, and we are also collecting a few files as an output artifact for the following stages.

The Snyk code scan stage

At this stage, we will push our code to the Snyk service to perform the vulnerability scan, and Snyk can provide us with a report, indicating vulnerabilities and possible solutions. We can configure our build to fail if any vulnerability is found, or we can decide to continue with the build.

The QA deployment stage

At this stage, we are going to configure the CodeDeploy service to perform the Spring Boot deployment of our application to EC2 instances, using the CodeDeploy agent. These EC2 instances will run behind a load balancer, `chapter-11-test-alb`. The CodeDeploy agent will use the `appspec.yml` file and `start.sh` and `stop.sh` scripts, available in the `scripts` directory, to perform deployment to a QA environment.

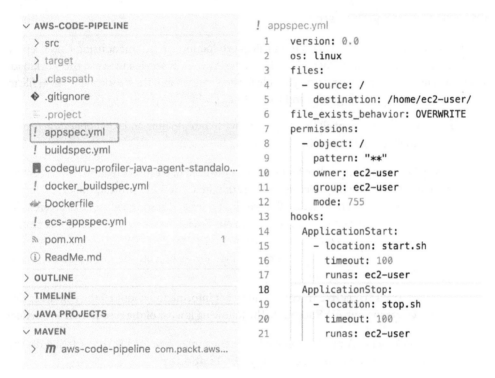

Figure 11.7 – appspec.yml for EC2 deployment to QA environment

The performance test stage

At this stage, we will be configuring a BlazeMeter test on our QA environment. After deployment to the QA environment, CodePipeline will trigger the BlazeMeter performance test to ensure that the application code is able to meet the performance expectations.

The production deployment approval stage

At this stage, we will configure the pipeline to require manual approval before it can be moved to the next stage of the flow. This stage will require someone to manually click on a link or a button to approve the paused pipeline. A pipeline can wait up to 7 days for approval before it gets a timeout. If someone didn't approve the pipeline within the time limit or rejects the approval request, then the pipeline build will fail at its current stage.

The production deployment stage

At this stage, we will deploy code changes to our ECS production environment using CodeDeploy with a blue/green strategy. At this stage, we will use the AWS ECS service to serve the workload. CodeDeploy will use the `ecs-appspec.yml` file for deployment and the `taskdef.json` file to provision the containers from our generated Docker image, which we pushed to ECR.

Now that you have a fair overview of all the stages we have in our pipeline and what operation occurs at each stage, let's start creating the pipeline by following these steps:

1. As a first step in creating the pipeline for our project, we need to create the required infrastructure and resources required by the pipeline, such as the the CodeBuild project, the CodeDeploy project, and the ECS cluster. To create the required roles and resources, we have created the Terraform template, which you can download from our GitHub repository: `https://github.com/PacktPublishing/Building-and-Delivering-Microservices-on-AWS/tree/main/chapter_11/terraform`. Please execute the `terraform init` command to initialize the Terraform.

2. Now, run the `terraform plan` command to get information about all the resources this template will create in your AWS account. The following is a list of the resources it will create:

    *   `chapter-11_alb`: A production environment application load balancer to handle the traffic and distribute it to `chapter-11_alb_listener`.

    *   `chapter_11_test_env_alb`: A QA environment application load balancer to handle the traffic and distribute it to `chapter-11_test_env_alb_listner`.

    *   `chapter-11_alb_listner`: A production environment application load balancer listener to transfer traffic to the `chapter-11_blue_alb_tgt_group` and `chapter-11_green_alb_tgt_group` target groups.

    *   `chapter_11_test_env_alb_listner`: A QA environment application load balancer listener to transfer traffic to the `chapter_11_test_env_alb_tgt_group` target group.

    *   `chapter-11_blue_alb_tgt_group`: A production load balancer target group to attach traffic on port 80 for existing task sets. Traffic will be switched from this blue target group to green when we deploy a newer version of the application.

- `chapter-11_green_alb_tgt_group`: A production load balancer target group to attach traffic on port 80 for a replacement task set. Traffic will be switched to this green target group from blue when we deploy a newer version of the application.

- `chapter_11_test_env_alb_tgt_group`: A QA load balancer target group to attach traffic on port 80 and transfer it to `chapter_11_test_env_asg`.

- `chapter_11_test_env_asg`: An auto-scaling group for QA environment EC2 instances so that a minimum capacity of the instances can be made available all the time.

- `chapter-11_code_maven_build`: A CodeBuild project to compile the Java source code and generate a Java artifact.

- `chapter-11_docker_build`: A CodeBuild project to generate a Docker image and push it to an ECR repository.

- `chapter-11_code_deploy_app`: An AWS CodeDeploy application for the production environment to create deployment groups, for use by CodeDeploy for the ECS cluster.

- `chapter_11_test_env_app`: An AWS CodeDeploy application for the QA environment to create deployment groups, for use by CodeDeploy for EC2 deployment.

- `chapter-11_code_deploy_group`: An AWS CodeDeploy deployment group for the `chapter-11_code_deploy_app` production application.

- `chapter_11_test_env_deploy_group`: An AWS CodeDeploy deployment group for the `chapter_11_test_env_app` QA application.

- `chapter-11_ecs_cluster`: A new ECS cluster with a Fargate capacity provider to deploy an application in the production environment.

- `chapter-11_ecs_capacity_providers`: An ECS cluster capacity providers list, which includes Fargate and Fargate SPOT instances. It defaults to Fargate, so you can add additional providers such as the EC2 auto-scaling group.

- `chapter-11_ecs_service`: An ECS service to run the desired capacity of the tasks for `aws-code-pipeline`. This service starts with the original task set, which will be replaced after another deployment later in this chapter.

- `chapter-11_task_definition`: A task definition with a container setting to deploy the `aws-code-pipeline` application from an ECR repository.

- `chapter_11_test_env_instance_profile`: An AWS EC2 instance profile to create the EC2 instance for a QA environment deployment.

- `chapter_11_test_env_deploy_policy`: An IAM policy that is attached to the `chapter_11_test_env_deploy_agent_role` role.

- `chapter_11_test_env_deploy_policy_attach`: Attached to the `chapter_11_test_env_deploy_policy` policy with the `chapter_11_test_env_deploy_agent_role` CodeDeploy agent role.

- `chapter-11_code_build_service_role`: The service role for the CodeBuild project to access code and push the image to ECR.

- `chapter-11_code_deploy_service_role`: The AWS IAM service role used by the CodeDeploy service to perform deployment to the ECS cluster; this role will be used later in this chapter while configuring the application.

- `chapter_11_test_env_deploy_agent_role`: This is the IAM role assumed by the EC2 instance, allowing the CodeDeploy agent to connect to the AWS CodeDeploy Service and download the required files from Amazon S3 for QA deployment.

- `chapter_11_test_env_ec2_launch_template`: The launch template to create EC2 instances with the same setting, and apply user data to download the CodeDeploy agent on instance launch.

- `chapter-11_lb_sg`: This is the security group used by the load balancer for both production and QA load balancers.

3. Creating these resources will incur a cost to you, so after you complete the exercise, run the terraform destroy commands to delete all the resources you created through terraform. For now, let's create the resources by running the `terraform apply -auto-approve` command.

4. Once the `apply` command is executed, you can review the resources in the AWS console. For now, let's start creating the pipeline. Search for `AWS CodePipeline` in the console and click on **Create pipeline** button.

5. Provide a name to the pipeline and click **Next**. In the **Advance** setting section, you can select a custom location and encryption for CodePipeline to store the artifacts generated during pipeline execution. We will keep the default values, so let's click on **Next**.

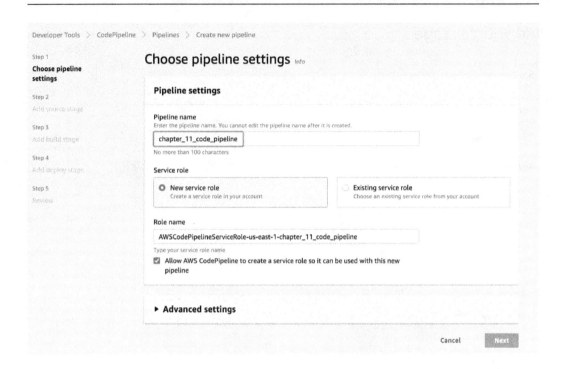

Figure 11.8 – The CodePipeline settings screen

6.   On the next screen, select the source code provider for the pipeline. We will use AWS CodeCommit and the `aws-code-pipeline` repository we created in *Chapter 5*. We will choose the master branch so that whenever a new code change is pushed to the branch, the pipeline will trigger a new build using the CloudWatch events. Click on the **Next** button to add the source stage.

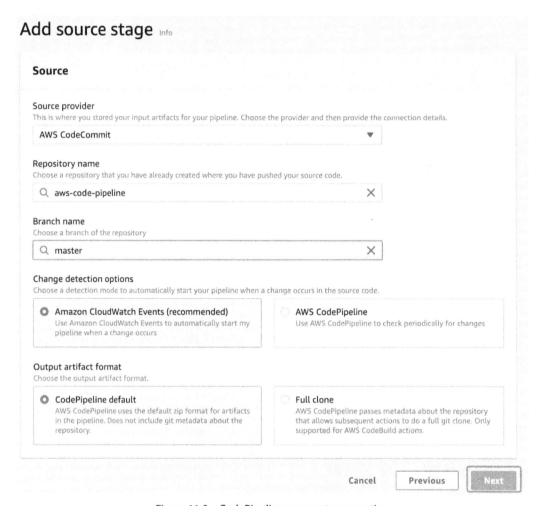

Figure 11.9 – CodePipeline source stage creation

7. On the next screen, we need to configure a build stage for the pipeline. Select **AWS CodeBuild** as the build provider and the `chapter-11_code_maven_build` CodeBuild project we created through Terraform. Then, click on the **Next** button.

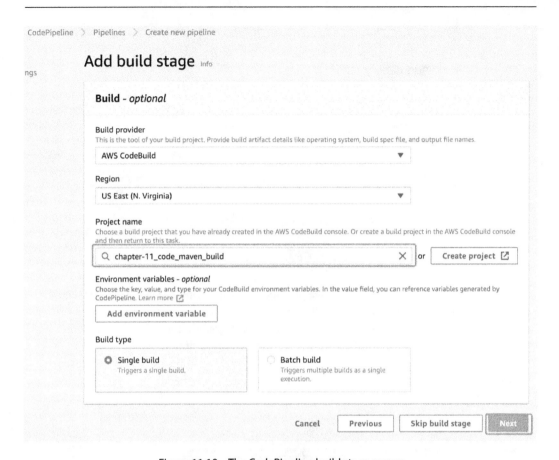

Figure 11.10 – The CodePipeline build stage screen

8. On the next screen, CodePipeline provides an option to add a **Deploy** stage. For now, we will skip this stage, as we will add these stages later in the pipeline. Click on the **Skip deploy stage** button and then the confirmation dialog's **Skip** button.

Figure 11.11 – The CodePipeline deploy stage screen

9. On the review screen, click on the **Create pipeline** button to create your first AWS CodePipeline. As soon as the pipeline is created, it will display a confirmation message and start a build. We are going to make changes to our pipeline, so for now, if you want to stop the current pipeline execution, you can do so by clicking on the **Stop execution** button. Now, we need to make changes to this pipeline to add more stages to align with the pipeline strategy we discussed earlier. In order to make changes to any of the existing stages or add new stages, click on the **Edit** button:

Figure 11.12 – The CodePipeline execution screen to edit

10. The following screen shows our chapter_11_code_pipeline pipeline in an editable state, where we can edit an existing stage or add new stages to our pipeline. We already have the source stage and build stage; we need to add a new stage Docker build stage to build Docker images, as we discussed earlier in this chapter. Click on the **Add stage** button next to the build stage in CodePipeline. You need to provide a name for this new stage – say, DockerBuild – and then click on the **Add stage** button to add this new stage to our pipeline:

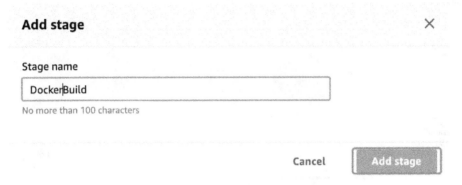

Figure 11.13 – The CodePipeline Add stage screen

11. Each stage in a pipeline can have one or more actions to perform a set of operations, so we need to add at least one action to this stage. Click on the + **Add action group** button to add an action to the `DockerBuild` stage.

Figure 11.14 – The CodePipeline Add action group screen

12. On the next screen, choose **AWS CodeBuild** as the action provider and select **BuildArtifact** as the input artifact from the previous stage, the `Build` stage, which has the generated JAR file. In the **Project name** field, select the `chapter-11_docker_build` CodeBuild project we created as part of the Terraform resources. Name the output artifact `DockerBuildArtifact`, which contains files such as `appsepc` and `taskdef.json`, required by subsequent stages.

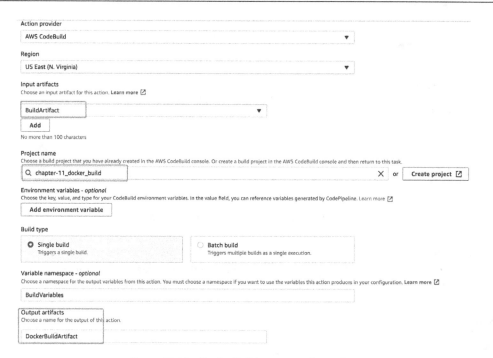

Figure 11.15 – DockerBuild stage configuration

13. Click on the **Done** button to finish the action creation, and you will see the new stage with the `DockerBuild` action. Now that we have our artifacts ready, let's add a new stage for the Snyk code scan. Click on **Add stage**, provide a name, and then click on **Add action group**, similar to the last step. Choose **Snyk** as the action provider, and select **SourceArtifact** for scanning.

Edit action

Action name
Choose a name for your action

SynkCodeScan

No more than 100 characters

Action provider

Snyk   ▼

Input artifacts
Choose an input artifact for this action. Learn more ☑

SourceArtifact   ▼

No more than 100 characters

Connect to Snyk

Connect to Snyk

Variable namespace - *optional*
Choose a namespace for the output variables from this action. You must choose a namespace if you want to use the variables this action produces in your configuration. Learn more ☑

Output artifacts
Choose a name for the output of this action.

No more than 100 characters

Figure 11.16 – Snyk Code Scan configuration

14. Now, we need to connect the Snyk service to our pipeline so that CodePipeline can push our code for scanning. When you click on the **Connect to Snyk** button, it will ask you to create an account; you can do so or log in using your GitHub credentials. Snyk will ask you to configure a project for the integration and ask for the pipeline behavior, and what needs to be done with the CodePipeline flow when a vulnerability is found. You can configure it to fail the deployment when a vulnerability of a configured severity is found. Since we are just demonstrating the capability, we will not fail our pipeline and let it continue, even if a critical vulnerability is found in our code. You need to provide a name for the project, and then click on the **Continue** button:

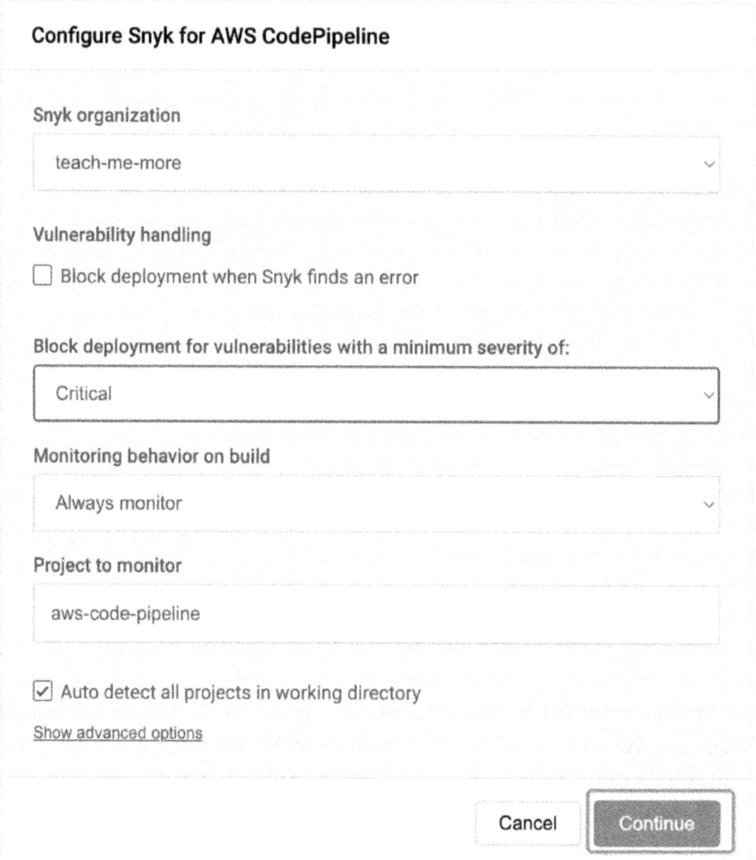

Configure Snyk for AWS CodePipeline
Snyk organization
teach-me-more
Vulnerability handling
Block deployment when Snyk finds an error
Block deployment for vulnerabilities with a minimum severity of:
Critical
Monitoring behavior on build
Always monitor
Project to monitor
aws-code-pipeline
Auto detect all projects in working directory
Show advanced options
Cancel
Continue

Figure 11.17 – Snyk project configuration

Confirm the Oauth request to connect to AWS CodePipeline and your Snyk account by clicking on the **Confirm** button:

aws
Services
N. Virginia
console_user @
Processing OAuth request
Choose Confirm to complete the connection
Cancel
Confirm

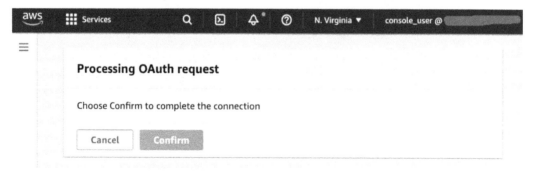

Figure 11.18 – Snyk OAuth integration

15. Now, you have successfully connected your pipeline stage to a Snyk account, so click on the **Done** button to finish the setup.

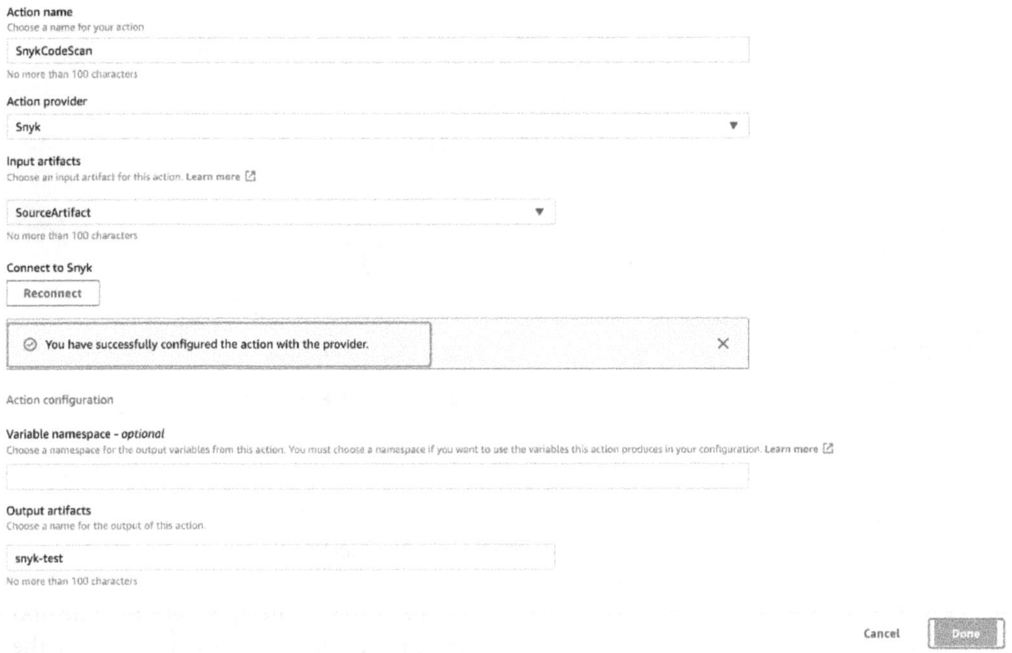

Action name
Choose a name for your action

SnykCodeScan

No more than 100 characters

Action provider

Snyk

Input artifacts
Choose an input artifact for this action. Learn more ☐

SourceArtifact

No more than 100 characters

Connect to Snyk

Reconnect

⊘ You have successfully configured the action with the provider.   ✕

Action configuration

Variable namespace - *optional*
Choose a namespace for the output variables from this action. You must choose a namespace if you want to use the variables this action produces in your configuration. Learn more ☐

Output artifacts
Choose a name for the output of this action.

snyk-test

No more than 100 characters

Cancel   Done

Figure 11.19 – Snyk integration confirmation

16. Next, we need to configure a new stage QA deployment to deploy our application to EC2 instances, using the CodeDeploy service. So, click on the **Add Stage** button, provide a name, and then click on **Add action group** and fill in the details, as shown in the following screenshot. We have to choose AWS CodeDeploy as an action provider and select `BuildArtifact` as input, since this contains our Java application JAR file and other artifact scripts used by the CodeDeploy agent, for deployment to an EC2 instance.

Figure 11.20 – QA deployment stage configuration

Select the `chapter_11_test_env_app` application in the application name and `chapter_11_test_env_deploy_group` as the deployment group, which are the resources we created using our terraform template. Then, click on the **Done** button, which will create the QA deployment pipeline stage.

17. As a next step, we need to add a BlazeMeter performance test as part of our pipeline, so let's click on the **Add Stage** button, provide a name, and then click on the **Add action group** button and fill out the following details. We need to choose **BlazeMeter** as an action provider and connect it to our BlazeMeter account. Click on the **Connect to BlazeMeter** button.

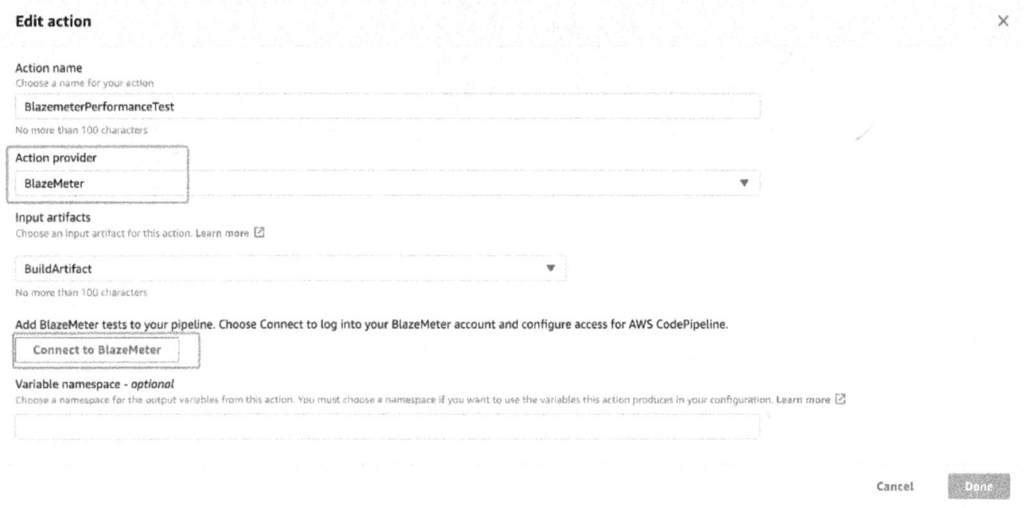

Figure 11.21 – Performance test stage configuration

18. When you click on the **Connect to BlazeMeter** button, it will open a new window to sign in to your BlazeMeter account (if you don't have one, you can create one for free). Once you sign in to your BlazeMeter account, you will see a screen similar to the following. Click on the **Create Test** button, and add a test case for our sample `aws-code-pipeline` microservice.

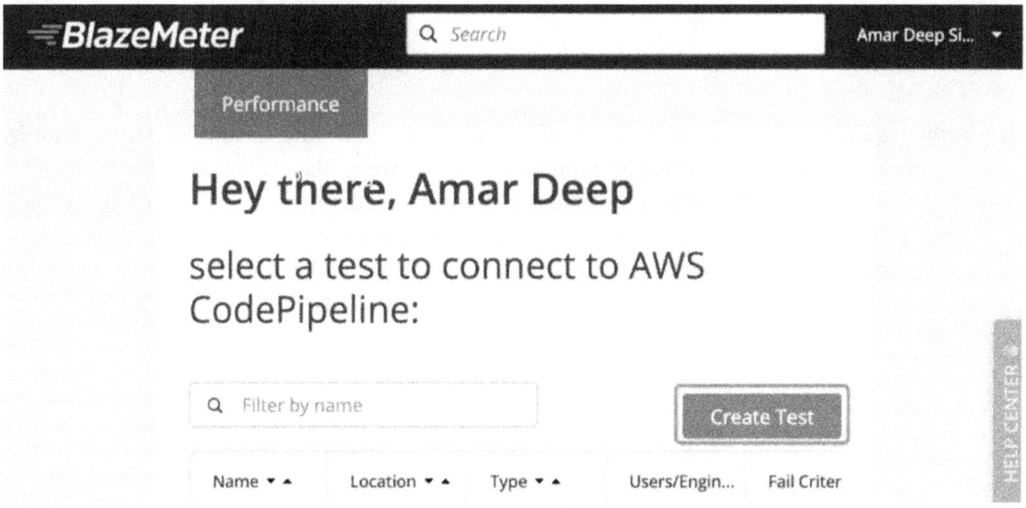

Figure 11.22 – The BlazeMeter home screen

19. We will create a basic test here by just pinging the URL of our QA environment load balancer, and if we get a 200 response, our test case will pass. Go to the EC2 dashboard and copy the DNS name of the test/QA environment URL, as shown in the following screenshot – in our case, the URL is `http://chapter-11-test-alb-2134268546.us-east-1.elb.amazonaws.com/`.

20. Now, add a test case in BlazeMeter and paste the `http://chapter-11-test-alb-2134268546.us-east-1.elb.amazonaws.com/` URL, and use the HTTP GET method, as our API responds through a GET request:

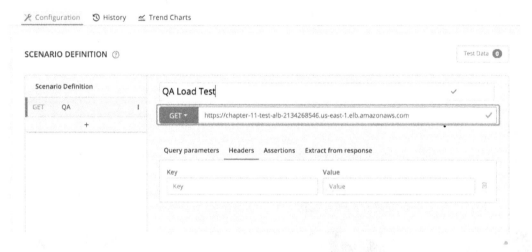

Figure 11.23 – BlazeMeter API test configuration

21. In the **Configuration** section, you can specify how long you want to run the performance test. We have configured the test for 5 minutes with 20 users. BlazeMeter will run simulated test cases against our QA environment application and report the different statistics for the performance of the application. Provide a name for your test suite, and then click on the **Save & Connect** button in the left panel to integrate with AWS CodePipeline.

Figure 11.24 – QA environment load test configuration

22. When you click the **Save & Connect** button, it will initiate an OAuth request to the AWS CodePipeline to confirm the connection. Click on the **Confirm** button, and once you confirm the request, CodePipeline will show a confirmation that BlazeMeter is connected. Click on the **Done** button and our performance stage is created.

Figure 11.25 – BlazeMeter connection confirmation

23. Next, we need to create a stage to get approval before deploying to the production environment. Let's create the deployment approval stage by clicking on **Add Stage**, providing a name, and then clicking on the **Add action group** button. You need to select **Manual approval** as an action provider, and then select the SNS topic on which you want to trigger a notification for approval. You can add a message and a URL to be sent out in the SNS notification. Click on the **Done** button, which will add the approval stage to the pipeline.

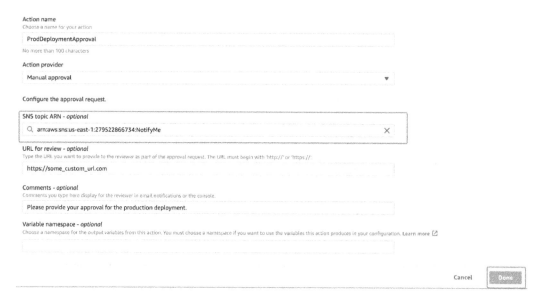

Figure 11.26 – Production deployment approval configuration

24. Now, we need to add the final stage to our pipeline, which is production deployment. To add this stage, click on **Add Stage**, provide a name, and then click on the **Add action group** button. Provide a name, `ProdDeployment`, and choose **Amazon ECS (Blue/Green)** as the action provider, since we are going to deploy to the ECS service in a production environment.

Figure 11.27 – ECS production deployment stage configuration

We need to select the Input artifact folder, from where CodePipeline can read the ECS task definition file, taskdef.json, and CodeDeploy the appespec file, ecs-appspec. yml. We need to select the AWS CodeDeploy application, chapter-11_code_deploy_ app, and the deployment group, chapter-11_code_deploy_group, which we created through Terraform for the production workload. Click on the **Done** button.

25. Now, we have all the stages ready for our deployment, so let's click on the **Save** button to save all the pipeline stages.

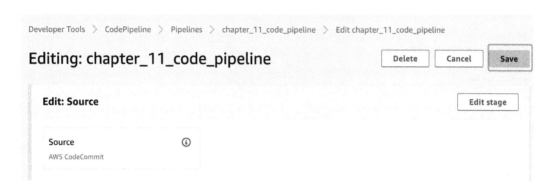

Developer Tools > CodePipeline > Pipelines > chapter_11_code_pipeline > Edit chapter_11_code_pipeline

Figure 11.28 – The pipeline editing screen

Our pipeline is now ready. In the next section, we need to make some changes to the source code, and then we will execute the pipeline.

Executing the pipeline

We need to make a couple of changes to the source code before we execute the pipeline specific to our environment:

1. We need to create a `ecs-appsepc.yml` file in our source code to specify how CodeDeploy will deploy the containers to the ECS cluster. We need to make sure that the task definition **Amazon Resource Names (ARN)** matches what we have in our environment:

```
! ecs-appspec.yml M  ✕

! ecs-appspec.yml
1    version: 0.0
2    Resources:
3      - TargetService:
4          Type: AWS::ECS::Service
5          Properties:
6            TaskDefinition: "arn:aws:ecs:us-east-1:▓▓▓▓▓▓▓:task-definition/chapter-11_task_definition:1"
7            LoadBalancerInfo:
8              ContainerName: "chapter-11_aws_code_pipeline_container"
9              ContainerPort: 80
```

Figure 11.29 – ecs-appspec.yml

2. Another file we need to create is the `taskdef.json` file, based on which ECS service will create the containers. This task definition is part of the GitHub repository source code; you just need to make sure that the task definition ARN and the image URL match your environment:

```
{} taskdef.json M ×

{} taskdef.json > [ ] containerDefinitions > {} 0 > [ ] portMappings > {} 0
   1    {
   2        "taskDefinitionArn": "arn:aws:ecs:us-east-1▮▮▮▮▮▮▮▮▮▮task-definition/chapter-11_task_definition:1",
   3        "containerDefinitions": [
   4            {
   5                "name": "chapter-11_aws_code_pipeline_container",
   6                "image": "▮▮▮▮▮▮▮▮▮dkr.ecr.us-east-1.amazonaws.com/packt-ecr-repo:aws-code-pipeline",
   7                "cpu": 512,
   8                "memory": 3072,
   9                "portMappings": [
  10                    {
  11                        "containerPort": 80,
  12                        "hostPort": 80,
  13                        "protocol": "tcp"
  14                    }
  15                ],
  16                "essential": true,
  17                "environment": [],
  18                "mountPoints": [],
  19                "volumesFrom": []
  20            }
  21        ],
  22        "family": "chapter-11_task_definition",
  23        "executionRoleArn": "arn:aws:iam::▮▮▮▮▮▮▮role/ecsTaskExecutionRole",
  24        "networkMode": "awsvpc",
```

Figure 11.30 – taskdef.json

3. Once you check these files into your CodeCommit repository, aws-code-pipeline, AWS CodePipeline will start the execution. Alternatively, you can start the build by clicking on the **Release change** button on your pipeline's home page:

Figure 11.31 – CodePipeline execution

4. Once the pipeline execution has started, you can see the execution progress stage by stage. After the pipeline execution passes the SnykCodeScan stage, you can log in to your Snyk account and validate the scanning results. The following screenshot shows the different vulnerabilities found in our code. Since we opted not to block the deployment while configuring our pipeline, the execution continued to the next stage:

Figure 11.32 – The Snyk vulnerability report

You can click on different issue types and find out which code part has a vulnerability, and Snyk will also provide you with a solution that how you can fix it.

5. Once you have finished executing the QA deployment stage in the pipeline, you can open up the browser and paste the test environment load balancer DNS URL to validate that the application is successfully deployed to the QA environment. The following screenshot shows a successful output:

aws-code-pipeline-1.0 ip-172-31-3-162.ec2.internal/172.31.3.162

Figure 11.33 – The QA environment service response

6.   After the QA deployment stage, the pipeline will enter the performance test phase, where BlazeMeter will start executing the load test in our QA environment by making requests to the QA environment load balancer URL: `http://chapter-11-test-alb-2134268546.us-east-1.elb.amazonaws.com/`. Log in to your BlazeMeter account to see the progress. Once the performance test is completed, you can review the report, as shown in the following screenshot, and it should confirm that our service has 0% errors:

Figure 11.34 – BlazeMeter QA environment performance test metrics

7.   BlazeMeter also sends a test execution to your email account, and you can review that. In our case, there were about 549,731 requests made by BlazeMeter within 5 minutes, and only two of them failed to respond, so the error rate was negligible.

8.   Once the pipeline completes the performance test stage, it will be paused for approval, before the source code can be deployed to the production environment. The following screenshot shows the pipeline waiting for approval.

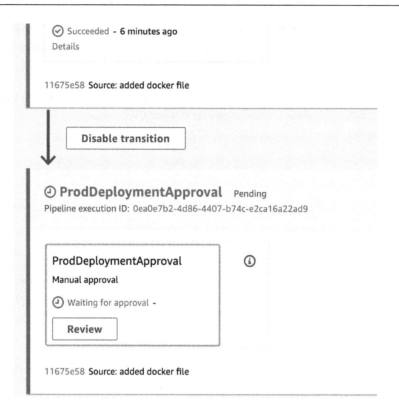

Figure 11.35 – The production deployment pending approval stage

9. While the pipeline waits for approval, it also triggers an SNS notification, as we configured earlier. You can click on the link in the SNS notification, or you can click on the **Review** button shown on the previous screen.

10. When you click on the **Review** button, it asks you to provide any comments, and you can either approve the deployment or reject it by clicking on the appropriate button. Let's go ahead and click on the **Approve** button to approve the production deployment.

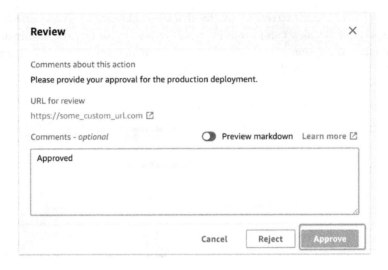

Figure 11.36 – The production deployment approval review

11. The following screenshot shows that the pipeline is approved for production rollout, and production deployment is in progress. To see the deployment progress, click on the **Details** link in the **ProdDeployment** action in the stage.

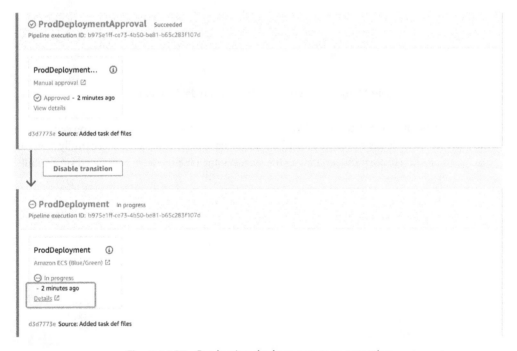

Figure 11.37 – Production deployment stage execution

12. In the CodeDeploy details screen, we can see that our application version is already deployed to the **Replacement** task set and is waiting for the original task set to be terminated. At this point, you have the option to immediately stop the original task set or perform a rollback if you notice any problems with the application deployment.

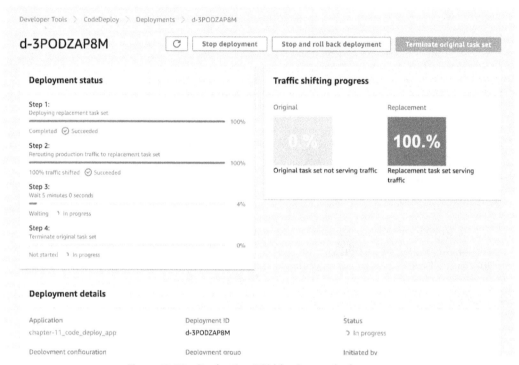

Figure 11.38 – Production ECS blue/green deployment

13. Once deployment is completed and CodeDeploy terminates the original task set, you can see that the pipeline execution is complete and all the stages have been completed successfully. You can also go to pipeline history and see that all the stages of the pipeline were completed successfully.

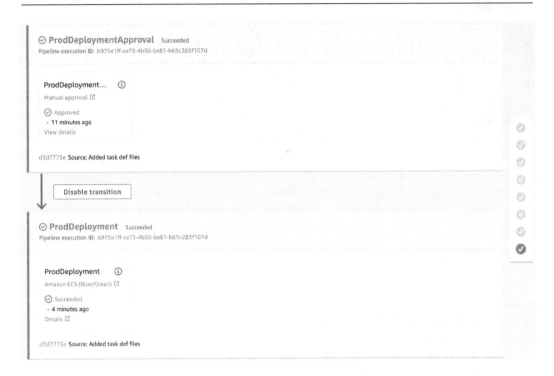

Figure 11.39 – CodePipeline end-to-end successful execution

14. As a final verification step, let's open up the web browser and paste the production load balancer DNS URL to confirm that the application is live in the production environment: `http://chapter-11-alb-1075133995.us-east-1.elb.amazonaws.com/`.

aws-code-pipeline-1.0 ip-172-31-3-64.ec2.internal/172.31.3.64

Figure 11.40 – Production environment service response

15. At this point, our end-to-end pipeline is ready and can deliver any CodeCommit to production without any manual intervention, except the approval step. The approval step is an optional step, for your use case; if you don't require approval, you can remove that stage from the pipeline. Now, you can go ahead and terminate your environment by executing the `terraform destroy` command to avoid extra charges.

16. Any active pipeline in your account will incur a fixed cost to your account, so if you are done with the exercise, you can delete the pipeline by selecting and clicking on the **Delete pipeline** button.

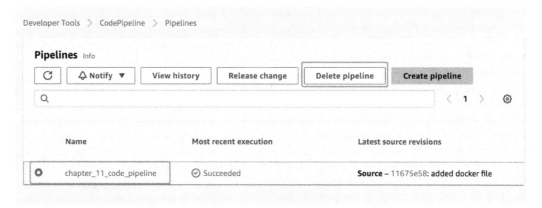

Figure 11.41 – Deleting the pipeline

It is very easy to create an AWS CodePipeline and add different stages one by one to make complex pipelines. Different integration services help to automate and deliver end-to-end automated software.

Summary

In this chapter, we learned about the AWS CodePipeline service, and its limitations and benefits. We created a pipeline for our sample `aws-code-pipeline` microservice and deployed that to QA and production environments. In addition to this, we also expanded our knowledge of how the CodePipeline service integrates with other third-party services, enhancing our pipeline to automate the end-to-end release process.

In the next chapter, we will learn about AWS Lambda functions, and we will expand our knowledge of AWS CodePipeline to the automated delivery of serverless deployments.

12

Setting Up an Automated Serverless Deployment

In this chapter, we will be focused on setting up CodePipeline for serverless deployment. This chapter will introduce you to serverless ecosystems and explain how AWS provides scalable solutions through the Lambda service. We will develop a sample Lambda function and set up automated CodePipeline for Lambda deployment. In this chapter, we will be covering the following topics:

- What is a serverless ecosystem?
- What is AWS Lambda?
- AWS Lambda development
- AWS Lambda manual deployment
- AWS Lambda pipeline setup

In the previous chapter, we learned about the AWS CodePipeline service and showed you how you can deploy a microservice to **Elastic Container Service (EC2)** and EC2-based instances in an automated way. In this chapter, we will discuss how you can automate the delivery of AWS Lambda functions using AWS CodePipeline. But before we start talking about Lambda functions, we need to understand what serverless is in general and how AWS uses this pattern.

What is a serverless ecosystem?

In *Chapter 1*, we talked about **serverless** architecture patterns and the popularity these patterns have gained in the last few years. Let's revise our knowledge and see what serverless means in software development.

A serverless application is not actually serverless, but it is a term used to signify that, as an application developer or owner, you are not responsible for maintaining the servers required for your application to run.

Your application is still deployed on servers, but instead of you owning those servers, a service provider is responsible for maintaining the hardware, OS patching, and other operational aspects. You, as a consumer, just request how much memory or CPU is needed and providers handle the provisioning of the required capacity from a pool of resources. You just upload your code and it gets executed on the infrastructure provided. You only pay for the duration of the runtime environment for which your code is executed; you don't pay for idle time. Since you don't own and maintain the server, from a consumer's perspective, it is referred to as serverless architecture.

In traditional computing, you are responsible for buying the hardware, installing it, patching the OS, and maintaining the application runtime. In traditional application environments, you incur costs all the time your service is up and running, but in serverless style, you are not responsible for buying and maintaining the servers and you are not charged for the idle time when your code is not executing. In serverless architecture, your provider gives you compute and memory to execute your code, and you just upload the code and only pay for the execution time.

In *Chapter 1*, we also learned that serverless architecture is divided into two categories – **Backend as a Service (BaaS)** and **Function as a Service (FaaS)**.

In BaaS style, you use the backend created by someone else as a service, such as a third-party authentication system (for example, Amazon Cognito), while in FaaS style, you write your own code but use third-party infrastructure to execute your code. The major difference between these two styles is that BaaS is not very customizable and you have to use what a service provider is offering, whereas FaaS is very customizable and you can get exactly what you need.

AWS supports many serverless services, including AWS Lambda, AWS Fargate, Amazon EventBridge, API Gateway, AWS Step Functions, SQL, SNS, S3, DynamoDB, Amazon Aurora Serverless, and many more. AWS Lambda is one of the popular ones and an example of FaaS-style serverless architecture, in which you execute your code on AWS infrastructure. Let's learn in depth about AWS Lambda functions in the next section.

What is AWS Lambda?

AWS Lambda is a serverless compute service to let you execute your code to do a specialized task. You can upload your code to AWS and define the runtime configuration, CPU and memory requirements, and an event on which your function should be executed. AWS is responsible for booting up servers and required runtime environments to execute the code you provided as soon as the triggering event takes place. You don't pay for the time AWS takes to provision the servers for you or idle time for the servers when you aren't using the service; you only pay for the execution time of your code.

AWS Lambda can be triggered from a majority of AWS services and SaaS applications to provide an entry point for your code. Before we go a bit deeper, let's first understand different concepts related to Lambda to help us understand how it works internally:

- **Function**: A Lambda function is an AWS resource that receives an event as a parameter from the invocation and executes the code you uploaded.

- **Event**: An event is a document in JSON format that is passed to your code; it contains information required by your function. Event structure and data are defined by the invoker, and they will be converted and passed as objects to your code.

- **Execution environment**: The code you upload as part of your Lambda function gets executed in a secure and isolated runtime environment based on the language runtime and configuration you provided. That environment is called the execution environment.

  The execution environment is responsible for managing the process and resources that are required to execute the code. It provides lifecycle support for the function and any required extensions.

- **Extensions**: You can extend your Lambda functions by using extensions. Extensions allow you to integrate your Lambda function with your choice of monitoring, security, compliance, or observability tools supported by AWS, or you can write a custom extension. AWS Lambda supports two types of extensions: internal and external. Internal extensions run as part of your runtime within the same process and terminate along with the runtime. External extensions run as a separate process in the execution environment and are initialized before the Lambda function invocation. External extensions run in parallel with the Lambda function runtime.

- **Trigger**: In the context of a Lambda function, a trigger is an AWS resource that is used to invoke the Lambda function. A trigger can be an AWS service that you can configure to invoke the Lambda function. You can add a trigger in the Lambda configuration section and specify the event in which your Lambda function should be invoked.

- **Concurrency**: In a Lambda function, concurrency defines how many requests your Lambda can handle in parallel. When you invoke a Lambda function, Lambda creates a runtime environment for it to handle that request, and any other requests remain in a *wait* condition until Lambda finishes executing the first request. Concurrency defines how many instances of your function can be created at a time to handle simultaneous requests. You can configure two types of concurrency for your functions: reserved concurrency and provisioned concurrency. When you use reserved concurrency, it guarantees that the maximum number of concurrent instances of the function will be available for you and no other function will use that concurrency. In provisioned concurrency, the configured number of execution environments will be initialized and ready to process your request, so you will be charged additionally for that.

- **Layer:** A layer in a Lambda context is a `.zip` archive that contains additional libraries or contents that are needed by your Lambda function code, which you want to separate from the main Lambda function code due to modularity or some other reason.

  A layer can contain a library, data, configuration file, or a customized runtime for the Lambda function. Layers allow you to speed up the Lambda deployment and share the layers across multiple Lambda functions. The contents of the layers are extracted in the `/opt` directory of the execution environment. You can use the layers across AWS accounts and your layers will remain accessible to your Lambda deployment, even if it is deleted from the published account, unless you create a new version of your Lambda function.

- **Destination:** A destination is a place or service where the Lambda function can send your event after processing. A Lambda function can be invoked in a synchronous or asynchronous fashion. In a synchronous invocation, if any error occurred during execution, the caller would immediately get an error response, but in the case of asynchronous invocation, you can configure AWS Lambda to transfer events to a destination, where you can take any additional action on these events. You can configure destinations only for asynchronous invocations.

Now that you have understood the Lambda concepts, let's focus on how Lambda actually works. As a developer, you create a Lambda function in one of the supported languages and upload your code as a ZIP file or from Amazon S3 to the Lambda service. The following diagram explains how AWS Lambda works:

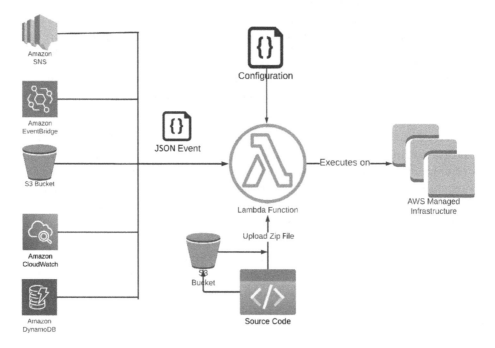

Figure 12.1 – Lambda function working style

As part of the Lambda function, you have to define the configuration for your code, such as the language, platform architecture, memory, storage, concurrency, SnapStart, trigger, destination, and so on.

Once you upload your source code or define a container image, the Lambda service will maintain this configuration and create an ARN to uniquely identify your Lambda function. The Lambda service maintains a pool of EC2 instances and it will create a Lambda execution environment for your function when you first time invoke the Lambda function. The Lambda service may keep the execution environment for a certain time to serve subsequent requests, but if there is no request for some time, Lambda function containers will be terminated and returned to the pool:

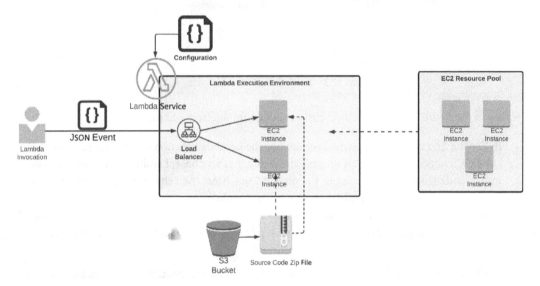

Figure 12.2 – Lambda function architecture

Based on the need and the number of requests, more resources can be pulled from the resource pool, and requests are served through a load balancer. The load balancer will handle the Lambda invocation and distribute traffic to appropriate EC2 instances. The process of pulling up an instance from the pool, downloading Lambda function code to the instance, and initializing and loading up any extension is known as a **cold start**.

On the first Lambda invocation, there will not be any existing execution environment available, so the Lambda function will have to go through a cold start. For that reason, the request will take additional time to complete. Since there is no guarantee of how long the execution environment is kept after processing a request and whether the Lambda function is scaled out due to more requests, a cold start can be faced by any request. The following figure shows the Lambda initialization process:

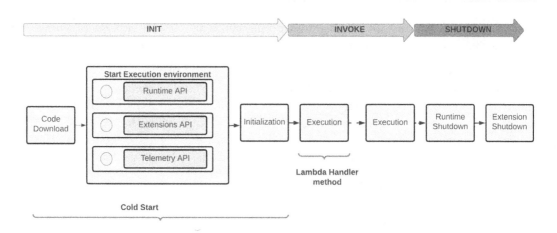

Figure 12.3 – Lambda function lifecycle

As shown in the figure, the Lambda lifecycle is divided into different phases:

- **INIT**: The init phase starts when you invoke the Lambda function for the first time, or your Lambda function is scaling out to serve the traffic. In this phase, Lambda first downloads the source code from the S3 bucket to the physical machine, then the runtime environment is bootstrapped, all extensions are started, and the Lambda function initialization code (static code) is executed. The init phase times out after 10 seconds. If it doesn't complete within that time, then it gets retried. Once the init phase is ready, it sends a Next API request to signal its completion.

- **INVOKE**: The Lambda function sends an Invoke event to runtime and to each extension part of it in response to the Next API request. The Invoke function executes the handler method within your code and its timeout if it doesn't complete the execution within the specified configured time limit. At the time of writing this book, the maximum time you can configure for a Lambda function to execute is 15 minutes, which includes the time taken by the runtime and extensions.

  The Invoke phase ends once Lambda receives the Next API request from all extensions and runtime.

- **SHUTDOWN**: The Lambda environment is shut down as a response to the Next API request from the Invoke phase. When the Lambda runtime is about to shut down, it sends a Shutdown event to all extensions to shut down and do any final cleanup.

When there is no event to process, the Lambda execution environment is maintained for a certain time after the Shutdown phase for any subsequent Lambda invocation. Maintaining the same runtime environment is called a **Lambda freeze**, and during this freeze period, your /tmp directory is maintained and can be used as a transient cache for multiple invocations.

When the freeze runtime environment is reused, any objects declared outside your handler method remain initialized and just execute your handler method. So, if you have declared a database connection in your code and opened up a database connection in your handler method code, then it will remain initialized and you should check whether it already exists; if so, you shouldn't be creating it again.

When creating a Lambda function, we should always consider that it is stateless and the Lambda service may or may not be using the same runtime environment on subsequent invocations, depending on scaling needs and freeze time. You can create Lambda functions in Java, Go, PowerShell, Node.js, C#, Python, and the Ruby language out of the box. AWS Lambda also provides a runtime API to support any other language of your choice. Now that we have a fair idea about Lambda functions, let's look at the benefits they provide to us.

The benefits of using Lambda functions

Lambda functions inherit all the features supported by a FaaS architecture. We will summarize a few here to refresh your memory:

- One of the greatest advantages that Lambda provides is that we don't have to provision or manage any infrastructure; we can just focus on writing a function and upload the code as a .zip file or as a container image and the Lambda service will help us to do the magic on a scalable platform.

- Amazon does provide an EC2 service, which is **Infrastructure as a Service (IaaS)**, but AWS Lambda goes beyond that and you only pay for the time that your code is executing, not for the idle time. So, it is a great upgrade on IaaS.

- Lambda functions support auto-scaling, and you can scale from a workload of a few requests a day to thousands of requests per second without manual configuration.

- With Lambda functions, you pay only for the compute time, not for the entire time to provision infrastructure or for the idle time, so you can bring your costs down.

- AWS Lambda provides easy integration with other AWS services, and you can use it to execute your code on a certain event and send data to a different destination.

- AWS Lambda supports integration with the database proxy, which helps to manage a pool of connections; this helps Lambda functions to get the highest level of concurrency without exhausting all database connections.

The limitations of using Lambda functions

AWS Lambda provides a lot of flexibility and cost optimizations, but we can't compare it directly with a dedicated infrastructure, and it has its limitations. Let's look at some of them:

- Lambda functions run on a shared infrastructure, so if you have a compliance requirement to not use a shared infrastructure, then Lambda functions are not the correct choice for that kind of workload.

- The Lambda function execution is limited to 15 minutes. If your process is going to take longer than that, then it is not the right use case for your workload.

- Each Lambda execution environment is limited to serve 10 requests per second, so if you have a concurrency of 5 setups in configuration, then your Lambda function can take up to 50 requests per second.

- A Lambda function can have only a maximum of five layers in each function.

- Your function can have a maximum of 6 MB of request and response allocation for synchronous invocation, while this limit is set to 256 KB for asynchronous invocations.

- Lambda function concurrency is limited to 1,000 requests at a time across all your functions within your account, but this can be increased with an additional quota limit increase.

Now you have a fair idea of the different benefits and limitations that Lambda functions have, let's start creating a Lambda function.

AWS Lambda development

Before we get into writing the code, we need to understand the handler method in Lambda functions. The handler method is a specialized method in Lambda function code that processes the event when the Lambda function is invoked. On invocation, the Lambda service will execute this handler method and the Lambda execution environment remains busy until this method finishes execution or errors out.

You can write Lambda functions in different supported languages, but for our example, we are going to use Java 11. In a Lambda function, any code written within the handler method gets executed each time Lambda is invoked, but class-level code is initialized once when your runtime initializes, and it may or may not be executed every time depending on the Lambda service, and whether it is using the same execution environment or creating a new one due to scaling needs or idle timeout.

AWS provides libraries to each supported language for creating handler functions and easily writing code. AWS provides the `aws-lambda-java-core` library for Java to handle common handler types and configurations along with a context with useful information about the Lambda function itself. The library provides useful request handler interfaces to be implemented by your handler function.

RequestHandler

This interface is a generic interface and takes your Lambda input and output as parameters. This handler deserializes an incoming request event into your input object and serializes the handler return object to text. The following is an example of a `RequestHandler` implementation where we are deserializing the incoming event to an `EmployeeEntity` object and in response, we are sending a confirmation message as a `String` object:

```
public class CreateEmployeeHandler implements
RequestHandler<EmployeeEntity,String>{
@Override
public String handleRequest(EmployeeEntity event, Context context){
return "Employee is created successfully !";
}
}
```

RequestStreamHandler

`RequestStreamHandler` is another handler interface that is provided by the runtime with an input stream and output stream, so you can use your custom serializer and deserializer. This handler comes with a `void` return type so you can directly read from the stream and output to the stream:

```
void handleRequest(InputStream input, OutputStream output, Context
context) throws IOException;
```

For most of the Lambda function deployments, `RequestHandler` should be sufficient. Amazon additionally provides `aws-lambda-java-events` and a few other libraries to cover the different event types generated by different systems, which you can use in conjunction with the `aws-lambda-java-core` library to handle different types of requests. Now, let's begin to write Lambda function code for deployment.

Sample Lambda development

For our sample Lambda function development, we will be using the `RequestHandler` interface to process the incoming request and respond to the invoker. To begin the development, let's create a Java Maven project:

1.   Open the VS Code Editor, click on **Create Java Project**, and select the **Maven** project type:

Figure 12.4 – Visual Studio Code Java project creation

2. Select **maven-archetype-quickstart** and press *Enter*, and then choose the latest version, as shown in the following screenshot:

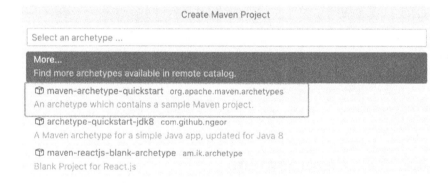

Figure 12.5 – Maven archetype selection

3. Provide the Maven artifact group ID as com.packt.aws.book.pipeline and press *Enter*:

Figure 12.6 – Maven group ID

4. Provide a name for the project, `aws-lambda-pipeline`, and press *Enter*:

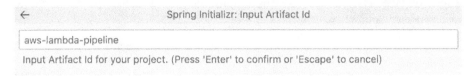

Figure 12.7 – Maven artifact ID

5. Select a destination location on your system, and the terminal will ask for a confirmation and will create your project:

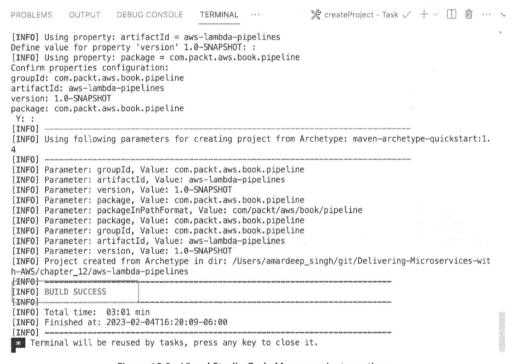

Figure 12.8 – Visual Studio Code Maven project creation

6. Open the `pom.xml` file in your project and update the `dependencies` section to match the following `pom.xml` file. Here, we have added the `aws-lambda-java-core` library as a dependency, and we are using Java 11 for development:

```
pom.xml
1   <?xml version="1.0" encoding="UTF-8"?>
2   <project xmlns="http://maven.apache.org/POM/4.0.0" xmlns:xsi="http://www.w3.org/2001/XMLSchema-instance"
3     xsi:schemaLocation="http://maven.apache.org/POM/4.0.0 http://maven.apache.org/xsd/maven-4.0.0.xsd">
4     <modelVersion>4.0.0</modelVersion>
5     <groupId>com.packt.aws.book.pipeline</groupId>
6     <artifactId>aws-lambda-pipeline</artifactId>
7     <version>1.0-SNAPSHOT</version>
8     <name>aws-lambda-pipeline</name>
9     <packaging>jar</packaging>
10    <properties>
11      <project.build.sourceEncoding>UTF-8</project.build.sourceEncoding>
12      <maven.compiler.source>11</maven.compiler.source>
13      <maven.compiler.target>11</maven.compiler.target>
14    </properties>
15    <dependencies>
16      <dependency>
17        <groupId>com.amazonaws</groupId>
18        <artifactId>aws-lambda-java-core</artifactId>
19        <version>1.2.2</version>
20      </dependency>
21    </dependencies>
22  </project>
```

Figure 12.9 – The pom.xml file

7. As a next step, let's add the `EmployeeEntity` class to our project with the `employeeName`, `department`, and `managerName` properties and setter and getter methods for these properties, along with the `toString()` method:

J EmployeeEntity.java ●

src > main > java > com > packt > aws > books > pipeline > awslambdapipeline > J EmployeeEntity.java >

```java
 5    public class EmployeeEntity {
 6        private String employeeName;
 7        private String department;
 8        private String managerName;
 9
10        public String getEmployeeName() {
11            return employeeName;
12        }
13        public void setEmployeeName(String employeeName) {
14            this.employeeName = employeeName;
15        }
16        public String getDepartment() {
17            return department;
18        }
19        public void setDepartment(String department) {
20            this.department = department;
21        }
22        public String getManagerName() {
23            return managerName;
24        }
25        public void setManagerName(String managerName) {
26            this.managerName = managerName;
27        }
28        @Override
29        public String toString() {
30            return "EmployeeEntity [employeeName=" + employeeName +
31            ", department=" + department + ", managerName="    + managerName + "]";
32        }
33    }
```

Figure 12.10 – EmployeeEntity.java

8. Now, as a final step, we need to add a CreateEmployeeHandler.java class, which implements the RequestHandler interface and takes the previously created EmployeeEntity object as input:

J CreateEmployeeHandler.java ●

src > main > java > com > packt > aws > books > pipeline > awslambdapipeline > J CreateEmployeeHandler.java > ...

```java
1    package com.packt.aws.books.pipeline.awslambdapipeline;
2    import com.amazonaws.services.lambda.runtime.Context;
3    import com.amazonaws.services.lambda.runtime.RequestHandler;
4    import java.util.logging.Logger;
5
6
7    public class CreateEmployeeHandler implements RequestHandler<EmployeeEntity,String>{
8        private static final Logger LOGGER =Logger.getLogger(name: "CreateEmployeeHandler");
9
10       @Override
11       public String handleRequest(EmployeeEntity event, Context context){
12           LOGGER.info("Received a request in your Lambda function "
13           +context.getFunctionName()
14           +" with details " + event);
15
16           return "Employee is created successfully  !";
17       }
18   }
```

Figure 12.11 – CreateEmployeeHandler.java

Here, a few things to observe are that we have implemented the `RequestHandler` interface with an input type of `Employee.java`, and the function is returning a `String` type of object. So, when our handler receives the request, it will convert the incoming event to an `EmployeeEntity` object. In our handler method, we are just logging the incoming request, along with the Lambda function name retrieved from the `Context` object, and sending a confirmation message back saying `Employee is created successfully!`. In a real production application, you can have your business logic here to connect with a database and create an employee in the database.

9.   Now we have our code ready, let's run the `mvn clean install` Maven command to build the required JAR file for the AWS Lambda deployment.

We now have our Lambda package ready to be deployed to AWS. In the next section, we need to create the Lambda function in the AWS console and upload the JAR file to test it.

Creating a Lambda function

Before you can deploy code to your Lambda function, we need to create the Lambda function in our account. So, let's log in to the AWS console and search for the Lambda service and perform the following steps:

1.   In the left panel, click on **Functions**, then click on the **Create function** button on the right to start creating the Lambda function:

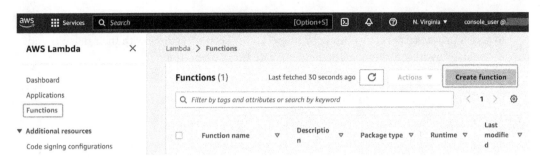

Figure 12.12 – Lambda functions home screen

2. Lambda provides an option to create a function from existing code templates (blueprints), from a container image, or from scratch. Let's select the **Author from scratch** option and provide the name and runtime information. We are using Java 11 for this Lambda development:

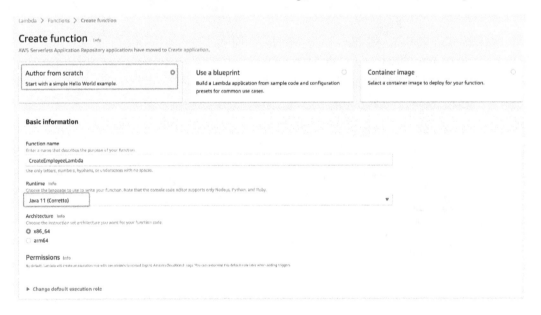

Figure 12.13 – Create function screen

3. In the **Advance settings** section, you can enable different settings. **Code signing** enables the Lambda service to ensure that only a verified and trusted code package is being deployed. **Enable Function URL** setting assigns an HTTPS endpoint to the Lambda function, so you can invoke your Lambda function as an HTTPS endpoint. Once you enable this setting, it will give you an option to select an authentication mechanism and the **cross-origin resource sharing** (**CORS**) policy. In the function URL configuration, you make requests to the HTTPS URL assigned to your function using the GET or POST method. The GET method takes a String parameter called message, and the POST method takes a message body and converts it into

an event before handing it over to the Lambda function. The **Enable VPC** option allows you to access private resources within your subnets. Let's keep everything with their defaults and not enable anything for now. Click on the **Create function** button:

Figure 12.14 – Lambda advanced settings

4. The following screenshots show a confirmation that the Lambda function has been created. Now, we need to click on the **Upload from** button, as shown in the following screenshot, to upload the source code as a `.zip` or `.jar` file or as an Amazon S3 URL:

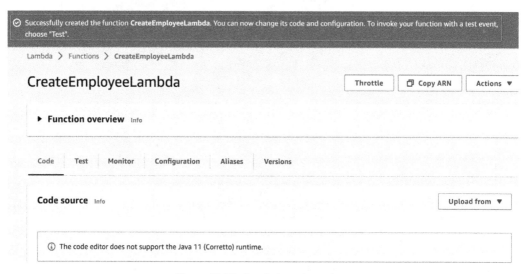

Figure 12.15 – Lambda code update

5. Click on the **Upload** button and select the `aws-lambda-pipeline-x.x.x.jar` file, then click on the **Save** button:

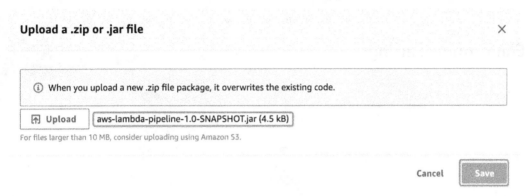

Figure 12.16 – Lambda Java code upload

6. Once you upload the source code, click on the **Edit** button, as shown in the following screenshot, as we have to change the default handler information:

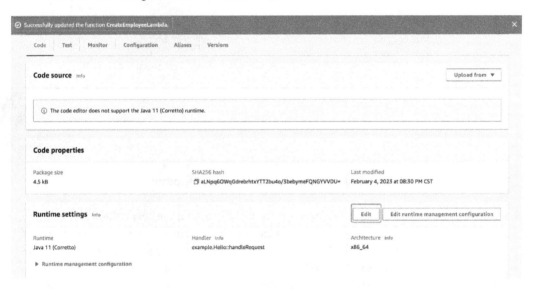

Figure 12.17 – Lambda runtime updated

7. **Handler** is the method in your source code that gets executed on a function invocation. It follows the pattern `packageName.ClassName::methodName`, so in our case, we created the `handlerRequest` method in our `CreateEmployeeHandler` class inside the `com.packt.aws.books.pipeline.awslambdapipeline` package, so let's change the **Handler** information to `com.packt.aws.books.pipeline.awslambdapipeline.CreateEmployeeHandler::handleRequest` and click on the **Save** button:

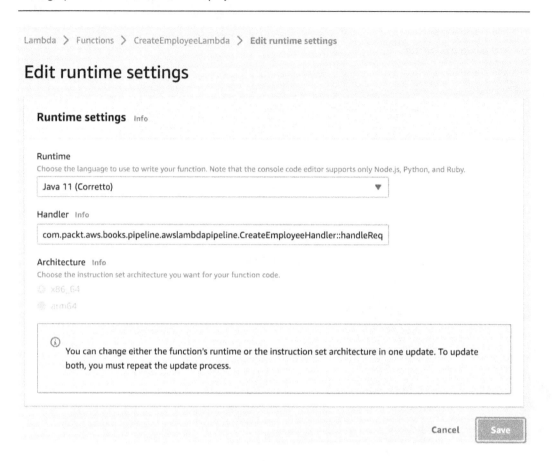

Figure 12.18 – Lambda function handler info update

8.   Now that we have saved the **Handler** information, let's click on the **Test** tab and start creating a test event to test our Lambda function. Click on the **Create new event** radio button and provide a name and employee event information, as shown in the following screenshot. We are basically passing an `Employee` JSON object, which our Lambda function is expecting, as we defined it in our `handleRequest(Employee event, Context context)` method. Click on the **Save** button to save this test event and then click on the **Test** button to test the Lambda function with `TestEvent`:

Test event Info

Save   Test

To invoke your function without saving an event, configure the JSON event, then choose Test.

Test event action

◉ Create new event

Edit saved event

Event name

TestEvent

Maximum of 25 characters consisting of letters, numbers, dots, hyphens and underscores.

Event sharing settings

◉ Private

This event is only available in the Lambda console and to the event creator. You can configure a total of 10. Learn more 🗗

○ Shareable

This event is available to IAM users within the same account who have permissions to access and use shareable events. Learn more 🗗

Template - *optional*

hello-world   ▼

Event JSON   Format JSON

```
1  {
2      "employeeName": "John Doe",
3      "department": "Information Technology",
4      "managerName": "Chris Veal"
5  }
```

Figure 12.19 – Lambda test event

9. The following screen shows that our Lambda function execution result is successful. To see more details about the invocation, click on the **Details** link, as shown here:

Code   Test   Monitor   Configuration   Aliases   Versions

⊘
Execution result: succeeded (logs)   ✕

▶ Details

Test event Info   Delete   Save   Test

Figure 12.20 – Lambda execution result

10. The following screenshot shows the return value from the Lambda function and also shows the log message we added in which you can see the entire test event we passed. The screenshot also shows the Lambda execution time for which you have been billed:

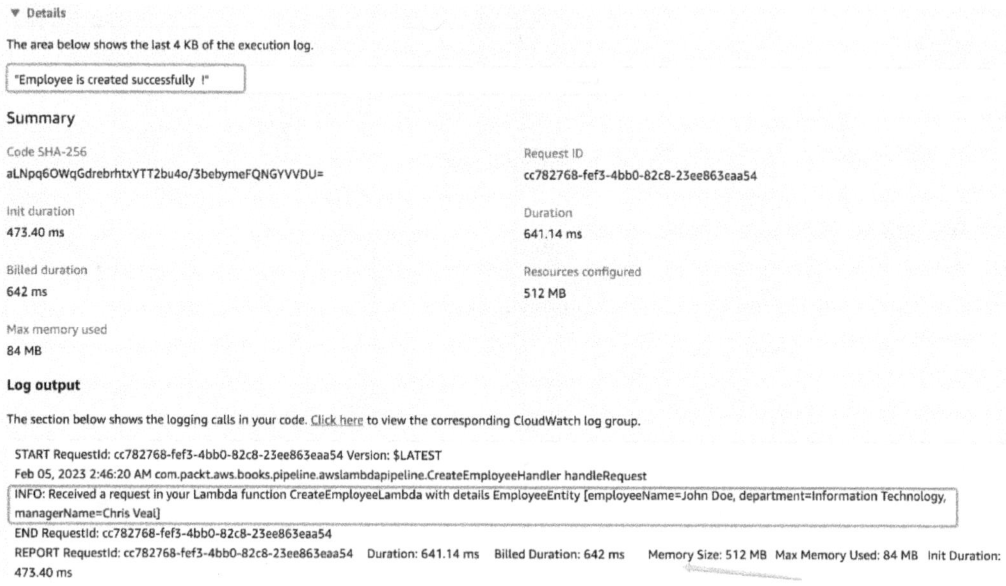

▼ Details

The area below shows the last 4 KB of the execution log.

"Employee is created successfully !"

Summary

Code SHA-256

aLNpq6OWqGdrebrhtxYTT2bu4o/3bebymeFQNGYVVDU=

Request ID

cc782768-fef3-4bb0-82c8-23ee863eaa54

Init duration

473.40 ms

Duration

641.14 ms

Billed duration

642 ms

Resources configured

512 MB

Max memory used

84 MB

Log output

The section below shows the logging calls in your code. Click here to view the corresponding CloudWatch log group.

START RequestId: cc782768-fef3-4bb0-82c8-23ee863eaa54 Version: $LATEST
Feb 05, 2023 2:46:20 AM com.packt.aws.books.pipeline.awslambdapipeline.CreateEmployeeHandler handleRequest
INFO: Received a request in your Lambda function CreateEmployeeLambda with details EmployeeEntity [employeeName=John Doe, department=Information Technology, managerName=Chris Veal]
END RequestId: cc782768-fef3-4bb0-82c8-23ee863eaa54
REPORT RequestId: cc782768-fef3-4bb0-82c8-23ee863eaa54   Duration: 641.14 ms   Billed Duration: 642 ms   Memory Size: 512 MB Max Memory Used: 84 MB Init Duration: 473.40 ms

Figure 12.21 – Lambda detailed execution result

Now, we are done developing our Lambda function, deploying it, and testing it out. In the next section, we need to automate the deployment of our Lambda function and also learn how we can auto-deploy other serverless services. Let's set up CodePipeline to deploy the Lambda function whenever you make any changes to the Lambda function code.

AWS Lambda pipeline setup

There are different ways you can set up a code pipeline to deploy Lambda functions for continuous delivery. In this example, we are going to use the AWS CLI to do so. In the last chapter, we learned how we can add several stages to our pipeline and use it for code scanning, testing, manual approval, and so on. Here, we will keep things simple and will have only two stages in the pipeline: one is for the source and the other one is for CodeBuild and the deployment of the Lambda function. Let's start creating the pipeline by following these steps:

1. As a first step to creating the pipeline for our project, we need to create the infrastructure and resources needed by the pipeline, such as the CodeBuild project, Lambda function, required roles for deployment, and so on. To create the required roles and resources, we have created a Terraform template. You can download the template from our GitHub repository here: `https://github.com/PacktPublishing/Building-and-Delivering-Microservices-on-AWS/tree/main/chapter_12/terraform`. Please execute the `terraform init` command to initialize Terraform.

2. Now, run the `terraform plan` command to get the information about all the resources this template is going to create in your AWS account. The following is a list of resources it is going to create:

    I.   `chapter-12_code_build`: CodeBuild project to compile the Java source code, generate the Java artifact, and run the `aws lambda` command to deploy the changes.

    II. `chapter_12_code_build_policy`: IAM policy to be attached to the `chapter-12_code_build_service_role` role for Lambda code updates by CodeBuild.

    III. `chapter-12_code_build_service_role`: Service role for the CodeBuild project to access the code and push the image to ECR.

    IV. `chapter-12_lambda_service_role`: Execution role used by the `chapter_12_lambda_function` Lambda function.

    V.   `chapter_12_lambda_function`: Lambda function to deploy and execute our code.

3. Run the `terraform apply --auto-approve` command to apply the Terraform changes. The following screenshot confirms that the Terraform execution is successful:

```
aws_iam_role.chapter-12_code_build_service_role: Creation complete after 1s [id=chapter-12_co
de_build_service_role]
aws_codebuild_project.chapter-12_code_build: Creating...
aws_lambda_function.chapter_12_lambda_function: Still creating... [10s elapsed]
aws_codebuild_project.chapter-12_code_build: Still creating... [10s elapsed]
aws_codebuild_project.chapter-12_code_build: Creation complete after 17s [id=arn:aws:codebuil
d:us-east-1:279522866734:project/chapter-12_code_maven_build]
aws_lambda_function.chapter_12_lambda_function: Still creating... [20s elapsed]
aws_lambda_function.chapter_12_lambda_function: Creation complete after 23s [id=chapter_12_la
mbda_function]

Apply complete! Resources: 5 added, 0 changed, 0 destroyed.
```

Figure 12.22 – The terraform apply result

4. Now, let's log in to the AWS Console, and go to the **Lambda** section to see that we have our new Lambda function created. As you can see, `chapter_12_lambda_function` has been created and the **Handler** section is also updated to the name we created in previous sections:

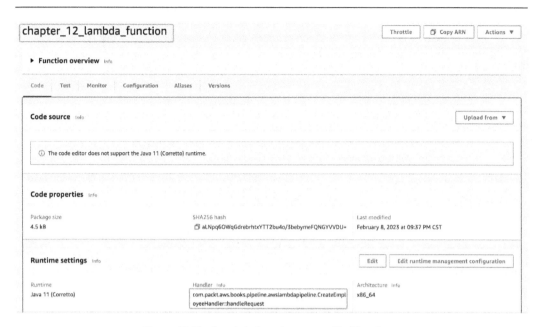

Figure 12.23 – Lambda function created by Terraform

5. For CodePipeline to work, we need to create a CodeCommit source code repository called `aws-lambda-pipeline`, so that we can use that as a trigger pipeline whenever there is a code change. So, open the **CodeCommit** service and click on the **Create repository** button:

Figure 12.24 – CodeCommit repository

6. Provide the name for the repository, `aws-lambda-pipeline`, and click on the **Create** button:

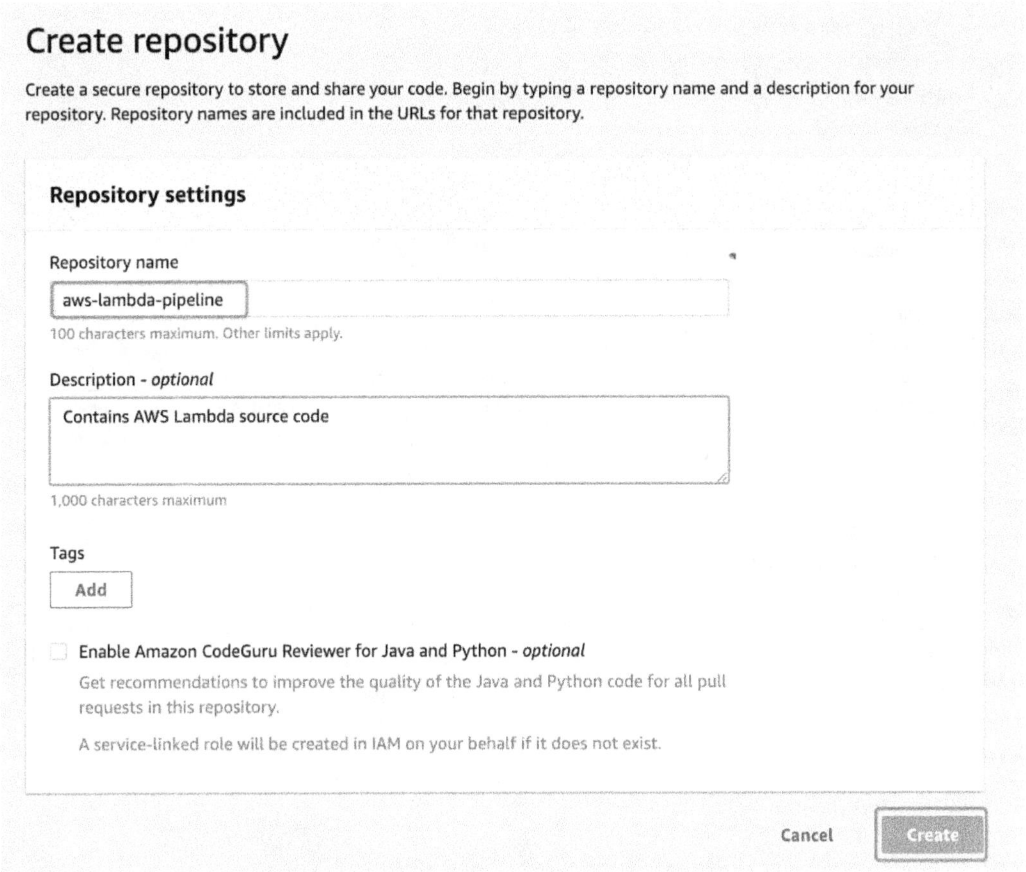

Create repository

Create a secure repository to store and share your code. Begin by typing a repository name and a description for your repository. Repository names are included in the URLs for that repository.

Repository settings

Repository name

```
aws-lambda-pipeline
```

100 characters maximum. Other limits apply.

Description - *optional*

```
Contains AWS Lambda source code
```

1,000 characters maximum

Tags

Add

☐ Enable Amazon CodeGuru Reviewer for Java and Python - *optional*

Get recommendations to improve the quality of the Java and Python code for all pull requests in this repository.

A service-linked role will be created in IAM on your behalf if it does not exist.

Cancel   Create

Figure 12.25 – CodeCommit new repository creation

7. Now we have the repository, as a next step, we need to create a `buildspec.yml` file for the CodeBuild service to use for building the project and update the `chapter_12_lambda_function` code. We can use the source code we created earlier in this chapter for the `CreateEmployeeHandler` function, or you can download it from here: `https://github.com/PacktPublishing/Building-and-Delivering-Microservices-on-AWS/tree/main/chapter_12/aws-lambda-pipeline`. Add the following code to the `buildspec.yml` file:

```
version: 0.2
phases:
install:
runtime-versions:
java: corretto11
pre_build:
commands:
```

```
- echo Starting pre-build phase
- java -version
build:
commands:
- echo Build started on `date`
- mvn test
post_build:
commands:
- echo Build completed on `date`
- mvn package
- ls
- aws lambda update-function-code --function-name chapter_12_
lambda_function --zip-file fileb://target/aws-lambda-pipeline-
1.0-SNAPSHOT.jar
artifacts:
files:
- target/aws-lambda-pipeline*.jar
discard-paths: yes
```

8. Now, push all of this Lambda code and the `buildspec.yml` file to the CodeCommit `aws-lambda-pipeline` repository we created in the previous steps. Once you push everything, the CodeCommit repository will have your latest code:

Figure 12.26 – The aws-lambda-pipeline repository after code push

9. Now, let's start creating the pipeline. Go to the **CodePipeline** service and click on the **Pipelines** link in the left panel, then click on the **Create pipeline** button on the right:

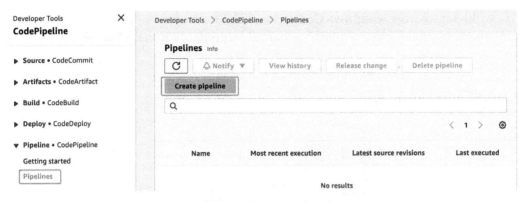

Figure 12.27 – Create pipeline home screen

10. Provide a name for the pipeline and click on the **Next** button:

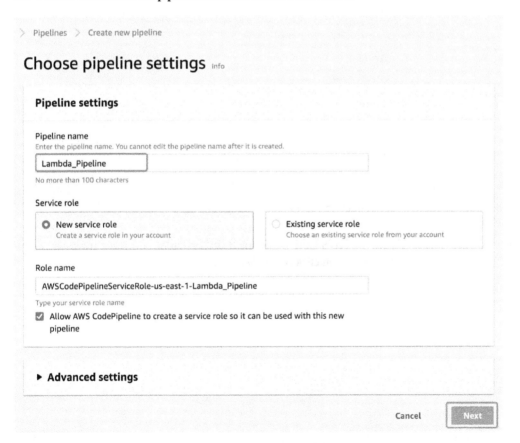

Figure 12.28 – Create pipeline screen

11. In the **Source** stage, select **AWS CodeCommit** and select the `aws-lambda-pipeline` repository and the branch to which you pushed your code. Click on the **Next** button:

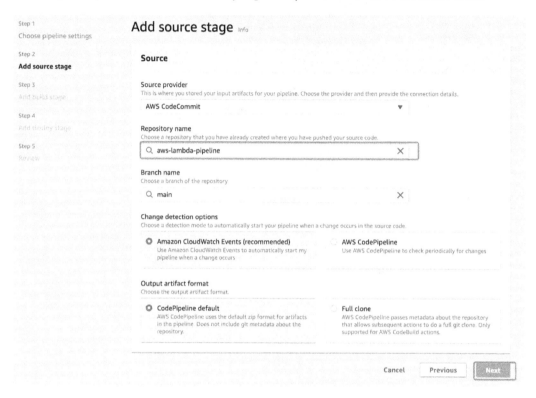

Figure 12.29 – CodePipeline source stage

12. In this **Build** stage, choose **AWS CodeBuild** for **Build provider** and select the `chapter-12_code_build` project, then click on the **Next** button:

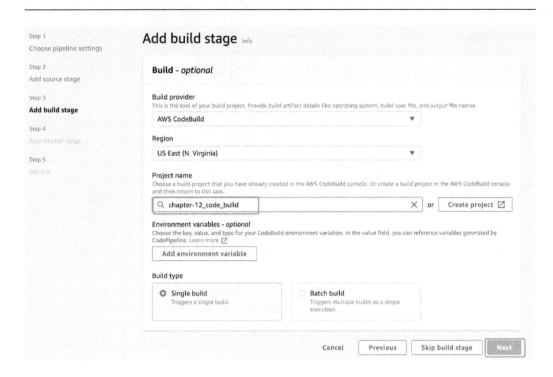

Figure 12.30 – CodePipeline build stage

13. In the next step, CodePipeline asks us to add a deploy stage. We are going to skip that here as we are deploying code changes to our Lambda function in the CodeBuild stage in the post-build action phase. So, go ahead and click on the **Skip deploy stage** button and proceed to the **Review** screen.

14. Confirm the information on the **Review** screen and click on the **Create pipeline** button. Once the pipeline is created, it will start the release process and start deploying the code from the CodeCommit repository to the Lambda service using CodeBuild:

Figure 12.31 – CodePipeline execution

15. Once the pipeline is finished, let's go to the Lambda console and test the function by providing the test event. As you can see, the Lambda function executed successfully:

Figure 12.32 – Lambda function test result

Now, your Lambda pipeline is ready, and when you make a change to the Lambda function, that code is pushed to the code repository, and it will be automatically deployed to the Lambda function. You can do more with CodePipeline and add more stages to create a new alias or create a new version of the Lambda function or add code scanning and tests for Lambda.

Summary

In this chapter, we learned about the AWS serverless offering and learned details about the Lambda functions, and their limitations and benefits. We deployed our Lambda function manually and then created a pipeline to deploy the changes to the function. In the next chapter, we will be learning more about CodePipeline and taking our knowledge beyond AWS infrastructure to integrate with other tools and deploy code to on-premises systems using CodePipeline.

13

Automated Deployment to an EKS Cluster

In this chapter, you will learn about Kubernetes and **Elastic Kubernetes Service** (EKS) provided by AWS for managing Kubernetes workload. First, you will get an overview of the Kubernetes platform and its components and architecture, and then we will learn how EKS service helps in managing the Kubernetes cluster. In this chapter, we will create an EKS cluster and then deploy our sample aws-code-pipeline application to EKS using CodePipeline.

In this chapter, we will cover the following topics:

- Kubernetes – an overview

- Kubernetes architecture

- Kubernetes objects

- What is EKS?

- Deployment to an EKS cluster

In previous chapters, we learned about Docker containers and how **Elastic Container Service** (ECS) helps orchestrate and deploy Docker containers.

Similar to the ECS service, Kubernetes is another competing product to manage the complexity involving thousands of containers. Kubernetes works as an orchestrator to manage container lifecycles and container placement across multiple servers in multiple data centers. Let's start by learning about Kubernetes and its components.

Kubernetes – an overview

Kubernetes is an open source, portable, and extensible platform for managing and orchestrating container workloads. Kubernetes helps to automate the creation, configuration, and placement of the containers across hundreds of container hosts. Kubernetes also orchestrates resource assignments such as memory, disks, CPU usage, health checks, and so on, on containers to easily manage the workload in a production environment.

> **Note**
> Kubernetes is platform neutral, and you can turn a set of servers running on-prem or in a third-party data center into a private cloud for deploying containers.

Kubernetes provides an abstraction on top of your servers so you aren't interacting with any specific server; instead, you use the standard APIs to deploy and manage containers. The following are some of the features Kubernetes provides:

- It provides a standard way to way to start/stop and manage the containers
- It handles the placement of the containers on specific servers in your cluster depending on the workload on each server and the resource
- It auto-scales the containers and handles the traffic demand accordingly by scaling up or down based on the need
- It provides the service discovery and load balancing feature to manage the traffic to the container instances and how they communicate with each other
- It allows you to orchestrate the storage needs of a container to automatically mount storage locally or from a cloud provider
- It provides features to self-heal containers and replaces any failed health check containers with healthy ones
- It makes it easy to configure logging and collect logs from these containers
- It makes it easy to configure and manage sensitive information such as passwords, OAuth tokens, **Secure Shell** (**SSH**) keys, and so on without rebuilding the image and exposing the information
- It helps to enforce common rules and security policies across the cluster and all containers, so you don't have to manage them individually

Let's get into the Kubernetes architecture now.

Kubernetes architecture

Kubernetes is a client-server architecture in which the control plane server is responsible for managing the cluster and provisioning containers on the worker nodes. Kubernetes architecture comprises several components that work together to form a Kubernetes cluster. Each component is responsible for certain tasks within clusters, so let's learn about different Kubernetes components to understand their functions.

Pod

The Pod is the smallest unit within the Kubernetes ecosystem. Kubernetes manages the containers using Pods, and when you must deploy or scale an application, Kubernetes creates Pods. A Pod contains one or more application containers, and they are considered short-lived. Kubernetes also creates a replica in case any Pod failed or was unreachable.

Worker nodes

A worker node is a server within the Kubernetes cluster on which Kubernetes places your Pods to support your workload. A worker node can be a physical machine or a virtual machine provisioned in your data center or through some cloud provider such as AWS. These worker nodes work as infrastructure for Kubernetes to create more Pods. You can add more worker nodes to your cluster to increase its capacity. Each worker node must have Kubelet and Kube-proxy components so that the control plane can manage the nodes and containers running on these worker nodes. A control plane is made of one or more nodes and is the engine for your Kubernetes cluster.

Control plane

In a Kubernetes cluster, the control plane manages the worker nodes and the Pods within a cluster. The control plane can be a single machine, or it can be a set of multiple machines providing fault resilience and high availability in a production environment. The control plane manages the cluster information and distribution of Pods to different worker machines. The control plane comprises of several components, as shown in the following diagram. Control plane components, by default, start on the same machine as the control plane, except for user containers, but they can be started on any machine in the cluster:

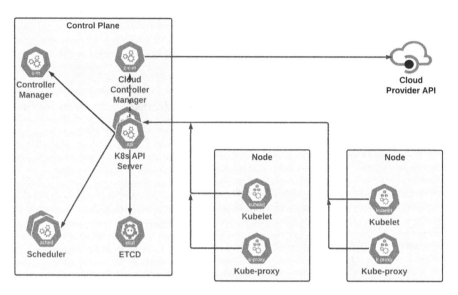

Figure 13.1 – Kubernetes architecture

Let's understand each of the control plane components and their functions in a little more detail:

- **kube-apiserver**: All communication between the Kubernetes cluster and from outside to Kubernetes happens through APIs, through the **command-line interface (CLI)**, **user interface (UI)**, or from the control plane to worker nodes. kube-apiserver works as a frontend to the control plane and is responsible for exposing and managing these APIs. kube-apiserver can scale horizontally and across different instances to handle the traffic.

- **etcd**: etcd is a highly available, strongly consistent key-value distributed data store. Kubernetes uses etcd to store all cluster-related information. Access to etcd happens through the kube-apiserver whenever there is a change in a cluster, or a Pod's etcd gets updated.

- **kube-scheduler**: When a new Pod is created using the kube-apiserver with no assigned node, the kube-scheduler component is responsible for placing the Pod on a worker node. kube-scheduler makes the decision based on several factors such as resource requirements, hardware and software constraints, data locality, affinity, and so on.

- **kube-controller-manager**: The controller manager is a daemon process that consists of core control loops. The controller manager is considered the brain of Kubernetes, which manages the cluster using the different controllers. Each controller watches and manages the shared state of the cluster using the API server. There are several controllers within Kubernetes, some of them are listed as follows, and each one is responsible for a specific task:

- **Replication controller**: A replication controller is responsible for ensuring that a specified number of Pods are always available at any time in a cluster. If a Pod goes offline, the Replication controller will create another Pod.

- **Deployment controller**: A deployment controller manages the Kubernetes deployments within your cluster. A deployment runs multiple replicas of the application and ensures that one or more instances of the application are always available.

- **Namespace controller**: A namespace is a way to subdivide a cluster into virtual sub-clusters for the isolation of resources within the cluster. A namespace controller is responsible for managing these namespaces within a cluster.

- **Node controller**: The node controller monitors nodes and responds to unhealthy ones. The node controller is primarily responsible for assigning a **Classless Inter-Domain Routing (CIDR)** block to a node, maintaining a list of healthy nodes, monitoring the node's health, and initiating Pod eviction from unhealthy nodes.

- **Job controller**: The Job controller monitors Job objects and manages Pods to complete those jobs. A Job is a short-lived task within Kubernetes.

- **ServiceAccount controller**: A ServiceAccount controller is responsible for managing the ServiceAccounts inside a namespace and ensuring that a default ServiceAccount is created for each active namespace.

> ServiceAccount
>
> A ServiceAccount serves as an identity to the process that runs within a Pod, so when a process connects to the Kubernetes cluster's API server, the ServiceAccount's identity can be used for authentication.

There are many more controllers within the Kubernetes controller manager, but we covered only a few here. In reality, all controllers are part of the single binary and run in the same process, but logically, they can be treated as different processes.

- **cloud-controller-manager**: A Kubernetes cluster can be extended to any cloud provider. Cloud controller manager is a control plane component to manage the link between your cluster and the cloud provider's API. Cloud controller manager is an optional component; it only runs your cloud provider controllers when your cluster has an extension to the cloud. If you are running Kubernetes only on on-prem infrastructure or on your local PC, then the cloud controller manager won't run a controller. Route controller, Node controller, and service controller are some examples of cloud-controller-manager that can have a dependency on cloud providers.

As we saw in the Kubernetes architecture diagram, the API server is the core of the control plane. The API server exposes a REST API, through which end users or different cluster components and external tools can communicate with each other. The API server lets you view and manipulate the state of the API objects such as Pods, ConfigMaps, Secrets, and so on. You can directly call this API using the REST API endpoint, or connect to the API server using command-line tools such as kubectl. Let's understand more about the kubectl utility, which we will use later in this chapter to communicate to the EKS cluster.

kubectl

kubectl is a command line tool provided by Kubernetes to interact with the Kubernetes control plane using the Kubernetes API. You need to install the kubectl utility to manage the cluster or perform deployments to the cluster. The kubectl utility needs the required credentials to connect with the cluster. By default, it looks for the credentials located in your system's `$HOME/.kube` directory. You can also provide credentials using the `KUBECONFIG` environment variable. The following are some of the frequently used commands:

- `kubectl get pods`: This gets the list of Pods available
- `kubectl get pod pod1`: This gets the details of a specific Pod
- `kubectl get services`: This gets the list of available services
- `kubectl get deployment deployment1`: This gets the details of a specific deployment service
- `kubectl get deployments`: This gets the list of available deployments
- `kubectl apply -f deployment.yaml`: This performs a deployment to a Kubernetes cluster

You can learn more about different commands related to kubectl by visiting the following page: `https://kubernetes.io/docs/reference/generated/kubectl/kubectl-commands`. Before we start creating Kubernetes applications, we need to first understand some of the important Kubernetes objects provided by the API server and how we manipulate those objects.

Kubernetes objects

You can interact with the Kubernetes cluster to view or change the stage of these objects using the REST API or a command utility such as kubectl to express these objects in `yaml` format. Kubernetes objects are persistent entities in the Kubernetes system, and Kubernetes use these entities to represent the state of your cluster.

When you create an object in Kubernetes, you tell the Kubernetes system what your desired state of the cluster is, then the Kubernetes system will constantly work to get to the desired state.

Kubernetes objects define what containerized applications are running and on which nodes, what resources are allocated to these applications, and how these applications will behave based on the defined policies such as restart, upgrades, and fault tolerance.

To create the object, you need to call the Kubernetes API server either using the REST API and defining the object specification and desired state in the JSON format or using the kubectl CLI then you need to provide the information in YAML format. Almost every Kubernetes object has two sections: the object's spec section, to specify the desired state, and the status section, which defines the object's current state in the Kubernetes environment. As a user, you specify the desired state, and the

Kubernetes system will manage the status section and update it when it ensures that the Kubernetes cluster is in the desired state.

Let's learn about some of the major object types in Kubernetes.

ReplicaSet

In a Kubernetes ecosystem, a ReplicaSet defines how many Pods you want to maintain at a time. When you specify a ReplicaSet, Kubernetes will try to maintain the specified number of Pods created from the specified template in the configuration. The following is a sample YAML configuration file to create a ReplicaSet. Here, you can specify how many copies of the Pod template you want to create. To identify the Pod template, ReplicaSet uses a tier label with the my_Pod value and then creates that many instances of the Pod:

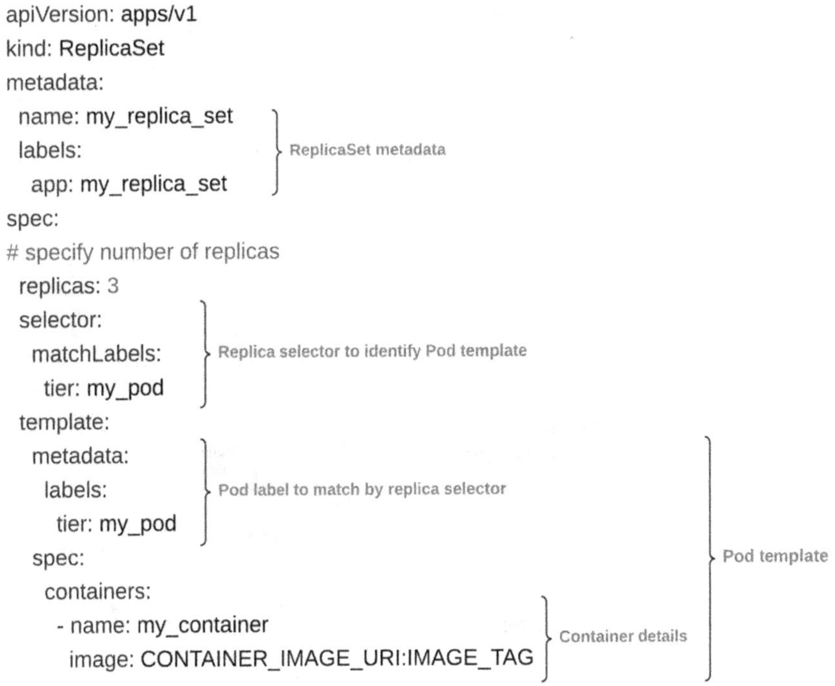

```
apiVersion: apps/v1
kind: ReplicaSet
metadata:
  name: my_replica_set
  labels:                          } ReplicaSet metadata
    app: my_replica_set
spec:
# specify number of replicas
  replicas: 3
  selector:
    matchLabels:                   } Replica selector to identify Pod template
      tier: my_pod
  template:
    metadata:
      labels:                      } Pod label to match by replica selector
        tier: my_pod
    spec:                                                                      } Pod template
      containers:
        - name: my_container       } Container details
          image: CONTAINER_IMAGE_URI:IMAGE_TAG
```

Figure 13.2 – A Kubernetes ReplicaSet configuration file

Pod templates have a spec section that specifies the container details along with the image to create containers. To create the ReplicaSet with this configuration file, you can run the kubectl apply -f FILE_NAME.yaml command, and once execution is successful, we can see the created ReplicaSets by running the kubectl get rs command.

Deployment

In a Kubernetes ecosystem, you specify the desired state of the application using a deployment object, and the Deployment controller will change the state at a controlled rate. Deployment provides a declarative way to create Pods and replicas; you can manage the deployments using the `deployment.yaml` file defining the deployment details, replica, Pod template, and container details. You can apply the deployment configuration using the `kubectl apply -f deployment.yaml` file and you can get the deployment status using the `kubectl rollout status deployment/MY_DEPLOYMENT_NAME` command. The following is an example `deployment.yaml` file:

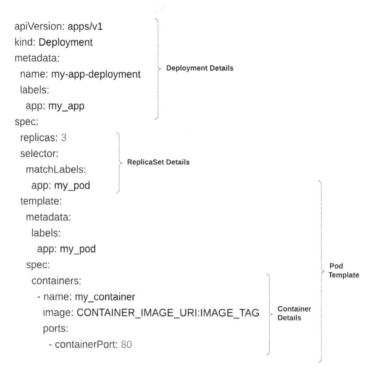

Figure 13.3 – A Kubernetes deployment sample configuration file

Once a deployment is completed and you have to update the container version or change anything in the Pod, you can run the command and update the deployment as follows:

```
kubectl set image deployment/MY_DEPLOYMENT_NAME label=CONTAINER_IMAGE_
URI:CONTAINER_TAG
```

While deployment is in progress, you can get the status by running the following command:

```
kubectl rollout status deployment/MY_DEPLOYMENT_NAME
```

If the deployment is stuck or has failed for some reason and you have to roll back to the previous version, you can run the following command:

```
kubectl rollout undo deployment/MY_DEPLOYMENT_NAME -to-revision=2
```

You need to scale up or down the replicas to support the load when running a deployment. The following command allows you to do that:

```
kubectl scale deployment/MY_DEPLOYMENT_NAME --replicas=15
```

Scaling up or down workload is not always feasible so you can setup Kubernetes to auto-scale replicas horizontally when your CPU usage reaches a certain threshold using the following command:

```
kubectl autoscale deployment/MY_DEPLOYMENT_NAME --min=5 --max=20
--cpu-percent=90.
```

This will start with 5 Pods for the application, and when CPU usage goes to 90% usage, it will scale up the number of Pods until the maximum Pod count reaches 20.

StatefulSet

The StatefulSet object helps to manage stateful applications such as databases in the Kubernetes environments. In non-stateful applications created by the Deployment object type, you can create or delete the Pods any time, and new Pods will start serving requests as they don't care about the application's state.

In a StatefulSet, Pods are created from the same specification, but they are not identical and interchangeable. Each Pod is assigned a persistent sticky identity. In a StatefulSet, you can use a storage volume to provide persistence to the workload, and in the case of a failure and a Pod needing to be created, the same persistent volume can be reattached to the Pod with the same sticky identifier.

Service

When you create a Pod, each Pod in a cluster gets its own IP address, so you don't have to create an explicit link between Pods or map the container ports and host ports. Pods are considered temporary by nature, and based on scaling, they can be created or destroyed dynamically, which brings us to the problem of how to refer these Pods from outside as a single entity so that no matter what IP address is assigned to each one of them and how many Pods are running, outside services can still connect to these.

A Service is an API server object that works as an abstraction on a group of Pods to expose them over the network. So, basically, a Service is a mechanism to expose the application over the network, which runs one or more Pods in your cluster. When you define a Service in Kubernetes, you can define a selector, which identifies what Pods need to be targeted for transferring the traffic. The following is a sample `service.yaml` file:

```
apiVersion: v1
kind: Service
metadata:
 labels:
   app.kubernetes.io/name: aws-code-pipeline
 name: aws-code-pipeline
spec:
 selector:
   app: aws-code-pipeline          ⎫ Application label to forward
 type: LoadBalancer                ⎬ traffic to
 ports:                            ⎭
  - protocol: TCP
    port: 80
    targetPort: 80
```

Figure 13.4 – A Kubernetes Service configuration file

You can create a Service by executing the `kubectl apply -f service.yaml` command. There are four types of Services you can create. Let's learn about each of them.

ClusterIP

ClusterIP is the default type when you don't specify one while creating a Service. This Service type exposes the service on a cluster internal IP address, so the Service is only reachable within the cluster. You can still expose the Service to the public using the Gateway API or Ingress, which we will learn about later.

NodePort

In the NodePort Service type, Kubernetes exposes the Service on a static port of each node within the cluster. In this configuration, Kubernetes allocates a port from the `30000-32767` range by default, and each node proxies the same port. You can access the Service by each node's IP using the same port.

LoadBalancer

In the LoadBalancer Service type, Kubernetes exposes the Service externally using the cloud provider's load balancer. Provisioning of the load balancer happens asynchronously, and traffic from the provider load balancer is routed to the backend Pods.

ExternalName

In the ExternalName Service type, the Service is mapped to a **Domain Name System (DNS)** name, and the DNS name is specified using the `spec.externalName` parameter.

Services are internal entities to a Kubernetes cluster, and how traffic is managed and forwarded to a Service is managed by Ingress, so let's learn about the Ingress object.

Ingress

Ingress is a Kubernetes API object to manage the external HTTP/HTTPS traffic to the Services in a cluster. Ingress helps expose the traffic routes from outside of the cluster to the Services and it is controlled by the rules defined in the Ingress configuration. The following diagram shows how Ingress, Service, Deployment, Pods, and so on relate to each other:

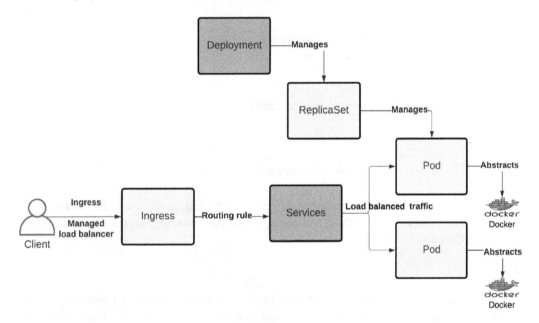

Figure 13.5 – Kubernetes object relationships

As you can see, Deployment is a way to create the Pods, and ReplicaSet manages the number of Pods needed. Each Pod is an abstraction created on top of a container. Service and Ingress help transfer traffic from outside the cluster to a particular Pod to respond to a client request. You can create an Ingress configuration as shown in the following command block and define the traffic rules based on a **Uniform Resource Locator** (**URL**) prefix and handover traffic to a service:

```
apiVersion: networking.k8s.io/v1
kind: Ingress
metadata:
  name: aws-code-pipeline-ingress
  annotations:
    nginx.ingress.kubernetes.io/rewrite-target: /
spec:
  ingressClassName: aws-code-pipeline
  rules:
  - http:
      paths:
      - path: /aws-code-pipeline
        pathType: Prefix
        backend:
          service:
            name: aws-code-pipeline
            port:
              number: 80
```

Ingress Rule

Figure 13.6 – A Kubernetes Ingress config file

This Ingress can be created using the `kubectl apply -f ingress.yaml` file. Other than networking, configuring the Pods without hardcoding within the container image is important; that's where ConfigMaps come in handy.

ConfigMap

ConfigMap is another very important object used in the Kubernetes ecosystem. ConfigMap helps to define the configuration outside your environment. You can store any non-confidential data as a key-value pair in a ConfigMap to decouple data from your containers. You don't need to hardcode values in your Pod, and Pods can consume ConfigMap data as an environment variable, command line input, or configuration file.

Secrets

A Secret is another object used in Kubernetes to hold information similar to ConfigMap, but it is specialized in storing confidential data such as passwords. Secrets provide a way to store sensitive data, so we don't have to store it in the application code.

We have covered most of the important Kubernetes objects; there are a lot more, but covering every object type is outside the scope of the book. You can read more about these objects on the Kubernetes documentation website: `https://kubernetes.io/docs/concepts/`.

As we learned, Kubernetes has multiple moving parts in its architecture, and managing the cluster can be challenging. Amazon has ECS to support container workloads, but customers who already have their workloads running on a Kubernetes platform usually choose to set up their clusters using AWS **Elastic Compute Cloud (EC2)** instances. To support the customer base that prefers to run their workload on Kubernetes, Amazon launched EKS to manage the Kubernetes cluster and to scale the worker nodes as needed. In the next section, let's learn about EKS in detail.

What is EKS?

EKS is a Kubernetes service that manages Kubernetes in the AWS cloud and on-premises data centers. EKS manages the Kubernetes control plane nodes for you, which are responsible for scheduling containers, storing cluster data, maintaining application availability, and providing up-to-date information about the cluster.

EKS automatically maintains and scales cluster nodes to effectively run Kubernetes on AWS infrastructure and your infrastructure using AWS Outposts. EKS integrates well with AWS security services, logging, and networking and you can quickly launch Kubernetes worker nodes using EC2 instances or the serverless Fargate offering.

> **AWS Outposts**
>
> AWS Outposts is a suite of fully managed AWS services that allow you to run AWS native services to virtually any on-premises or edge location. With the help of AWS Outposts, you run AWS services in your data center and connect to your local AWS Region. AWS Outposts supports running application workloads locally, which requires low latency to on-premises systems or needs local data processing. AWS Outposts provides a true hybrid experience and lets you expand to your on-premises data centers with low latency using the native cloud APIs and services.

EKS runs the Kubernetes control plane services across multiple availability zones, ensuring that the cluster is highly available and scalable. EKS ensures that if any of the nodes within a cluster is unhealthy, it is replaced by a healthy node without any degradation in the performance.

Kubernetes comes with **role-based access control (RBAC)** to ensure that cluster users and workloads running on the cluster have the appropriate access to the resources they need. EKS provides integration between AWS **Identity and Access Management (IAM)** users and Kubernetes RBAC. In EKS, when a new cluster is created, by default, the IAM user gets added to the Kubernetes RBAC systems, and that user becomes the admin for the cluster. That user has access to add other IAM users to the Kubernetes RBAC systems so that they can execute commands against the Kubernetes cluster.

EKS allows you to connect and visualize to any compatible Kubernetes cluster running on-premises using EKS anywhere, a self-managed cluster on Amazon EC2, or running outside AWS infrastructure. The Kubernetes version running on EKS is an open source version and fully compatible with Kubernetes community tools and add-ons such as kubectl.

The following diagram shows the architecture of the EKS cluster, where EKS is responsible for managing the Kubernetes control plane, which includes the components such as kube-apiserver, etcd, kube-scheduler, kube-controller-manager, and other controllers. AWS is responsible for scaling and managing the master nodes of the control plane:

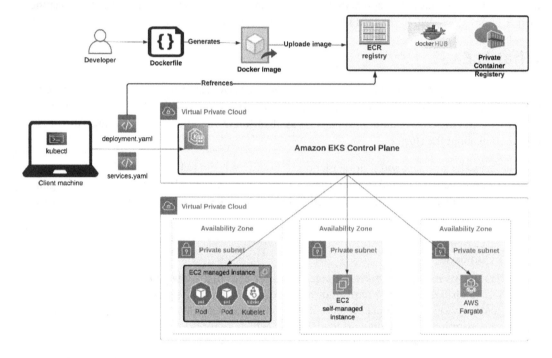

Figure 13.7 – EKS cluster architecture

As a customer, you are responsible for launching and attaching the worker nodes required to run the Kubernetes workload. EKS lets you run the control plane in a different **virtual private cloud** (**VPC**), and you can have your worker nodes in a different VPC. You can run worker nodes using serverless Fargate or you can have EC2 managed or self-managed instances.

As a developer, you can develop your Docker files and push your images to the Docker registry such as **Elastic Container Registry** (**ECR**), and then you define your Kubernetes in the form of a deployment.yaml file, which you can deploy to Kubernetes using the kubectl command utility, and Kubernetes will create the required Pods and place them in the worker nodes. You can also create a Kubernetes service to distribute the load between Pods, like a load balancer. The following is a summary of the benefits EKS provides:

- EKS provides a managed Kubernetes cluster, so you don't have to worry about managing the control plane and persistence of the cluster; EKS takes care of all this.

- EKS provides an integrated console to visualize and troubleshoot through the Kubernetes cluster and applications running on Kubernetes.

- EKS lets you create, update, and delete nodes, so you can run the workload on these nodes. EKS lets you create node groups using spot instances to save operating costs.

- EKS makes it easy to upgrade the running EKS cluster to the latest version of Kubernetes.

- EKS supports running Kubernetes workload on managed EC2 instances such as Fargate, self-managed EC2 instances, and on-premises instances using EKS Anywhere, so you get a choice about where you want to run your workload with the same set of tools and interface.

- EKS supports Kubecost and cost allocation tags to simplify the billing and understand the usage of Kubernetes at the cluster level and on an individual application level.

- EKS delivers Kubernetes control plane logs to CloudWatch for easy access and debugging.

- All EKS API calls are logged, and the history can be viewed in CloudTrail for auditing purposes.

- EKS provides **AWS Controllers for Kubernetes** (ACK), which lets Kubernetes control AWS services from the Kubernetes environment to simplify integration.

- EKS provides a lot more secure Kubernetes; it integrates well with AWS fine-grained IAM controls on top of Kubernetes RBAC and isolates traffic within your own Kubernetes-specific VPC.

Now that we have learned about the EKS service and its benefits, let's learn how to create an EKS cluster, add working nodes, and how to deploy an application to this service.

Deploying application on EKS cluster

We will deploy the aws-code-pipeline application that we created in previous chapters. This time, we will deploy it to the EKS Kubernetes cluster using CodePipeline. The following diagram shows the overall deployment structure; we will use the CodeBuild service to create artifacts and Docker images and push that container image to the ECR repository:

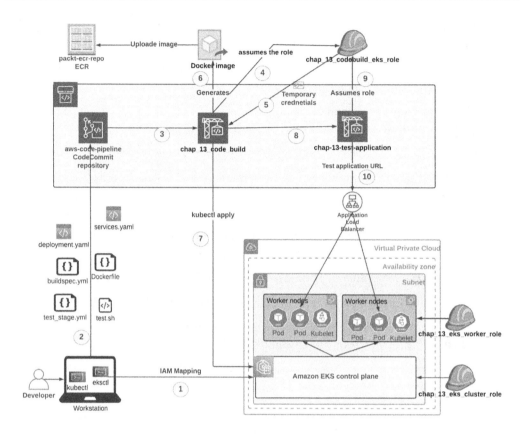

Figure 13.8 – EKS Pipeline deployment architecture

This container image is referenced by the `deployment.yaml` file and deployed to the Kubernetes cluster. The CodeBuild service needs appropriate access to connect with ECR and the Kubernetes cluster to deploy the code. Once the application is deployed, it passes the control to another `chap-13-test-application` CodeBuild project, which is responsible for connecting to the application using a load balancer to test the success of the deployment. Let's go through this journey step by step in the following sections.

Creating an EKS cluster

To deploy an application to an EKS cluster, we first need to create one. We can create a cluster using the AWS console, or we can create one using Terraform. As we discussed earlier in this chapter, the user who creates the EKS cluster becomes the default owner, and that user needs to create the identity mapping and assign access to other users using the commands through eksctl (a utility provided by AWS). In this exercise, we will create the cluster using the Terraform `command_user` user, then we will change permissions and assign access to our `console_user` user. Follow these steps to continue with the exercise:

1. As a first step to creating the EKS cluster for our project, we need to execute the Terraform template, which will create the EKS cluster, CodeBuild project, and other different roles needed. You can download the template from our GitHub repository: `https://github.com/PacktPublishing/Building-and-Delivering-Microservices-on-AWS/tree/main/chapter_13/terraform`. Please execute the `terraform init` command to initialize Terraform.

2. Now run the `terraform plan` command to get the information about all the resources this template will create in your AWS account. The following is an overview of the resources it is going to create:

    I.   `chap_13_code_build_policy`: This IAM role policy provides access to describe an EKS cluster so that the CodeBuild service can request information about the cluster.

    II.   `chap_13_eks_worker_role`: This is the IAM role to be assumed by the EC2 instances part of the EKS worker node group. This role gives the EC2 instances access to ECR to download the container images, and the EKS cluster can perform Pod placement and manage these worker nodes.

    III. `chap_13_codebuild_eks_role`: This is the IAM role used by the CodeBuild service to connect to ECR to push and download Docker images to connect to the EKS cluster for deployment.

    IV.   `chap_13_eks_cluster_role`: This is the IAM role to be assumed by the EKS cluster to manage the Kubernetes cluster.

    V.   `chap_13_eks_cluster`: This is the EKS cluster to perform the application deployment to.

    VI.   `chap_13_code_build`: This is the CodeBuild project that compiles the Java source code and generates Java artifacts, uploads them to the ECR repository, and deploys to the Kubernetes `chap_13_eks_cluster` cluster.

    VII. `chap-13-test-application`: This is the CodeBuild project that tests the deployed Kubernetes application by pinging the service URL.

3. Creating these resources will incur a cost, so after you complete the exercise, run the `terraform destroy` command to delete all the resources you created through Terraform. For now, let's create the resources by running the `terraform apply -auto-approve` command.

4. Once the `apply` command is executed, you can review the created resources in the AWS Console. For now, let's search for EKS to see the newly created cluster. The following screenshot shows our newly created EKS `chap-13-eks-cluster` cluster:

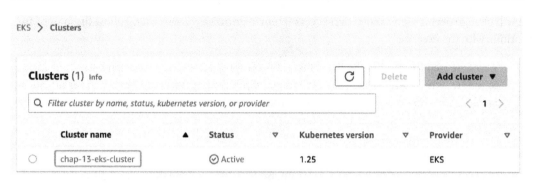

Figure 13.9 – The EKS Clusters home screen

Now, we have created the EKS cluster that manages the Kubernetes control plane. For the next step, we need to add the worker nodes on which the control plane can deploy the workload.

Adding worker nodes

Worker nodes are the nodes on which Kubernetes creates the Pods and runs your containers. EKS allows adding the worker node as worker node groups and as a serverless compute through Fargate worker nodes. We will use EC2 worker nodes in the following exercise:

1. To add worker nodes on the EKS cluster, click on the **Compute** tab, as shown in the following screenshot, and click on the **Add node group** button:

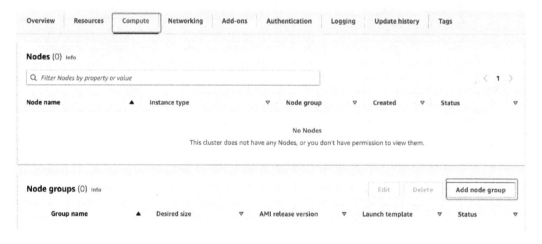

Figure 13.10 – Adding a worker node group

2. Provide the `chap-13-eks-worker-nodes` name to the worker node group and select the **chap-13-eks-worker-role** role, which we created using Terraform. EKS provides the option to create worker nodes using the EC2 launch template, and you can select an existing one, but here, we aren't going to use one, so click on the **Next** button at the bottom of the screen:

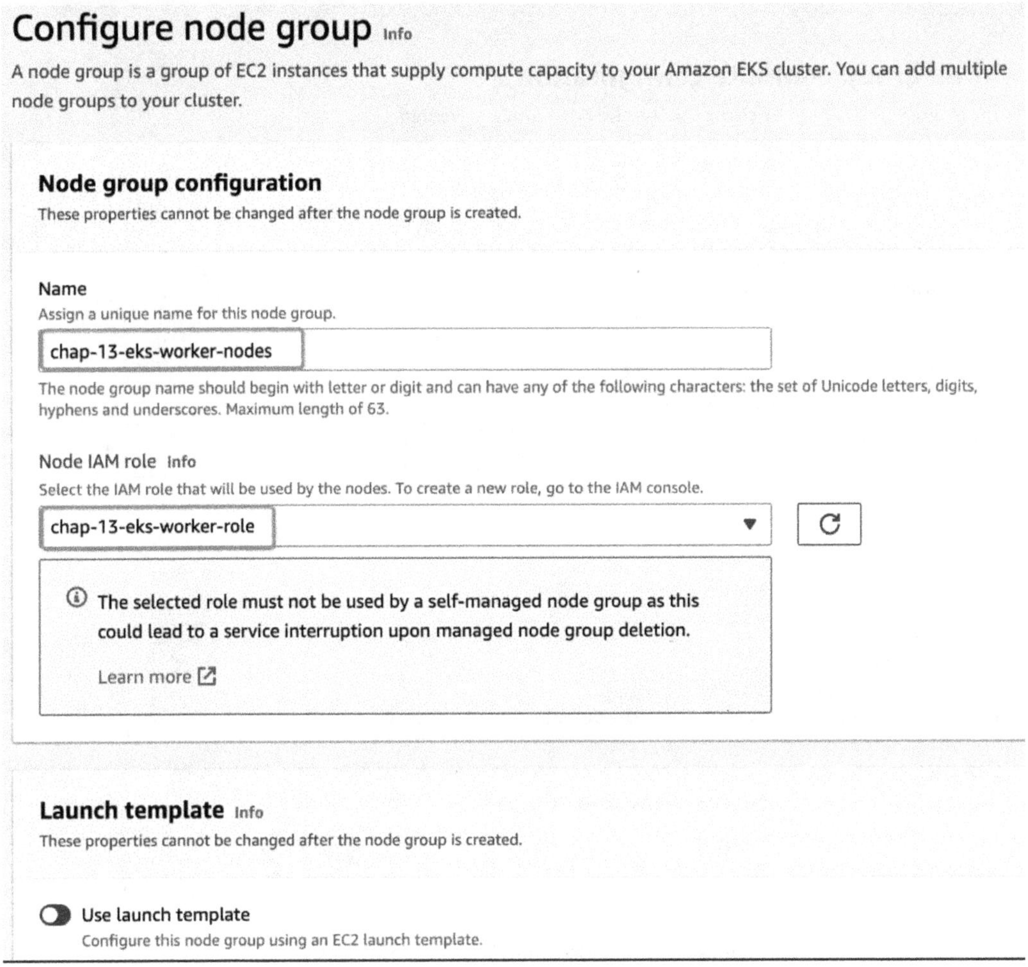

Configure node group Info

A node group is a group of EC2 instances that supply compute capacity to your Amazon EKS cluster. You can add multiple node groups to your cluster.

Node group configuration

These properties cannot be changed after the node group is created.

Name

Assign a unique name for this node group.

chap-13-eks-worker-nodes

The node group name should begin with letter or digit and can have any of the following characters: the set of Unicode letters, digits, hyphens and underscores. Maximum length of 63.

Node IAM role Info

Select the IAM role that will be used by the nodes. To create a new role, go to the IAM console.

chap-13-eks-worker-role	▼		C

ⓘ The selected role must not be used by a self-managed node group as this could lead to a service interruption upon managed node group deletion.

Learn more ☑

Launch template Info

These properties cannot be changed after the node group is created.

⬤ **Use launch template**

Configure this node group using an EC2 launch template.

Figure 13.11 – Node group configuration

3. On the next screen, AWS allows us to select the compute resource type. We are keeping the defaults suggested by EKS:

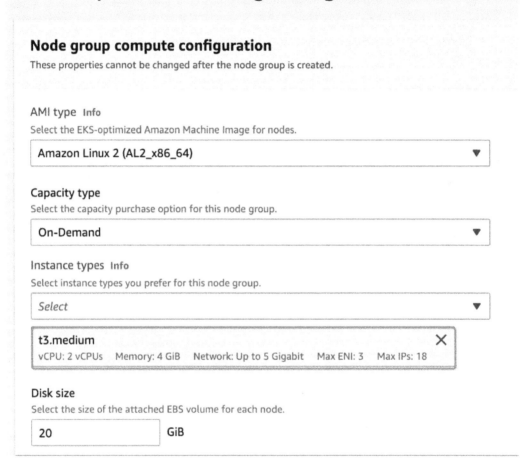

Figure 13.12 – Node group instance resources

4. If you scroll down, you can configure the auto-scaling of the worker nodes to how many instances you want to be running at a time and what will be the maximum number of nodes. At the bottom of the screen, you can configure how many instances or a percentage of instances can be offline during the upgrade process. Keep the default configuration and click on the **Next** button:

Node group scaling configuration

Desired size
Set the desired number of nodes that the group should launch with initially.

| 2 | nodes |

Minimum size
Set the minimum number of nodes that the group can scale in to.

| 2 | nodes |

Maximum size
Set the maximum number of nodes that the group can scale out to.

| 2 | nodes |

Node group update configuration Info

Maximum unavailable
Set the maximum number or percentage of unavailable nodes to be tolerated during the node group version update.

○ **Number**
    Enter a number

○ **Percentage**
    Specify a percentage

Value

| 1 | node |

Cancel   Previous   **Next**

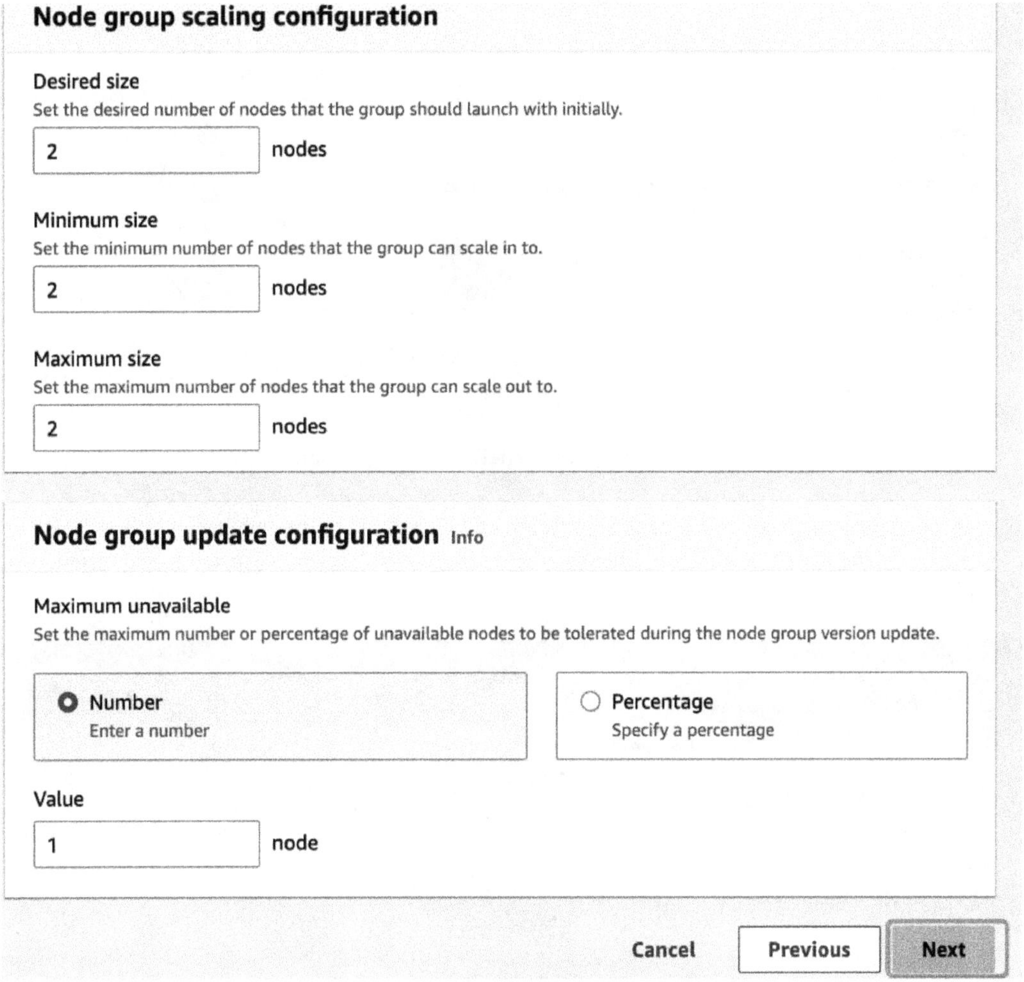

Figure 13.13 – Node group scaling configuration

5. On the next screen, select the subnets within your VPC, where you want to install these worker nodes and click on the **Next** button:

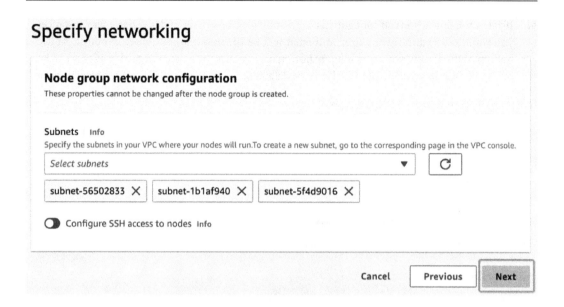

Figure 13.14 – Node group network configuration

6. On the review screen, verify the configuration and click on the **Create** button. AWS will start providing the worker nodes, and once they are ready, you will see newly provisioned instances in your cluster:

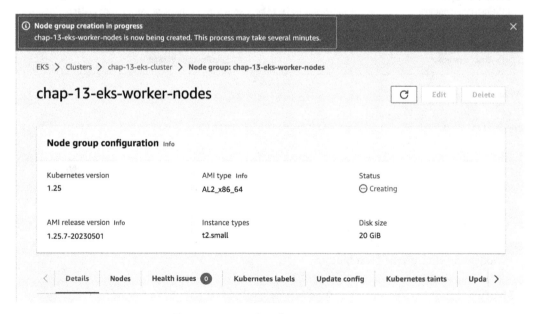

Figure 13.15 – EKS node group creation

If you look at the preceding screenshot, you can see that EKS displays a warning that the current `console_user` user doesn't have access to the Kubernetes objects, as we created this cluster using the Terraform `command_user` user. In the next section, we will learn how to provide access to other users.

EKS permission mapping

By default, EKS adds the user who creates the cluster to the Kubernetes RBAC system so that the user has permission to interact with the cluster. In our case, we have created the cluster using `command_user`, so that is already part of the Kubernetes RBAC system. Now we need to assign RBAC access on this cluster to our AWS console user named `console_user`, and we also need to assign access to the CodeBuild service `chap_13_codebuild_eks_role` role so that it can perform the deployment to EKS. Let's take the following steps to install the required utilities and assign appropriate permissions:

1. First, we need to install the kubectl utility to our system so that we can interact without the EKS `chap-13-eks-cluster` cluster. To download the kubectl utility, run the following command:

    ```
    curl -sS -o kubectl https://amazon-eks.s3.us-east-1.amazonaws.
    com/1.20.4/2021-04-12/bin/linux/amd64/kubectl
    ```

2. Assign the executable permission to the downloaded kubectl file by typing the following command:

    ```
    chmod +x ./kubectl
    ```

3. Now you have the kubectl utility on your system, let's add it to your path so you can run the kubectl commands from anywhere. Execute the following command:

    ```
    export PATH=$PWD/:$PATH
    ```

4. kubectl uses the credentials stored in the `.kube/config` directory in the user's home to connect to the Kubernetes cluster. To get these credentials, we need to run the following command. This command will connect to your EKS cluster and update the cluster information:

    ```
    aws eks update-kubeconfig --name chap-13-eks-cluster --region
    us-east-1
    ```

5. Now run the `kubectl version` command to confirm that the installation is successful and you are connected to the cluster.

6. AWS provides the eksctl utility to manage the EKS cluster and allow easier user mapping. To provide access, follow the instructions provided on the following link to install the eksctl utility: `https://docs.aws.amazon.com/eks/latest/userguide/eksctl.html`.

7. Once you have installed the eksctl utility, run the following command to ensure that it is correctly installed:

```
eksctl version
```

8. Now that we have everything ready, let's assign the chap-13-eks-cluster command user access to our EKS cluster. You need to provide the correct **Amazon Resource Name (ARN)** of the user to whom you are assigning access and use the correct EKS cluster name and AWS Region:

```
eksctl create iamidentitymapping \
--cluster chap-13-eks-cluster \
--region us-east-1 \
--arn arn:aws:iam::xxxxxxxxxxxx:user/console_user \
--group system:masters \
--no-duplicate-arns \
--username console_user
```

9. Once you run the command, console_user will be mapped to the Kubenetes RBAC system and it will be able to interact with the EKS cluster:

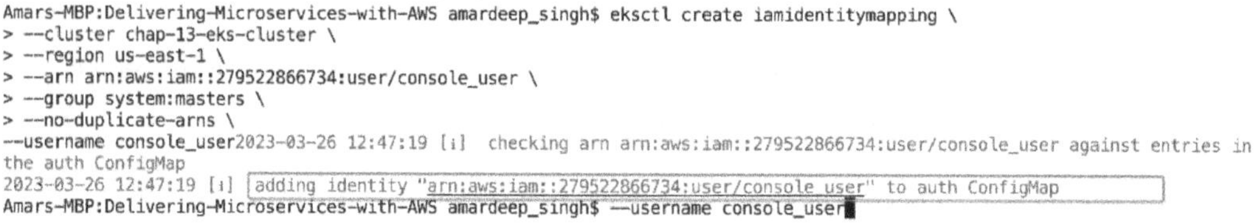

```
Amars-MBP:Delivering-Microservices-with-AWS amardeep_singh$ eksctl create iamidentitymapping \
> --cluster chap-13-eks-cluster \
> --region us-east-1 \
> --arn arn:aws:iam::279522866734:user/console_user \
> --group system:masters \
> --no-duplicate-arns \
--username console_user2023-03-26 12:47:19 [i]  checking arn arn:aws:iam::279522866734:user/console_user against entries in
the auth ConfigMap
2023-03-26 12:47:19 [i] adding identity "arn:aws:iam::279522866734:user/console_user" to auth ConfigMap
Amars-MBP:Delivering-Microservices-with-AWS amardeep_singh$ --username console_user
```

Figure 13.16 – EKS cluster IAM user permissions assignment

10. Now the console_user user has access to the EKS cluster, if you refresh the AWS console and go to the cluster, you can see the warning displayed earlier is gone, and the worker nodes are associated with the cluster. You can browse around the cluster by clicking on the **Resources** tab. You can see the different Kubernetes objects currently deployed:

Figure 13.17 – EKS Cluster worker nodes

11. Now we need to run the `iamidentitymapping` command one more time to assign access to the CodeBuild `chap-13-codebuild-eks-role` service role. You need to get the correct ARN of the role, cluster name, and AWS Region name. eksctl allows you to add users and roles to the Kubernetes RBAC:

```
eksctl create iamidentitymapping \
--cluster chap-13-eks-cluster \
--region us-east-1 \
--arn arn:aws:iam::xxxxxxxxxxxxx:role/chap-13-codebuild-eks-role \
--group system:masters \
--no-duplicate-arns \
--username chap-13-codebuild-eks-role
```

We now have everything set up from an access perspective and CodeBuild should be able to deploy applications to the EKS cluster. For the next step, we need to set up the source code for the deployment.

Code changes

In this section, we are going to focus on the code changes we need specifically for deployment to the EKS. You can download the full source code from the GitHub repository: `https://github.com/PacktPublishing/Building-and-Delivering-Microservices-on-AWS/tree/main/chapter_13/aws-code-pipeline`. In this exercise, we are going to use two CodeBuild projects: `chap-13-code-build` and `chap-13-test-application`. Let's review the core changes needed for this project:

1. We need to create a `deployment.yaml` file for Kubernetes to create the required Pods of the aws-code-pipeline application. In the containers section, we refer to the image from the `packt-ecr-repo` ECR repository created in *Chapter 10*. We have specified the tag of the image as a `TAG_VERSION` variable, which CodeBuild will replace with the latest version of the code during the build process:

```
aws-code-pipeline > ! deployment.yaml > ⊡ kind
 1    apiVersion: apps/v1
 2    kind: Deployment
 3    metadata:
 4      labels:
 5        app.kubernetes.io/name: aws-code-pipeline
 6        app.kubernetes.io/instance: aws-code-pipeline-instance
 7        app.kubernetes.io/version: "1.0.0"
 8        app.kubernetes.io/managed-by: kubectl
 9      name: aws-code-pipeline-deployment
10    spec:
11      replicas: 2
12      selector:
13        matchLabels:
14          app: aws-code-pipeline
15      template:
16        metadata:
17          labels:
18            app: aws-code-pipeline
19        spec:
20          containers:
21            - image:          .dkr.ecr.us-east-1.amazonaws.com/packt-ecr-repo:TAG_VERSION
22              imagePullPolicy: Always
23              name: aws-code-pipeline
24              ports:
25                - containerPort: 80
```

Figure 13.18 – The aws-code-pipeline application deployment.yaml

2. To expose the application outside the EKS cluster, we need to create a Kubernetes service using the `service.yaml` file. This service object will refer to the Pods created using the `deployment.yaml` file, and the service will create a load balancer to transfer traffic to the application Pods. The following is the configuration for the `service.yaml` file, which listens to the traffic on port `80` and transfers to Pods on the same port:

```
aws-code-pipeline > ! service.yaml > ᴬᴮ apiVersion
 1  apiVersion: v1
 2  kind: Service
 3  metadata:
 4    labels:
 5      app.kubernetes.io/name: aws-code-pipeline
 6    name: aws-code-pipeline
 7  spec:
 8    selector:
 9      app: aws-code-pipeline
10    type: LoadBalancer
11    ports:
12    -  protocol: TCP
13       port: 80
14       targetPort: 80
```

Figure 13.19 – The aws-code-pipeline service.yaml config file

3. We are going to create the `buildspec.yml` file, which is used by the CodeBuild `chap-13-code-build` project to build the project and create and push the Docker image to the `packt-ecr-repo` ECR repository. Once the image is published to the ECR repository, CodeBuild will change the image tag in the `deployment.yaml` file and create the deployment using the `kubectl apply` command and start the Pods. CodeBuild will also create and apply the `service.yaml` file to create the Kubernetes service, get the service URL, and change the current URL in the `test.sh` script file:

```
version: 0.2
phases:
 install:
  commands:
   - aws ecr get-login-password --region us-east-1 | docker login --username AWS --password-stdin    } Log in to ECR
279522866734.dkr.ecr.us-east-1.amazonaws.com
    - curl -sS -o aws-iam-authenticator https://amazon-eks.s3.us-west-2.amazonaws.com/1.19.6/2021-01-05/bin/linux/amd64/aws-iam-authenticator
    - curl -sS -o kubectl https://amazon-eks.s3.us-west-2.amazonaws.com/1.20.4/2021-04-12/bin/linux/amd64/kubectl    } Download and install
    - chmod +x ./kubectl ./aws-iam-authenticator                                                           aws-iam-authenticator
    - export PATH=$PWD/:$PATH                                                                              and kubectl'
 pre_build:
  commands:
   - TAG="$(date +%Y-%m-%d.%H.%M.%S)"      } Creating the TAG variable
   - export KUBECONFIG=$HOME/.kube/config
 build:
  commands:
   - aws eks update-kubeconfig --name chap-13-eks-cluster --region us-east-1   } Updating the EKS cluster configuration to connect to
   - mvn clean install                                                           the cluster for the CodeBuild service role
   - cp target/aws-code-pipeline*.jar .
   - docker build -t aws-code-pipeline:$TAG .                                   } Maven and Docker image
   - docker tag aws-code-pipeline:$TAG 279522866734.dkr.ecr.us-east-1.amazonaws.com/packt-ecr-repo:$TAG    build
 post_build:
  commands:
   - docker push 279522866734.dkr.ecr.us-east-1.amazonaws.com/packt-ecr-repo:$TAG  } Docker image push to ECR
   - sed -i "s/TAG_VERSION/$TAG/g" deployment.yaml
   - kubectl apply -f deployment.yaml
   - kubectl rollout restart -f deployment.yaml                                 } Kubernetes deployment and Service creation
   - kubectl apply -f service.yaml
   - SERVICE_URL=$(kubectl get service aws-code-pipeline -o jsonpath='{.status.loadBalancer.ingress[].hostname}')  } Service URL extraction
   - sed -i "s/CLUSTER_URL/$SERVICE_URL/g" test.sh                                                                    and replacement in test
artifacts:                                                                                                           script
 files:
  - test.sh
  - test_stage.yml
discard-paths: yes
```

Figure 13.20 – The buildspec.yaml file for CodeBuild

4. In this step, we need to create the buildspec file named test_stage.yml, which will be used by the CodeBuild chap-13-test-application project for testing the EKS deployment:

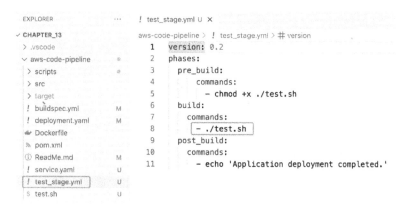

Figure 13.21 – The test_stage.yml file for CodeBuild

5. The test stage uses a shell script file to connect to the EKS service and check whether the service is up and running to confirm the successful deployment. The following is the code for the `test.sh` script. In this script file, the `CLUSTER_URL` string will be replaced by the CodeBuild `chap-13-code-build` project during the build process with the actual deployed service URL once the service is created. In this script, we are using the `curl` command to request the service endpoint, and if we have a `200` status code, we are considering the application deployment a success. In a real production environment, these scripts can be much more complicated:

```
CHAPTER_13                      aws-code-pipeline > $ test.sh
> .vscode                         1    #!/bin/bash
∨ aws-code-pipeline       ⊜       2    URL=CLUSTER_URL
  > scripts                ⊜       3    RESPONSE=$(curl --write-out %{http_code} --silent --output /dev/null ${URL})
  > src                            4    if [ $RESPONSE -ne 200]
  > target                         5  ∨ then
  ! buildspec.yml          M       6        echo 'Application deployed failed!'
  ! deployment.yaml        M       7        exit 1
  ◬ Dockerfile                     8    else
  ⅀ pom.xml                        9    echo 'Application deployed succesfully !'
  ⓘ ReadMe.md              M      10    fi
  ! service.yaml           U
  ! test_stage.yml         U
  $ test.sh                U
```

Figure 13.22 – The test.sh script

Now we have everything set up from a code perspective and CodeBuild should be able to deploy applications to the EKS cluster. Let's now check in the code to the CodeCommit `aws-code-pipeline` repository we created in *Chapter 5*. As the next step, let's create CodePipeline to deploy this code in an automated fashion.

Setting up CodePipeline

To perform a deployment to the EKS cluster, we need to create a CodePipeline. We will keep this pipeline simple enough and only have three stages: **Source**, **Build**, and **Test**. To start creating the pipeline, perform the following steps:

1. Search for `CodePipeline` in the AWS console and click on the **Create pipeline** button. Provide the `chap-13-eks-pipeline` name for the CodePipeline and click on the **Next** button to select the source:

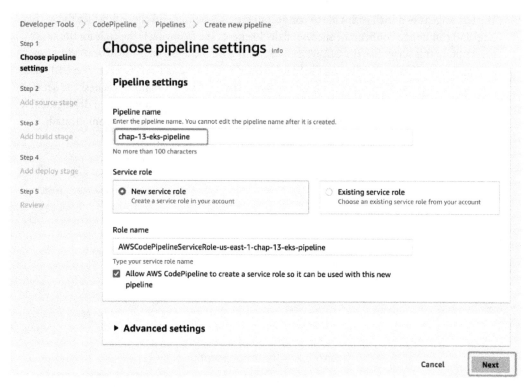

Figure 13.23 – Creating the pipeline

2. On the next screen, select **AWS CodeCommit** as the **Source provider** option and select the `aws-code-pipeline` repository, which we created in the previous chapters. Enter `master` for the **Branch name** field and click on the **Next** button:

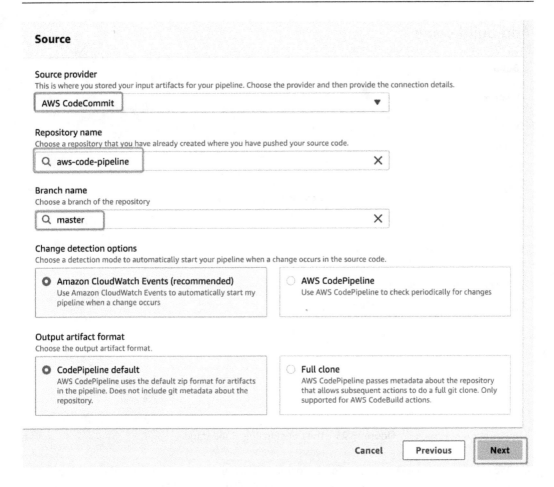

Source

Source provider
This is where you stored your input artifacts for your pipeline. Choose the provider and then provide the connection details.

AWS CodeCommit ▼

Repository name
Choose a repository that you have already created where you have pushed your source code.

🔍 aws-code-pipeline ✕

Branch name
Choose a branch of the repository

🔍 master ✕

Change detection options
Choose a detection mode to automatically start your pipeline when a change occurs in the source code.

◉ **Amazon CloudWatch Events (recommended)**
Use Amazon CloudWatch Events to automatically start my pipeline when a change occurs

○ AWS CodePipeline
Use AWS CodePipeline to check periodically for changes

Output artifact format
Choose the output artifact format.

◉ **CodePipeline default**
AWS CodePipeline uses the default zip format for artifacts in the pipeline. Does not include git metadata about the repository.

○ Full clone
AWS CodePipeline passes metadata about the repository that allows subsequent actions to do a full git clone. Only supported for AWS CodeBuild actions.

Cancel   Previous   **Next**

Figure 13.24 – The CodePipeline source repository

3.   On the next screen, select **AWS CodeBuild** for building the source code. In this step, we need to select the chap-13-code-build CodeBuild project, which we created using Terraform. This CodeBuild project uses the buildspec.yml file as configuration and is responsible for building the Maven project, creating a Docker image, pushing it to ECR, and then deploying this Docker image to our EKS cluster. Click on the **Next** button:

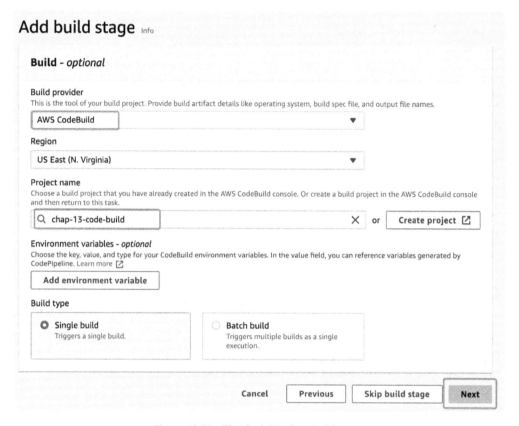

Figure 13.25 – The CodePipeline Build stage

4. We will not use CodeDeploy here for deployment. The build stage we configured in the previous step is going to take care of the deployment, so let's click on the **Skip deploy stage** button:

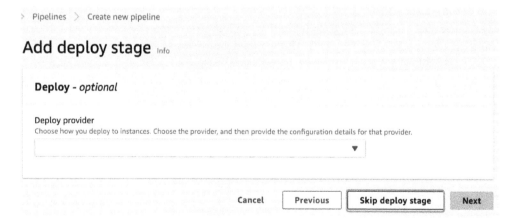

Figure 13.26 – Skip deploy stage

5. On the next screen, review the information and click on the **Create pipeline** button to create the pipeline. Now, we need to add an additional stage to this CodePipeline to perform a test after deployment, so let's click on the **Edit** button:

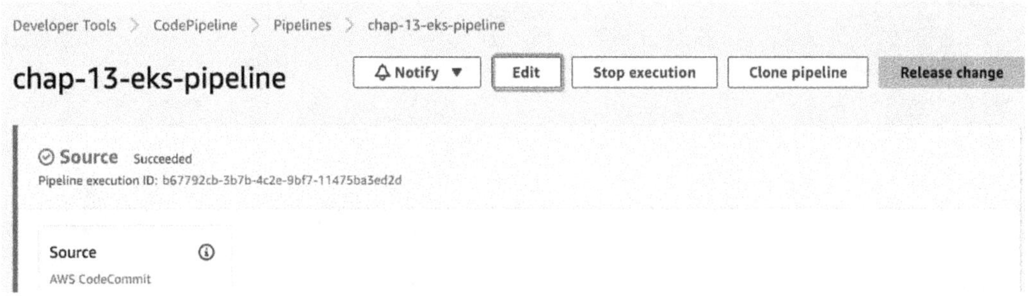

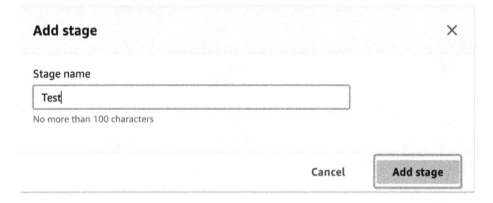

Figure 13.27 – Edit the CodePipeline

6. Click on the **Add stage** button towards the bottom of the screen, then provide the name of the stage as Test, and click on the **Add stage** button on the pop-up window:

Figure 13.28 – The Add stage screen

7. Click on the **Add action group** button in the newly created Test stage and provide details for it. We are using CodeBuild as a provider for the Test stage, so select the CodeBuild chap-13-test-application project, which we created using Terraform for testing the deployment. This stage uses the test.sh script to perform the health check with the Kubernetes deployed service. After selecting the CodeBuild chap-13-test-application project, click on the **Done** button:

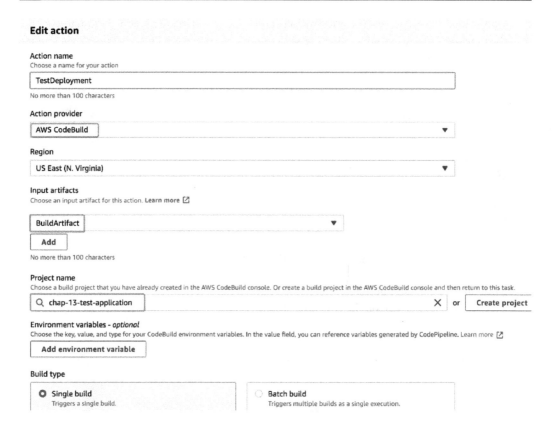

Figure 13.29 – The Test stage CodeBuild configuration

8. Click on the **Save** button to save the new Test stage changes and start the pipeline by clicking the **Release change** button:

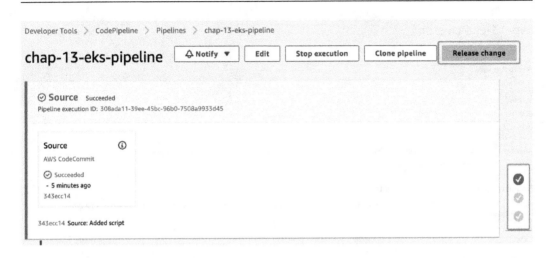

Figure 13.30 – CodePipeline execution

9. Once the pipeline is completed successfully, you can go to the load balancer's section to grab the DNS name of the load balancer to test the application. In our case, the URL is `http://aac97df331de0445fb98b33dba217f8e-372651673.us-east-1.elb.amazonaws.com`. Open that in a browser and you can see that we get a response from EKS:

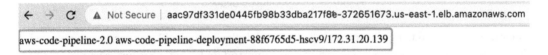

Figure 13.31 – EKS deployed application

10. To see the details of the deployed Pods, you can go to the EKS cluster and click on the **Resources** tab and you can click on different workload types and explore the details of Pods, ReplicaSets, Deployments, and other Kubernetes objects:

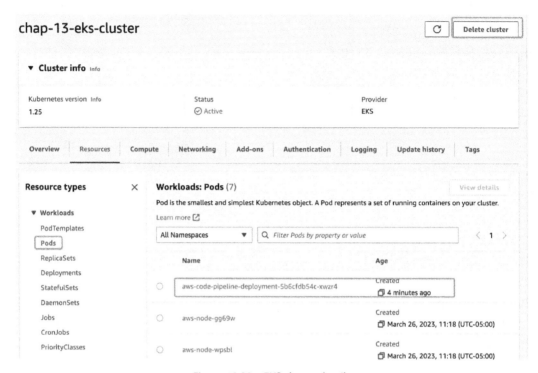

Figure 13.32 – EKS cluster details

We now have our aws-code-pipeline project all set up and deploying to the EKS cluster in an automated fashion. You can run the `terraform destroy` command to delete the resources you created to avoid the additional cost to your AWS account.

Summary

In this chapter, we learned about Kubernetes and EKS offered by AWS. We learned how to write Kubernetes configuration files and how to apply those to change the state of the cluster. We created an EKS cluster and explored EKS cluster architecture in more depth and added worker nodes to it. We updated our sample aws-code-pipeline microservices and added `deployment.yaml` and `service.yaml` files to deploy to EKS. We set up CodePipeline to automatically build and deploy the aws-code-pipeline microservice to the EKS cluster. In the next chapter, we will learn more about CodePipeline and take our knowledge beyond AWS infrastructure to integrate it with other tools and deploy code to on-prem systems using CodePipeline.

14

Extending CodePipeline Beyond AWS

In this chapter, we will be extending our knowledge of AWS CodePipeline and setting up a CodePipeline that integrates with code repositories hosted outside AWS. We will utilize AWS CodePipeline to integrate with the Jenkins build server and then deploy our application to an on-prem server, using AWS CodeDeploy. This chapter will guide you on how to configure CodePipeline to deploy applications on private servers or on-prem servers.

In this chapter, we will cover the following topics:

- What is GitHub?
- Integrating CodePipeline with GitHub
- What is Bitbucket?
- Extending CodePipeline

In the previous chapters, we created CodePipeline and used the CodeCommit repository as a source code provider. CodePipeline supports several version control systems, GitHub and Bitbucket being some of the most popular ones, so let's learn about them and then see how we can use them in our pipeline.

What is GitHub?

GitHub is an open source version control system that provides hosting of Git-based repositories. GitHub supports unlimited public and private repositories for free, although features are limited on private repositories. GitHub is a very popular source version control system within the developer community.

GitHub has an enterprise offering that is dedicated for the organization, and you can host it on your infrastructure. The following are some of the features that GitHub supports:

- GitHub provides an easy interface for creating and manage repositories, projects, and organizations
- GitHub allows contributors to easily notify a change pushed to the repository, and you can get notifications of changes to your subscribed activities

- GitHub supports easy code review with visual differences, so you can spot which files have been changed and what changes have been made

- You can configure rules and assign code to do peer code reviews in a `pull` request

- GitHub actions support the complete CI/CD, and you can design automatic software development workflows to build, test and deploy your code

- GitHub supports the hosting of packaging, and you can use packages as dependencies for building other projects

- You can create calls to get all the data and events you need within GitHub, and automatically kick off and advance your software workflows

- GitHub supports security scanning of code and finds vulnerabilities in code before they reach production

- GitHub supports integration with enterprise authentication and authorization mechanisms, such as SAML and LDAP, for quick onboarding

- GitHub supports the maintenance of secrets and sharing across multiple repositories to reduce workflow failures and increase security

Creating a GitHub repository

To get started with GitHub, you need an email address that you can use to register on `https://github.com/`. Perform the following steps to create a new repository:

1. To create a new repository, you can simply click on the + drop-down button and select the **New repository** option:

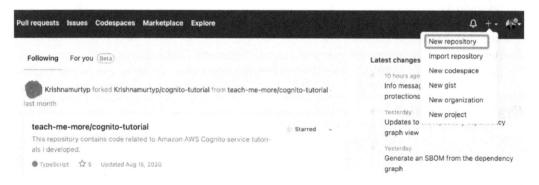

Figure 14.1 – The GitHub Home screen

2. On the **Create a new repository** screen, you can provide details about the repository, such as the repository name, a description, whether the repository is public or private, and the type of licenses attached to the repository. Once you have done so, click on the **Create repository** button:

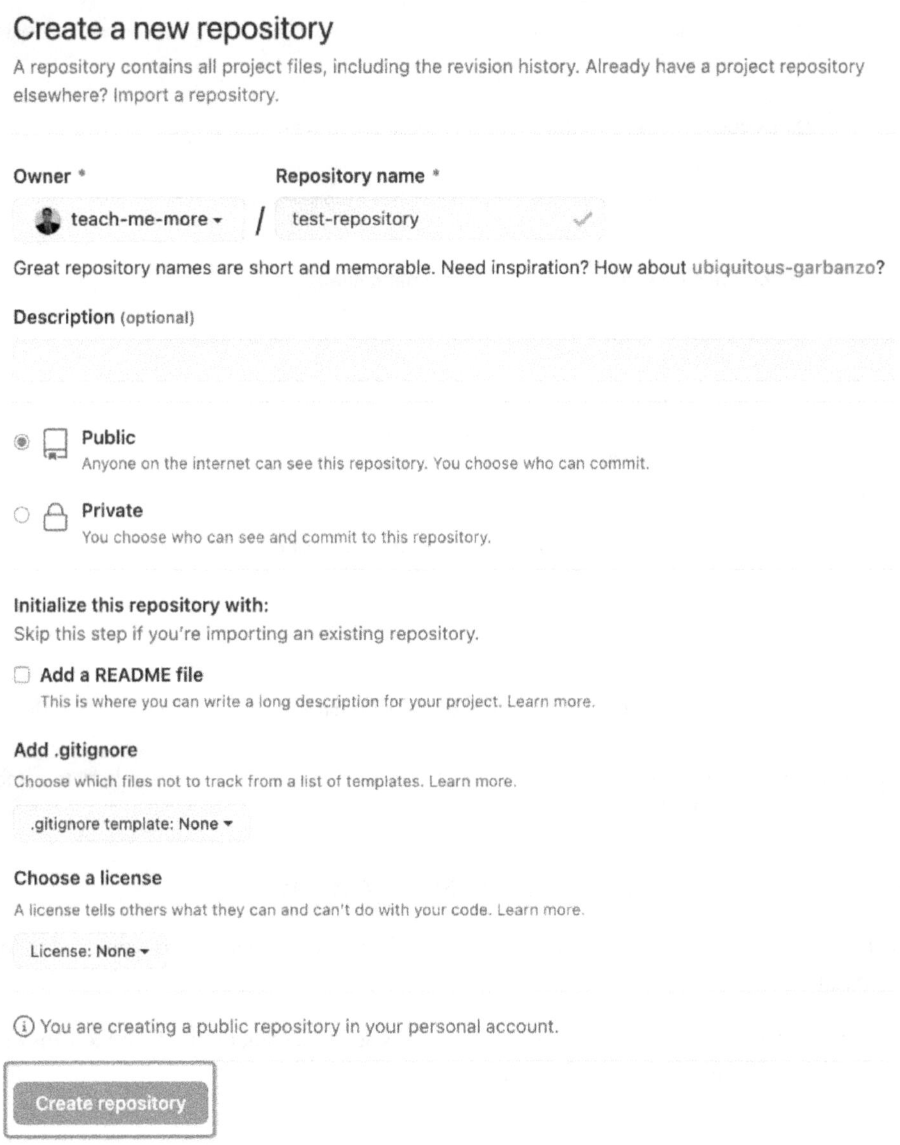

Create a new repository

A repository contains all project files, including the revision history. Already have a project repository elsewhere? Import a repository.

Owner *   **Repository name** *

👤 teach-me-more ▾   /   test-repository   ✓

Great repository names are short and memorable. Need inspiration? How about ubiquitous-garbanzo?

Description (optional)

◉ 🖥 **Public**
    Anyone on the internet can see this repository. You choose who can commit.

○ 🔒 **Private**
    You choose who can see and commit to this repository.

Initialize this repository with:
Skip this step if you're importing an existing repository.

☐ **Add a README file**
    This is where you can write a long description for your project. Learn more.

Add .gitignore
Choose which files not to track from a list of templates. Learn more.

.gitignore template: None ▾

Choose a license
A license tells others what they can and can't do with your code. Learn more.

License: None ▾

ⓘ You are creating a public repository in your personal account.

[Create repository]

Figure 14.2 – Creating a new repository on GitHub

3. Once the repository is created, you can push your code to it using the instructions provided on the screen:

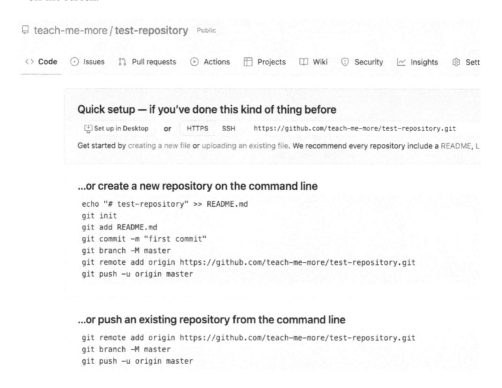

Figure 14.3 – A new GitHub repository

We now know how to create a GitHub repository. In the next section, you will learn how to establish a relationship between GitHub and AWS.

Connecting GitHub with AWS developer tools

To connect a GitHub repository to your AWS CodeBuild Service or CodePipeline, you need to follow the following instructions:

1. Go to the AWS console, and in the **Developer Tools** section, click on the **Connections** option and then the **Create connection** button:

Figure 14.4 – The AWS Developer Tools Connections page

2. On the next screen, you need to select the connection provider, such as **Bitbucket**, **GitHub**, or **GitHub Enterprise Server**. The process for all the providers is the same, but here, we will connect with GitHub, so select the **GitHub** option, provide a name for the connection, and then click on the **Connect to GitHub** button:

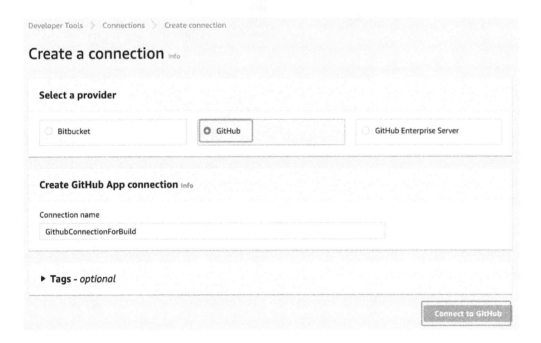

Figure 14.5 – The connection provider selection

3. On the next screen, GitHub will ask you to log in using your credentials, or if you are already logged in, it will ask you to authorize and confirm that you want to allow AWS access. Click on the **Authorize AWS Connector for GitHub** button:

Figure 14.6 – GitHub authorization confirmation

4. On the next screen, you have a connection listed, and you can link between AWS and GitHub. Click on the **Connect** button to complete the connection:

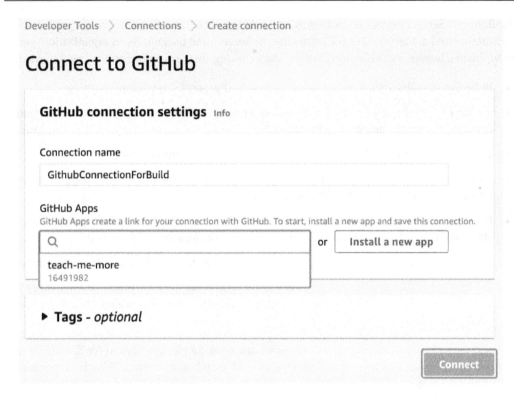

Figure 14.7 – The GitHub connection link

5. Now that you have a connection established between your GitHub and AWS developer tool, you can select this connection any time you configure a CodeBuild or CodePipeline project.

Later in this chapter, we will learn an alternative way to connect to a repository. For now, let's learn about Bitbucket, another popular version control system.

What is Bitbucket?

Bitbucket is a Git-based source code repository service owned by Atlassian. Atlassian provides a suite of tools for easy documentation, code management, and review processes. Jira, Confluence, and Bitbucket are some of the popular products from Atlassian. Bitbucket supports all Git-based operations and easy source code management. Bitbucket is a web-based application developed on top of the open source Git repository, so you can use the easy user interface to create and manage the source code. You can create unlimited public and private git repositories.

Bitbucket comes in two modes – cloud-hosted and privately hosted. Bitbucket Cloud is hosted by Atlassian as a SaaS offering, and as a consumer, you don't have to manage any infrastructure or worry about backup and resiliency.

In the Bitbucket Server offering, as a consumer, you host the Bitbucket software on your organization's infrastructure and are responsible for managing the servers and backup. As an organization, you have to obtain a license to use Bitbucket Server. The following are the benefits provided by Bitbucket:

- Bitbucket provides scaling architecture that you host privately on your infrastructure.

- Jira is a widely used issue-tracking system, and Bitbucket provides easy integration to track your commits against Jira items. You can track what changes were made for a particular Jira issue.

- You can easily set up code merge and code approver rules to enforce code quality.

- Bitbucket provides an easy interface for code reviewers, and you can see the changes side by side and provide necessary feedback to improve quality.

- Bitbucket supports CI/CD tools to design fully automated workflows, using Bitbucket Pipelines; alternatively, it connects with other CI/CD tools, such as Jenkins and Bamboo.

- Bitbucket provides built-in integration with security scanning tools, such as Snyk, for vulnerabilities and code issues.

- Bitbucket provides easy integration with other Atlassian products, such as Jira, Crucible, FishEye, and Bamboo.

In the next section, we will learn more about how we can expand a pipeline beyond AWS, and as well as that, we will integrate Bitbucket with CodePipeline and learn how to create a Bitbucket repository to push our source code.

Extending CodePipeline

CodePipeline integrates well with other AWS services such as CodeCommit, CodeBuild, and CodeDeploy to perform automated releases. In this section, we will learn how to extend the CodePipeline beyond AWS and perform deployment to non-AWS infrastructure. As shown in the following diagram, we will create an application, push the source code to the Bitbucket repository, and integrate Bitbucket into our CodePipeline project:

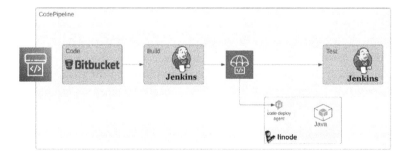

Figure 14.8 – CodePipeline extension beyond AWS

In this example, we use Jenkins to build the source code and integrate that with CodePipeline. In previous chapters, we learned about the CodeBuild service and understood its benefits, but at times, organizing it required the use of existing build tools such as Jenkins, so this chapter will help you understand how to plug in Jenkins in your CodePipeline.

> **Jenkins**
>
> Jenkins is an open source tool used to automate the build and deployment of software. Jenkins helps to automate development-related tasks, building, testing, and deployment and supports CI/CD. It also allows you to extend its functionality using plugins.

In previous chapters, you also learned about EC2 instances, ECS containers, and serverless deployments using the CodeDeploy service. In this section, we will deploy an application to an on-prem private server outside the AWS infrastructure. Once deployment is done, we will again utilize a separate Jenkins job to perform the validation of the deployment. CodePipeline will do all the orchestration work for us here and provide an end-to-end automated deployment.

As a first step, we need to create a new repository to push our code. We will use BitBucket to create the repository.

Creating a Bitbucket repository

In order to create a Bitbucket repository, you first need to sign up with Bitbucket Cloud by creating an account. Once you have a Bitbucket account created, perform the following steps to create a new Bitbucket repository:

1. Log in to your Bitbucket account, click on the **Create** button, and select **Repository** on your dashboard:

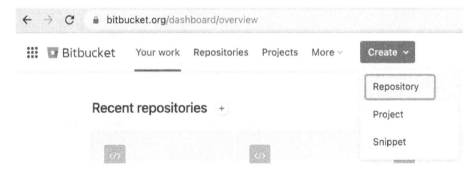

Figure 14.9 – The Bitbucket dashboard

2. When you create a repository, you need to select a project under which you want to create the repository. In our case, we don't have an existing project, so we can create the project on the fly while creating the repository. In the project name field, click on the **Create Project** option and provide `aws-code-pipeline` as a project name, as well as a repository name, and click on the **Create repository** button:

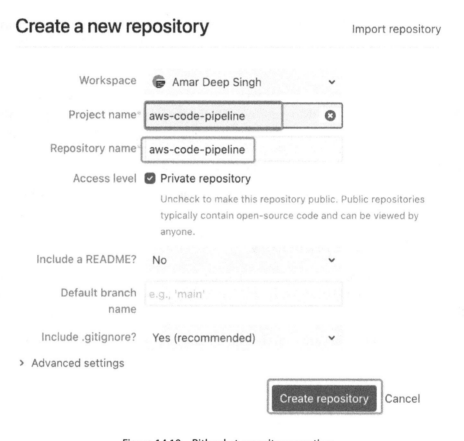

Figure 14.10 – Bitbucket repository creation

3. Once you click on the **Create repository** button, Bitbucket will create an empty repository for you. You can now click on the **Clone** button and copy the link to clone the empty repository:

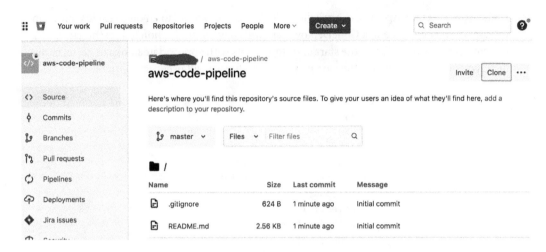

Figure 14.11 – Bitbucket repository confirmation

4. Run the following command to clone the empty repository:

    ```
    git clone https://bitbucket.org/{account_name}/aws-code-
    pipeline.git
    ```

5. The source code repository is ready, and as a next step, we need to change the code; you can download the source code from our GitHub repository at https://github.com/ PacktPublishing/Building-and-Delivering-Microservices-on-AWS/ tree/main/chapter_14/aws-code-pipeline and add it to the Bitbucket repository we just created. The application source code is the same as what we used in the last few chapters; we just need to make some minor changes to the start.sh, stop.sh, and appspec.yml files. Let's review those code changes:

```
∨ aws-code-pipeline          ⊛      9    import org.springframework.web.bind.annotation.GetMapping;
  ∨ scripts                   ⊛     10    import org.springframework.web.bind.annotation.RequestParam;
    $ build.sh            M          11    import org.springframework.web.bind.annotation.RestController;
    $ start.sh           M          12
    $ stop.sh            M          13    @RestController
    $ test.sh            U          14    public class WelcomeController {
  ∨ src                       ⊛     15     @Value("${app.name}")
    ∨ main                    ⊛     16        private String appName;
      > java/com/packt/aws/book... ⊛  17
      > resources                   18        @Value("${app.version}")
      > test                        19        private String appVersion;
    > target                        20        private static Logger LOGGGER=null;
    ◆ .gitignore                    21
    ! appspec.yml                   22        @GetMapping({ "/", "/info" })
    ℵ pom.xml              1         23        public String sayHello() throws UnknownHostException {
                                     24            LOGGGER =  LoggerFactory.getLogger(clazz: WelcomeController.class);
                                     25            InetAddress ip = InetAddress.getLocalHost();
                                     26            LOGGGER.info(appName+"-"+appVersion);
                                     27            return appName + "-" + appVersion+" \n "+ip;
```

Figure 14.12 – The aws-code-pipeline source code

6. CodeDeploy uses the `appspec.yml` file to deploy an application to an on-prem instance. On the application deployment, a CodeDeploy agent will copy the application to the `/home/root` directory, and then it will use the start and stop scripts with appropriate hooks to perform the deployment. The following is the `appspec.yaml` file we will use:

```
version: 0.0
os: linux
files:
  - source: /
    destination: /home/root/
file_exists_behavior: OVERWRITE
permissions:
  - object: /
    pattern: "**"
    owner: root
    group: root
    mode: 755
hooks:
  ApplicationStart:
    - location: start.sh
      timeout: 100
  ApplicationStop:
    - location: stop.sh
      timeout: 100
```

Figure 14.13 – The CodeDeploy appspec.yml configuration

7. The `start.sh` script starts the Spring Boot application on port `80`:

```
#!/usr/bin/env bash
cd /home/root
sudo /usr/bin/java -jar -Dserver.port=80 \
*.jar > /dev/null 2> /dev/null < /dev/null &
```

8. The `stop.sh` script is responsible to stop the running application and perform a cleanup of the previous version of the application:

```
#!/bin/bash
killall java
rm -f /home/root/aws-code-pipeline*.jar
rm -f /home/root/appspec.yml
rm -f /home/root/app.log
rm -f /home/root/*.sh
exit 0
```

9. Once we have reviewed the changes from the downloaded code, we need to push these changes to our Bitbucket repository. Copy all the files to the Bitbucket repository and run the `git add`, `commit`, and `push` commands to push the source code, as we learned in *Chapter 5, Creating Repositories with AWS CodeCommit*. Once you have pushed the changes, you can see them in the Bitbucket UI:

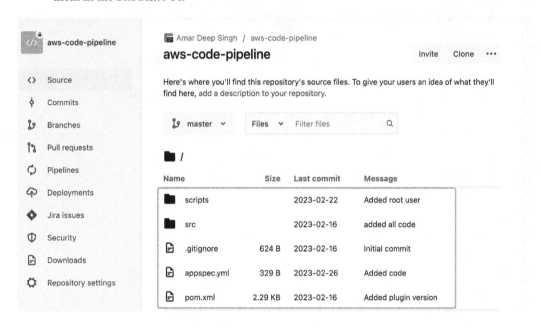

Figure 14.14 – The Bitbucket repository source code repository

Now, our source code and repository are ready to be built. As a next step, we need a build. In the next section, we will create a Jenkins server so that we can use it to build our source code.

Creating a Jenkins build server

You can get started with Jenkins either by running its Docker image or simply downloading the Java JAR file to run in an Apache Tomcat container. In this example, we will use an EC2 instance to create the Jenkins server, but you can use any on-prem or private server to do so. Let's start creating the Jenkins server by performing the following steps:

1. First, we need to create the EC2 instance and then download the Jenkins WAR file and start the Java process. We have packaged all of these in a Terraform template, which will create the EC2 instance, users, and roles for deployment. You can download the template from our GitHub repository: `https://github.com/PacktPublishing/Building-and-Delivering-Microservices-on-AWS/tree/main/chapter_14/terraform`. Execute the `terraform init` command to initialize Terraform.

2. Now, run the `terraform plan` command to get the information about all the resources this template will create in your AWS account. The following is a list of these resources:

    - `chap_14_ins_sg`: This is the security group used by EC2 instances to allow traffic on ports `80` and `8080` and allow remote connection to instance using `ssh`

    - `chap_14_instance_profile`: This is an instance profile to attach the IAM role to an instance type

    - `chap_14_ec2_launch_template`: This is a launch template to create EC2 instances with the same setting, apply user data to download the Jenkins JAR file from `https://get.jenkins.io/war-stable/2.375.3/jenkins.war`, and install Java 11 and Maven at the instance launch time

    - `chap_14_asg`: This is an autoscaling group for EC2 instances so that a minimum capacity of the instances can be made constantly available

    - `chap_14_pipeline_role`: This is the IAM role to be assumed by the Jenkins EC2 instance, to push code to S3 after the build

    - `chap-14-code-deploy-role`: This is the AWS IAM service role used by the CodeDeploy service to perform deployment; this role will be used later in this chapter while configuring the application

    - `chap_14_on_prem_user`: This is the newly created IAM user and will be used by the on-prem/private server outside the AWS infrastructure to register the instance and AWS CodeDeploy agent

    - `attach_code_deploy`: This is the CodeDeploy policy attachment to the `chap_14_on_prem_user` IAM user to allow the CodeDeploy agent access to the CodeDeploy service

    - `attach_s3_access`: This is the S3 policy attachment to the `chap_14_on_prem_user` IAM user to allow the CodeDeploy agent access to an S3 bucket

    - `chap-14-on-prem-code-deploy`: This is the CodeDeploy application of an EC2/on-premises compute platform type to deploy to a private server

    - `chap-14-on-prem-deploy-group`: This is the CodeDeploy deployment group to deploy the application to the on-prem Linode server, with matching `provider= linode_2` tags

3. Run the `terraform apply -auto-approve` command to apply the Terraform changes.

4. Now, let's log in to the AWS console and go to the EC2 section to look at the EC2 instance we have created using Terraform and deployed the Jenkins WAR file, along with Java and Maven. Copy the public IP address of the Jenkins instance and open that in the browser on port `8080`, as Jenkins started on port `8080`. Type `http://PUBLIC_IP_ADDRESS:8080/` in the address bar:

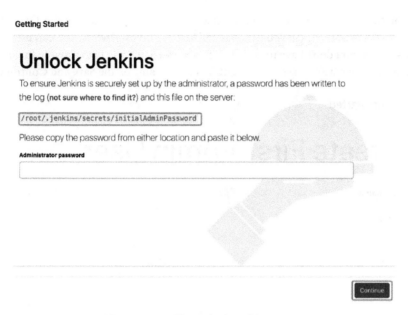

Figure 14.15 – The Unlock Jenkins screen

5.   As you can see, the Jenkins welcome screen is displayed and asks for the administrator password to unlock the server. You will see this screen only once, and the password to unlock the server can be found inside the `/root/.jenkins/secrets/initialAdminPassword` file on the EC2 instance. So, log into the EC2 instance and copy the password to log in to the Jenkins server, using the `sudo cat /root/.jenkins/secrets/initialAdminPassword` command. Once you enter the password and click the **Continue** button, Jenkins will ask for customizations and the installation of the required plugins. Go ahead and click on the first tile to install the suggested plugins:

Figure 14.16 – Jenkins plugin installation

6. Once you click on the suggested plugins, you will see them installed one by one. Once all plugins are installed, Jenkins will ask you to create the first admin user, which you can use to manage Jenkins so that you don't have to copy the password each time. Let's provide the details and create the first user, with a username of author, by clicking on the **Save and Continue** button:

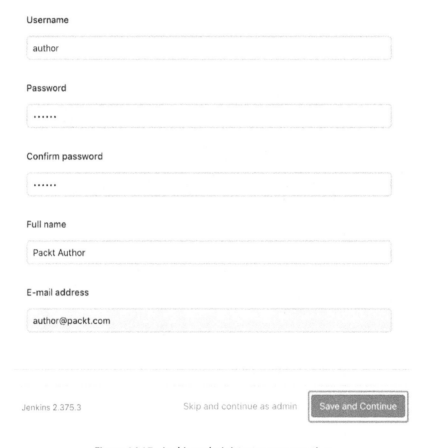

Figure 14.17– Jenkins administrator user creation

7. On the next screen, Jenkins asks to provide a URL to access Jenkins; you can customize it or leave it as is. If you have a load balancer in front of the EC2 instances, you can provide a load balancer DNS name here. Click on the **Save and Finish** button to proceed:

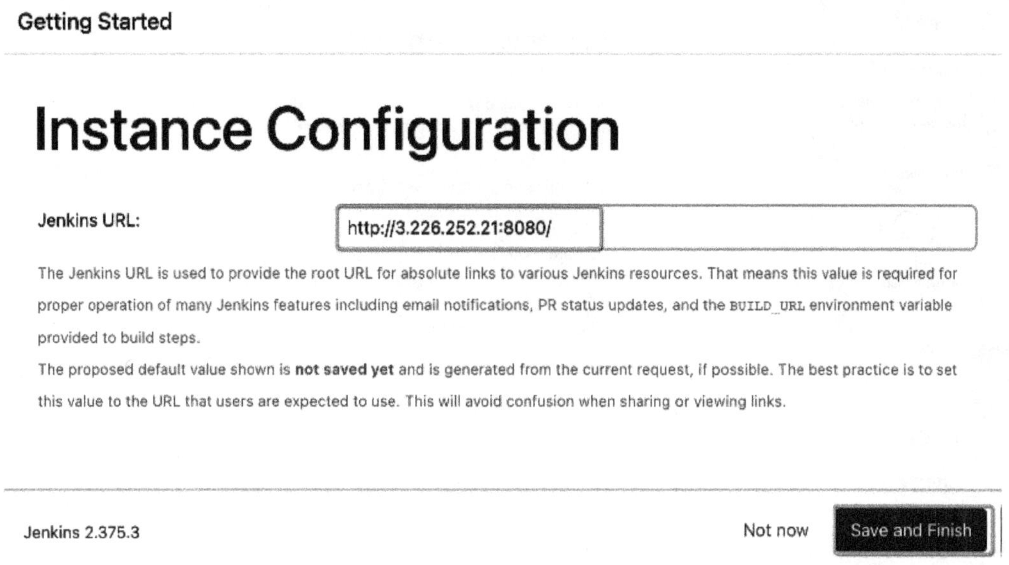

Getting Started

Instance Configuration

Jenkins URL: http://3.226.252.21:8080/

The Jenkins URL is used to provide the root URL for absolute links to various Jenkins resources. That means this value is required for proper operation of many Jenkins features including email notifications, PR status updates, and the BUILD_URL environment variable provided to build steps.

The proposed default value shown is **not saved yet** and is generated from the current request, if possible. The best practice is to set this value to the URL that users are expected to use. This will avoid confusion when sharing or viewing links.

Jenkins 2.375.3   Not now   Save and Finish

Figure 14.18 – Jenkins URL configuration

8. The next screen shows that Jenkins is ready to be used. Click on the **Start Using Jenkins** button to proceed. On the Jenkins home screen, you will see several options to create and manage a Jenkins job, users, and Jenkins itself. We still need to add a CodePipeline plugin to Jenkins, so let's click on the **Manage Jenkins** option on the left panel:

Jenkins
Search (⌘+K)
Dashboard
New Item
People
Build History
Manage Jenkins
My Views
Build Queue
No builds in the queue.
Build Executor Status
Welcome to Jenkins!
This page is where your Jenkins jobs will be displayed. To get started, you can set up distributed builds or start building a software project.
Start building your software project
Create a job
Set up a distributed build

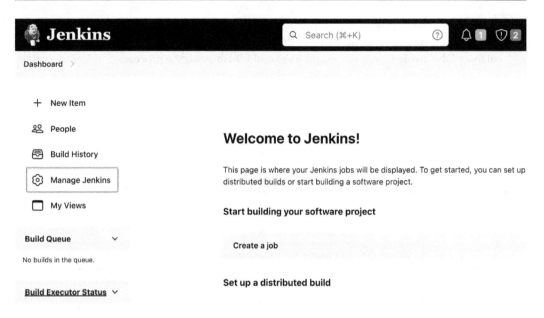

Figure 14.19 – The Jenkins home screen

9. Click on the **Manage Plugins** option, as shown in the following screenshot:

Manage Jenkins
Building on the built-in node can be a security issue. You should set up distributed builds. See the documentation.
Set up agent
Set up cloud
Dismiss
System Configuration
Configure System
Configure global settings and paths.
Global Tool Configuration
Configure tools, their locations and automatic installers.
Manage Plugins
Add, remove, disable or enable plugins that can extend the functionality of Jenkins.
Manage Nodes and Clouds
Add, remove, control and monitor the various nodes that Jenkins runs jobs on.

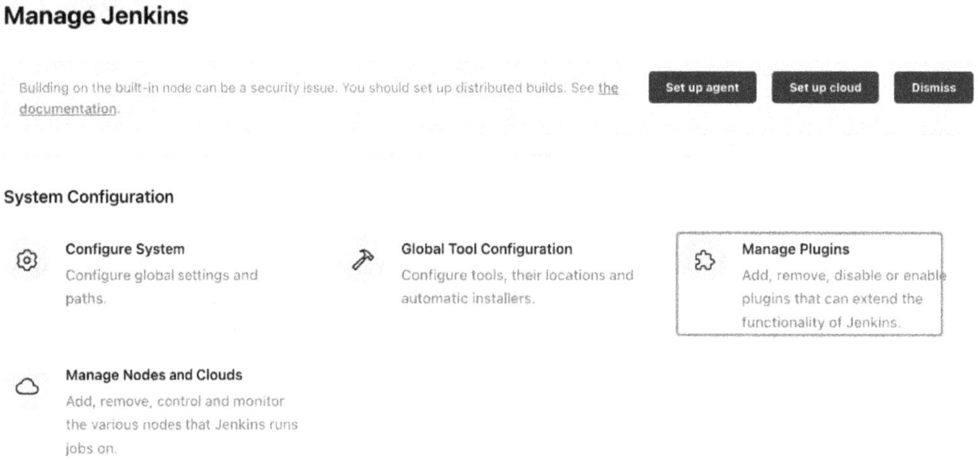

Figure 14.20 – The Manage Plugins option in Jenkins

10. Click on the **Available plugins** option on the left panel, search for `Aws CodePipeline`, and select the plugin, as shown in the following screenshot. Then, click on the **Install without restart** button to proceed with the installation:

Figure 14.21 – AWS CodePipeline plugin installation

11. The next screen will install the AWS CodePipeline plugin, show its progress, and confirm its successful installation.

At this point, we are done setting up our Jenkins server. Now, we are ready to create Jenkins jobs for the code build and tests.

Creating a Jenkins build job

To compile our source code and build the deployment artifact, we need to create a Jenkins job, which will be executed by the CodePipeline. Let's perform the following steps to create a build job:

1. Click on the **+ New Item** link on the Jenkins home page on the left panel:

Figure 14.22 – Creating a new job in Jenkins

2. Provide a name for the job, `aws-code-pipeline-build-job`, and select **Freestyle project** as the job type. Then, click on the **OK** button:

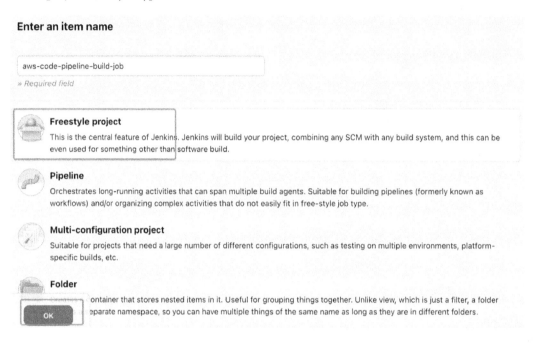

Figure 14.23 – A new build job creation in Jenkins

3. Provide a description of the project, and select the **Execute concurrent builds if necessary** checkbox:

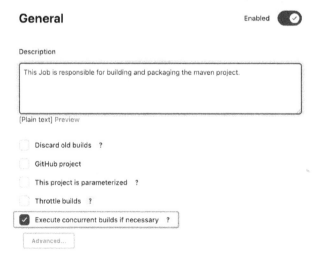

Figure 14.24 – Build job configuration in Jenkins

4. In the **Source Code Management** section, select the **AWS CodePipeline** radio button, as we want CodePipeline to start this Jenkins job. In the next section, for Jenkins to connect to the AWS CodePipeline and get the source code from an S3 bucket, you need to provide the AWS access key and secret key in the appropriate fields. We will leave these fields empty, as we have this Jenkins installed on the EC2 instance and attached the `chap_14_pipeline_role` IAM role to the instance, so AWS CodePipeline will assume the role, which has appropriate policies attached to it and provides the required access:

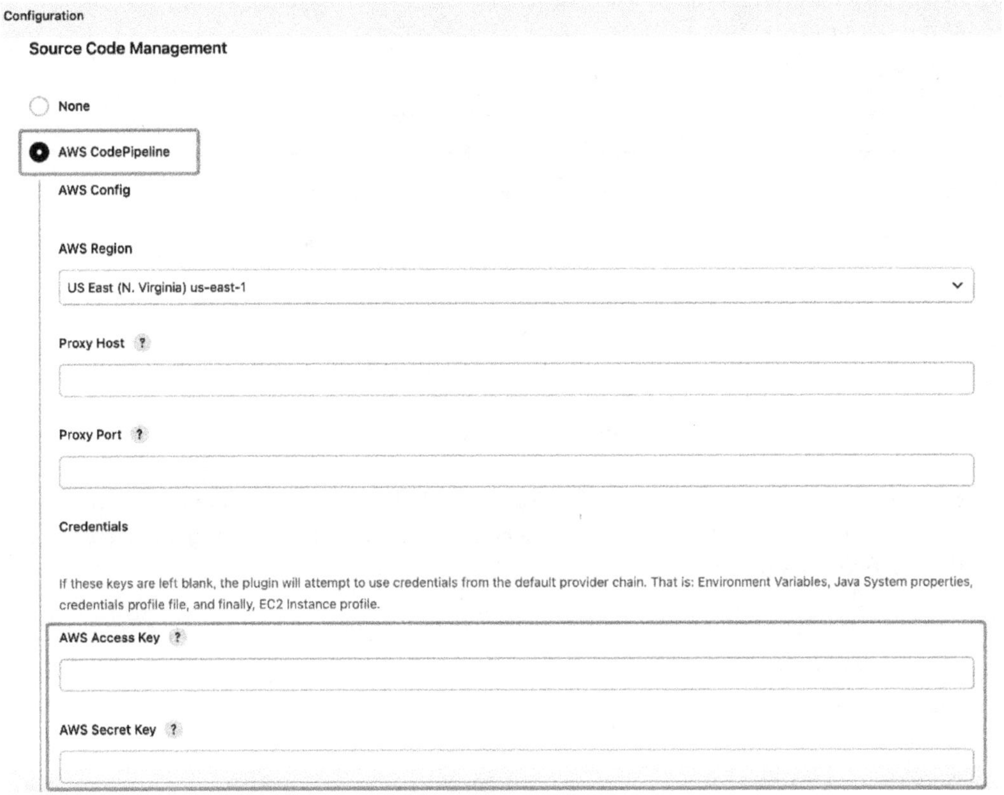

Figure 14.25 – Jenkins CodePipeline plugin configuration

5. When you scroll below, you will see that you need to provide the CodePipeline action type, which confirms whether you are using Jenkins to build the source code or for testing. We are configuring this job to build the source code, so let's select **Build** in the **Category** field and type `JenkinsBuildProvider` in the **Provider** input box. This text, `JenkinsBuildProvider`, needs to be matched with the **Provider name** field in the CodePipeline when we create a Jenkins provider, later in this chapter:

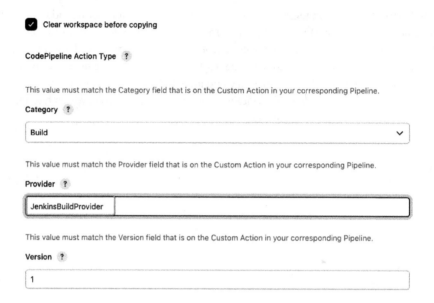

Figure 14.26 – Build job provider configuration in Jenkins

6.   The **Build Triggers** section specifies the ways to trigger this build job. You have multiple options to configure a trigger, but since there is no direct connection between the CodePipeline and the Jenkins server, here we need to force our CodePipeline plugin to pool to the AWS CodePipeline service each minute and see whether there is any build job request. Here, we set the poll frequency to each minute:

Figure 14.27 – Build trigger configuration in Jenkins

7. In the **Build Steps** section, we need to specify what we want to do when CodePipeline runs this Jenkins job. In this step, we will choose to execute a shell script. Once you choose that option, Jenkins will provide a text area to supply the commands to execute in this build step:

Build Steps

Figure 14.28 – Build step configuration in Jenkins

8. Enter the following script in the **Command text** area. In this script, we simply run the Maven clean install to compile and build our Java artifact, and then we copy that Java JAR file along with the script files and `appspec.yml` to the `artifact` directory:

```
mvn clean install
mkdir artifact
cp target/aws-code-pipeline*.jar artifact
cp scripts/* artifact/
cp appspec.yml artifact/appspec.yml
```

9. Post-build actions allow us to complete any post-build activity, such as cleaning up the directory, sending email notifications, or archiving artifacts. Here, we will use **AWS CodePipeline Publisher** as a post-build action so that we can return control to AWS CodePipeline and push the build artifacts to S3, allowing other pipeline steps to use the artifacts appropriately:

Aggregate downstream test results
Archive the artifacts
Build other projects
Publish JUnit test result report
Record fingerprints of files to track usage
Git Publisher
AWS CodePipeline Publisher
E-mail Notification
Editable Email Notification
Set GitHub commit status (universal)
Set build status on GitHub commit [deprecated]
Delete workspace when build is done

Add post-build action ▲

Save   Apply

Figure 14.29 – Post-build configuration in Jenkins

10. Once you have selected the post-build action type in the previous step, Jenkins will show a detailed section to provide the location in the Jenkins workspace to be pushed to S3, as shown in the following screenshot:

Post-build Actions

≡ **AWS CodePipeline Publisher**   ×

Output Locations

You can use the Location field to publish a file or folder in the workspace, or leave this field blank to publish everything in the workspace. Artifact name refers to the AWS CodePipeline action output artifact. You can leave this field blank or use it to match Jenkins locations and action output artifacts. If you leave the Artifact Name field blank for one artifact, you must leave all Artifact Name fields blank.

Location

/artifact

Artifact Name

|

Add more

Add

Add post-build action ▼

Save   Apply

Figure 14.30 – Post-build output configuration in Jenkins

11. If you leave the **Location** field empty, then AWS CodePipeline Plugin will push the entire workspace contents to an S3 bucket. In our case, we just want to publish the files we copied to the artifact directory. So, just click on the **Save** button, which will create our `aws-code-pipeline-build-job`.

We are done creating the Jenkins job that we will use to build the project. In the next step, we need to create a Jenkins job to test the project when deployment to our private server is complete. So, let's create the Jenkins test job.

Creating a Jenkins test job

In this section, we will create a new Jenkins job, similar to the previous one, but in this job, we will focus on testing the deployed service. In this job, we will be making an HTTP call to the deployed `aws-code-pipeline` service, and if we receive a `200` HTTP status code, then we will call this test successful. In a real-life application, you need to do a lot more in the test phase, but here, we will just validate a successful response. Let's go ahead and create a new Jenkins job:

1. Click on the **+ New Item** link on the Jenkins home page on the left panel, provide a name for the job, `aws-code-pipeline-test-job`, and select **Freestyle project** as the job type. Then, click on the **OK** button:

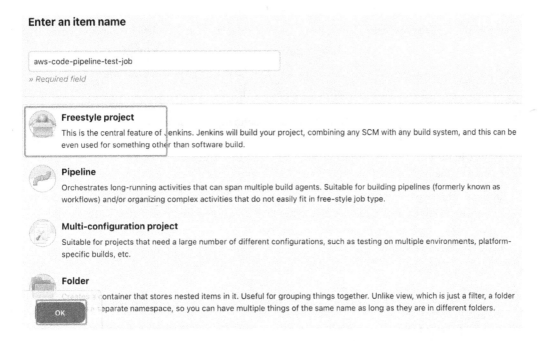

Enter an item name

aws-code-pipeline-test-job

» *Required field*

Freestyle project
This is the central feature of Jenkins. Jenkins will build your project, combining any SCM with any build system, and this can be even used for something other than software build.

Pipeline
Orchestrates long-running activities that can span multiple build agents. Suitable for building pipelines (formerly known as workflows) and/or organizing complex activities that do not easily fit in free-style job type.

Multi-configuration project
Suitable for projects that need a large number of different configurations, such as testing on multiple environments, platform-specific builds, etc.

Folder
Creates a container that stores nested items in it. Useful for grouping things together. Unlike view, which is just a filter, a folder creates a separate namespace, so you can have multiple things of the same name as long as they are in different folders.

OK

Figure 14.31 – Test job creation in Jenkins

2. Provide a description of the project, and select the **Execute concurrent builds if necessary** checkbox. In the **Source Code Management** section, select the **AWS CodePipeline** radio button, as we want CodePipeline to start this Jenkins job. When we execute the test step in our AWS CodePipeline, similar to the previous Jenkins job, this EC2 instance will assume the `chap_14_pipeline_role` role for any communication to AWS CodePipeline or S3. In the **CodePipeline Action Type** section, we need to select the category as **Test**, and in the provider field, enter `JenkinsTestProvider`. This text, `JenkinsTestProvider` needs to be matched with the **Provider name** field in the CodePipeline when we create a Jenkins provider for a test job, later in this chapter:

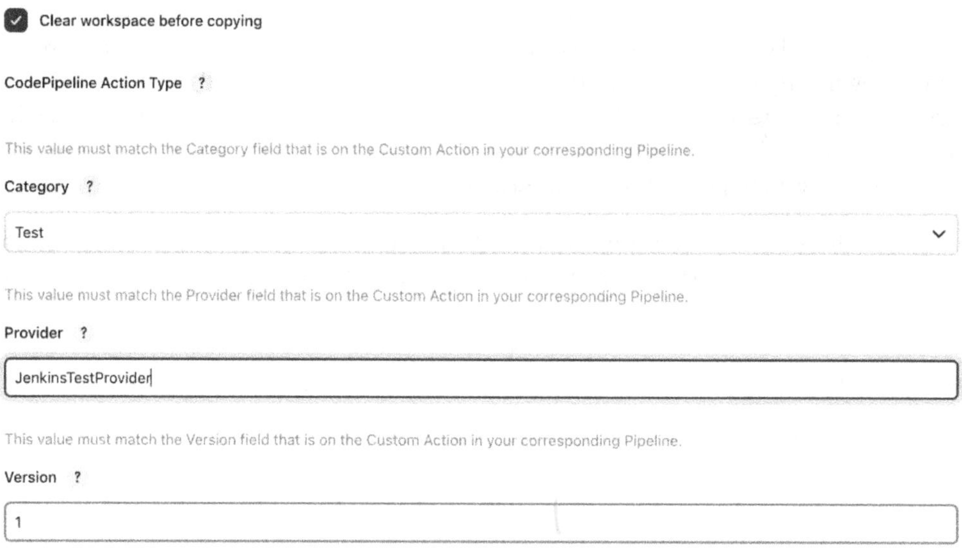

✅ Clear workspace before copying

CodePipeline Action Type   ?

This value must match the Category field that is on the Custom Action in your corresponding Pipeline.

Category   ?

```
Test                                                                            ⌄
```

This value must match the Provider field that is on the Custom Action in your corresponding Pipeline.

Provider   ?

```
JenkinsTestProvider
```

This value must match the Version field that is on the Custom Action in your corresponding Pipeline.

Version   ?

```
1
```

Figure 14.32 – Test job provider configuration in Jenkins

3. Similar to the build job, we need to configure the build trigger to force the CodePipeline plugin to pool to the AWS CodePipeline service each minute for any test job request.

4. In the **Build Steps** section, we need to execute the shell script that will perform the validation of the deployed service. Enter the following script in the command text area. In this script, we simply run a `curl` command to the private server IP address where our application is deployed. Click on the **Save** button, which will create our `aws-code-pipeline-test-job`:

```
#!/bin/bash
URL=http://45.33.82.63/
RESPONSE=$(curl --write-out %{http_code} -silent -output /dev/
null ${URL})
if [ $RESPONSE -ne 200]
then
    echo 'Application deployed failed!'
    exit 1
```

```
else
echo 'Application deployed succesfully !'
fi
```

We are done creating the test Jenkins job, which will be plugged into our pipeline at the test stage. Now, we need to create and set up a private server on which our pipeline will perform the deployment.

Creating the private server

In our pipeline, we will deploy our application to an on-prem instance, and in order to replicate it, we will create a **Virtual Private Server** (**VPS**). Linode is a VPS service provider company that provides private servers, so in this exercise, we will create a Linode server and use our CodePipeline to deploy the `aws-code-pipeline` microservice on this server. Let's follow these steps to create and set up the server:

1. Go to `https://cloud.linode.com/` and create an account in order to create a server. You can use a different VPS provider as well, but Linode provides a $100 free credit, so you can complete this exercise without a fee. Once you have logged in to your account, click on the **Create Linode** button, as shown in the following screenshot, or you can click on the **Create** button next to the left panel and choose **Linode**:

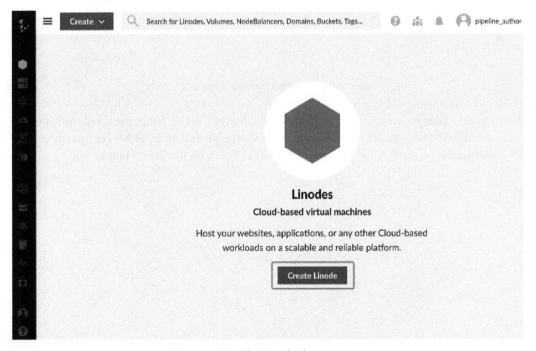

Figure 14.33 – The Linodes home screen

2. In order to create the server, you need to choose what kind of Linux image you are going to use; select a **Ubuntu 22.10 Linux** distribution and a region close to you. For our sample microservice, we don't need a dedicated CPU, so we can choose a shared CPU with 2 GB of memory, which should be sufficient:

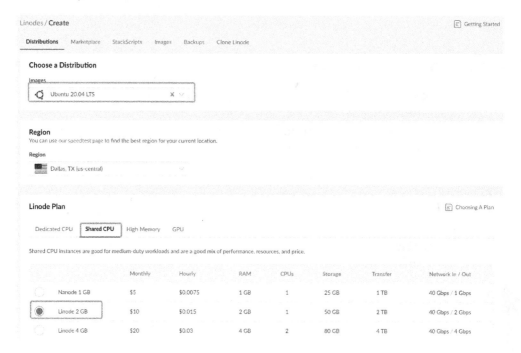

Figure 14.34 – Private server configuration

3. Provide a label for your server for easy identification and set up a root password, which you can use to log in to the server. We don't need to configure backup or VLAN configuration for this exercise, so scroll to the bottom of the page and click on the **Create Linode** button:

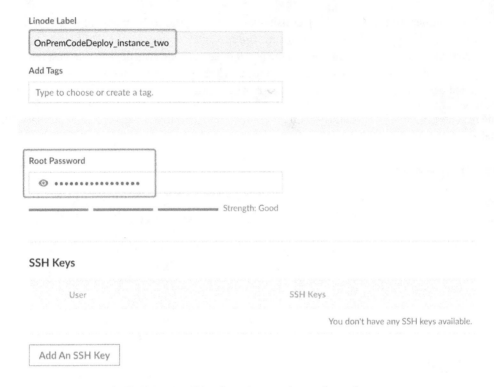

Figure 14.35 – Linode server security configuration

4. Once you click on the **Create Linode** button, it will provision a server for us and attach it to a public IP address, through which we can log in to this server to finish the setup:

Figure 14.36 – Linode server details

5. Now, connect to this Linode server using ssh by typing the `ssh root@45.33.82.63` command, and provide the password for the server. In your case, the IP address will be a different value, which you can find in the previous screenshot. As a deployment best practice, you should use a different user other than root, so let's create a new user on this Linode server called `ec2-user` by typing the `adduser ec2-user` command.

In the next step, we need to install AWS on this Linode server so that it can connect with the AWS CodeDeploy service.

Installing AWS CLI

In order for the on-prem server to run AWS commands and connect to AWS services, we need to configure the AWS CLI. Perform the following steps to install and configure the AWS CLI:

1. First of all, we need to install the latest updates to the Ubuntu server by executing the following command:

    ```
    sudo apt-get update
    ```

2. AWS CLI gets downloaded as a ZIP file, so in order to unzip that file, we need to install the `unzip` utility. Execute the following command to install the unzip utility to our server:

    ```
    sudo apt install unzip
    ```

3. Now, we need to download the AWS CLI from Amazon for Linux. Execute the following command:

    ```
    curl "https://awscli.amazonaws.com/awscli-exe-linux-x86_64.zip"
    -o "awscliv2.zip"
    ```

4. The preceding command will download the AWS CLI to the present directory. Now, we need to unzip the file by executing the following command:

    ```
    unzip awscliv2.zip
    ```

5. For the final step, we need to install the AWS CLI on this server by executing the following command:

    ```
    sudo ./aws/install
    ```

    Once the command is complete, you can check the version of the AWS CLI:

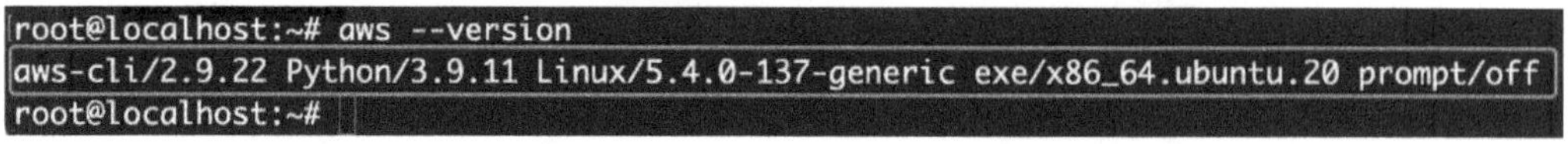

```
root@localhost:~# aws --version
aws-cli/2.9.22 Python/3.9.11 Linux/5.4.0-137-generic exe/x86_64.ubuntu.20 prompt/off
root@localhost:~#
```

Figure 14.37 – AWS CLI installation confirmation

Now, AWS CLI is installed successfully. Next, we need to register this Linode server with AWS as an on-prem instance so that the CodeDeploy service can perform the deployment and recognize this server.

Registering an on-prem server

In order to register the on-prem instance, we first need to configure the AWS credentials for the AWS CLI, which will be used to run the command to register the private instance within the CodeDeploy service. The following are the steps to do this:

1. Execute the aws configure command. You will then be asked for the access ID and the secret key. You can obtain these keys for the chap_14_on_prem_user user we created using Terraform earlier in this chapter. You can find instructions in the *Appendix* of this book on how to create an access key ID and secret access key:

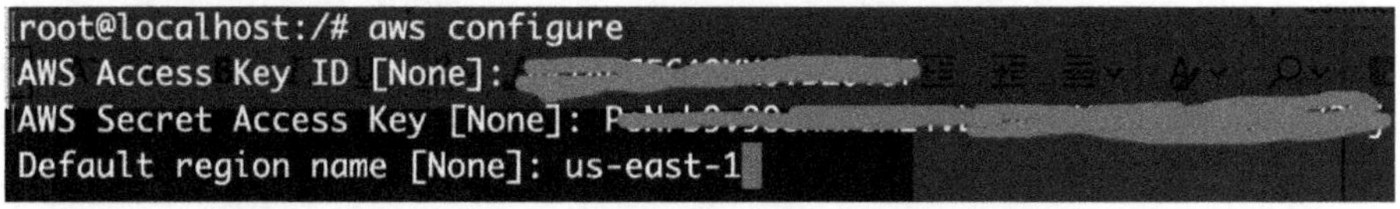

```
root@localhost:/# aws configure
AWS Access Key ID [None]:
AWS Secret Access Key [None]: P
Default region name [None]: us-east-1
```

Figure 14.38 – AWS CLI credential configuration

2. Once you have saved the credentials, you can execute the following command to register the Linode private server as an on-prem instance in AWS CodeDeploy. Here, you need to replace the ARN of chap-14-on-prem-user with what you have for your account:

```
aws deploy register-on-premises-instance --instance-name linode_
server_2 --iam-user-arn arn:aws:iam::xxxxxxxxxxxx:user/chap-14-
on-prem-user
```

3. The preceding command will register the Linode Server as the on-prem instance in the CodeDeploy service. You can validate that by navigating to the **CodeDeploy | On-premises instances** option in the AWS console, where you can see that a server with the name we provided in the preceding command linode_server_2 is listed. Following the same steps, you can create and register multiple on-prem instances with CodeDeploy. Note that you need a separate user for each on-prem instance; you can't share the same IAM user across multiple on-prem instances:

Figure 14.39 – On-premises instances

4.   Now, we have our instance configured, so let's go ahead and click on the server's name, as shown in the preceding screenshot, to provide a tag to this on-premises instance. These tags will later be used by the CodeDeploy service to target the deployment to specific servers, based on the tag value. Here, we set a new tag, named `provider`, with a `linode_2` value. Click on the **Submit** button to save it:

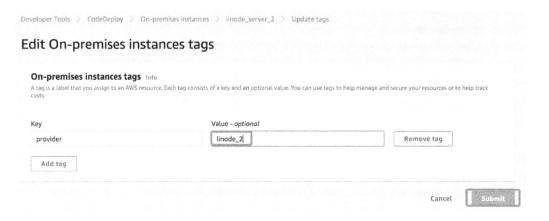

Figure 14.40 - On-prem instance tag configuration

Now, we have our on-prem instance registered with the CodeDeploy service, which will push the package to the instance for deployment. In previous chapters, we learned that in order for the CodeDeploy service to work on any instance, it needs the CodeDeploy agent to be available and running on the instance, which we haven't done yet, so let's go ahead and install the CodeDeploy agent on our Linode server.

Installing and configuring the CodeDeploy agent

To install the CodeDeploy agent on the Linode server, we need to log in to it and perform the following steps:

1. First of all, we need to install Java so that when we push our application to the server, it can execute the Java commands. Execute the following command:

    ```
    sudo apt install openjdk-11-jdk
    ```

2. Next, install Ruby support for the CodeDeploy agent to work by executing the following command:

    ```
    sudo apt install ruby-full
    ```

3. Now, we need to download the CodeDeploy agent from the S3 URL: https://aws-codedeploy-us-east-1.s3.amazonaws.com/latest/install. This URL can be different for different regions. Execute the following command:

    ```
    wget https://aws-codedeploy-us-east-1.s3.amazonaws.com/latest/install
    ```

4. Now, we need to provide appropriate permissions to the download binary so that it can be executed to install the CodeDeploy agent:

    ```
    chmod +x ./install
    ```

5. Install the CodeDeploy agent by executing the following command:

    ```
    sudo ./install auto
    ```

6. In order to start the CodeDeploy agent on the instance, run the following command:

    ```
    sudo service codedeploy-agent start
    ```

    To check the status of a running agent, you can use the following command:

    ```
    sudo service codedeploy-agent status
    ```

A response similar to the following confirms that the CodeDeploy agent is installed and running:

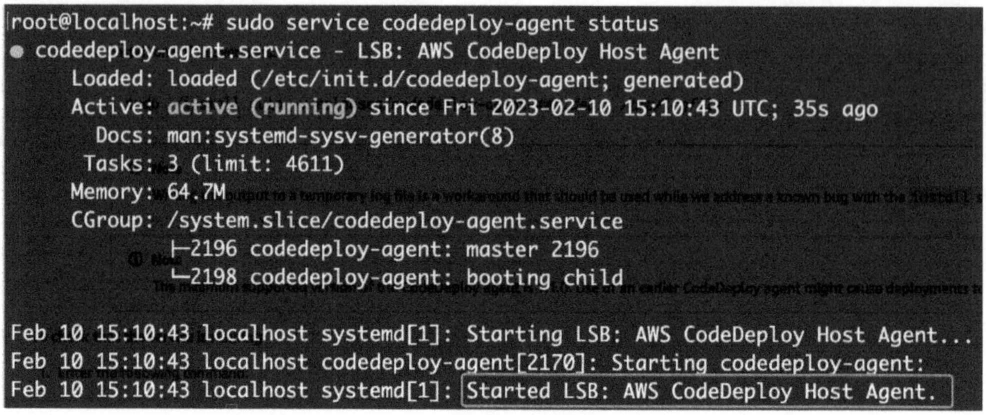

```
root@localhost:~# sudo service codedeploy-agent status
● codedeploy-agent.service - LSB: AWS CodeDeploy Host Agent
   Loaded: loaded (/etc/init.d/codedeploy-agent; generated)
   Active: active (running) since Fri 2023-02-10 15:10:43 UTC; 35s ago
     Docs: man:systemd-sysv-generator(8)
    Tasks: 3 (limit: 4611)
   Memory: 64.7M
   CGroup: /system.slice/codedeploy-agent.service
           ├─2196 codedeploy-agent: master 2196
           └─2198 codedeploy-agent: booting child

Feb 10 15:10:43 localhost systemd[1]: Starting LSB: AWS CodeDeploy Host Agent...
Feb 10 15:10:43 localhost codedeploy-agent[2170]: Starting codedeploy-agent:
Feb 10 15:10:43 localhost systemd[1]: Started LSB: AWS CodeDeploy Host Agent.
```

Figure 14.41 – The CodeDeploy agent startup

7. For the CodeDeploy agent to connect to the AWS services and provide life cycle events, it is necessary to configure credentials in a file called `codedeploy.onpremises.yml`. The CodeDeploy agent looks for this file in a specified location – `/etc/codedeploy-agent/conf/codedeploy.onpremises.yml`.

    In order to create the file, enter the following command:

    ```
    sudo nano /etc/codedeploy-agent/conf/codedeploy.onpremises.yml
    ```

    Enter the following details in the file and save it. Here, you need to replace the information related to the user ARN for the `chap-14-on-prem-user` user and your `ACCESS_KEY` and `SECRET_KEY` from the previous steps:

    ```
    aws_access_key_id: ACCESS_KEY
    aws_secret_access_key: SECRET_KEY
    iam_user_arn: arn:aws:iam::xxxxxxxxxxxx:user/chap-14-on-prem-
    user
    region: us-east-1
    ```

Now, we have our CodeDeploy agent configured and ready to perform deployment to the Linode server. In the next step, we need to create an AWS CodePipeline to put all these steps together.

Creating a CodePipeline

In the previous sections, we created a Bitbucket repository, a Jenkins job to build the code, and a Jenkins job to perform deployment validation. We also set up a private server to represent an on-prem instance for deployment. Now, we need to set up a CodePipeline project that will glue all this together and help us to automate deployment end to end. Let's log in to the AWS console and start setting up the CodePipeline:

1. After logging in to the AWS console, search for CodePipeline, and then click on the **Create pipeline** button to get started on a new CodePipeline. Provide a name for this pipeline, such as chap-14-on-prem-pipeline, and then click on the **Next** button.

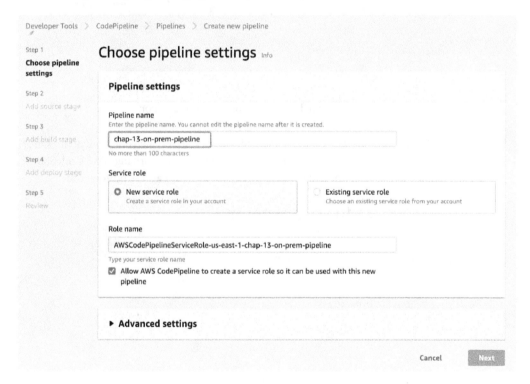

Figure 14.42 – CodePipeline configuration

2. Since we created our source code repository within Bitbucket, we need to select Bitbucket as the source provider. This is the alternative way to configure the source code provider connections. Once we have selected Bitbucket as the source provider, we need to establish a connection between CodePipeline and Bitbucket. To do so, click on the **Connect to Bitbucket** button next to the **Connection** field:

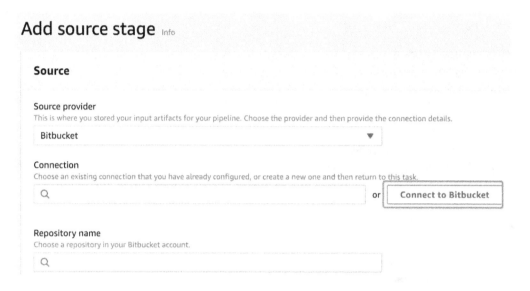

Figure 14.43 – CodePipeline source provider configuration

3. This will open up the Bitbucket OAuth login screen, where you provide your credentials. Once you are logged in, Bitbucket will ask you to confirm that you are authorizing AWS CodeStar to read data related to your account and project. Click on the **Grant access** button:

Confirm access to your account

AWS CodeStar is requesting access to the following:

Read your account information

Read your team membership information

This 3rd party vendor has not provided a privacy policy or terms of use. Atlassian's Privacy Policy is not applicable to the use of this App.

 Cancel

Figure 14.44 – Bitbucket OAuth confirmation

4. Once you grant access to AWS CodeStar, the AWS console connection screen will open up and show the connected Bitbucket apps. Click on the **Connect** button to finish the wizard:

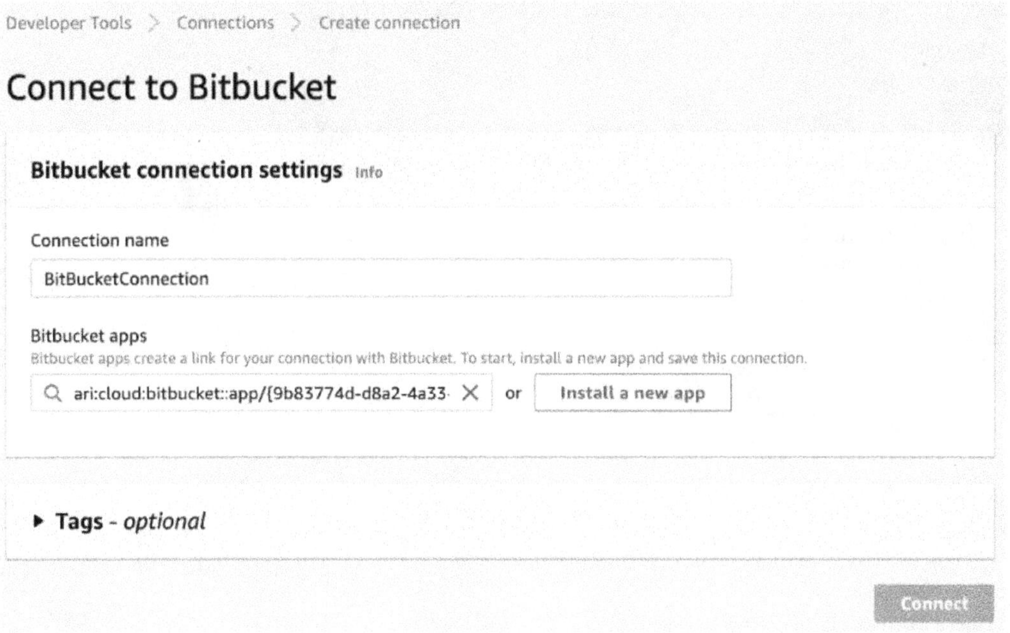

Developer Tools > Connections > Create connection

Connect to Bitbucket

Bitbucket connection settings Info

Connection name

BitBucketConnection

Bitbucket apps
Bitbucket apps create a link for your connection with Bitbucket. To start, install a new app and save this connection.

🔍 ari:cloud:bitbucket::app/{9b83774d-d8a2-4a33 ✕   or   Install a new app

▶ **Tags** - *optional*

Connect

Figure 14.45 – The Bitbucket connection link

5. Now, connection to Bitbucket is established, and you can select your repository and the branch name from which you want to build the source. Click on the **Next** button to finish the **Source** stage setup:

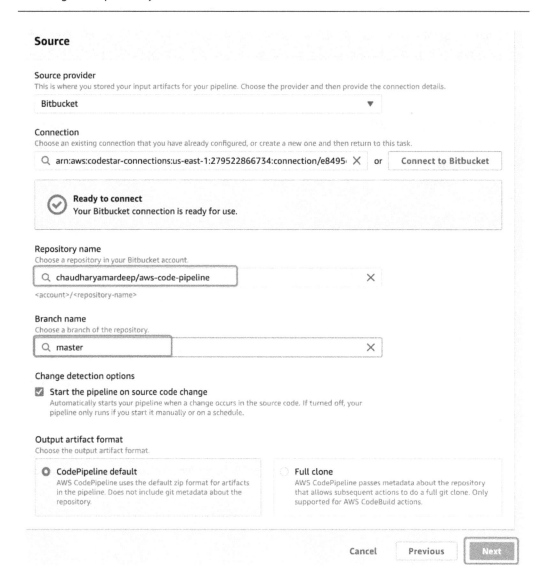

Figure 14.46 – CodePipeline source code provider configuration

6.   In the build stage, we need to select the build provider from the list. Here, we will connect to Jenkins and use it as a build provider instead of AWS CodeBuild, so select **Add Jenkins**, and then provide the details of the Jenkins server and job that we created for the build earlier in this chapter. The provider name is the same as the one we defined during the build job creation. Once you match all the details from the build job, click on the **Next** button.

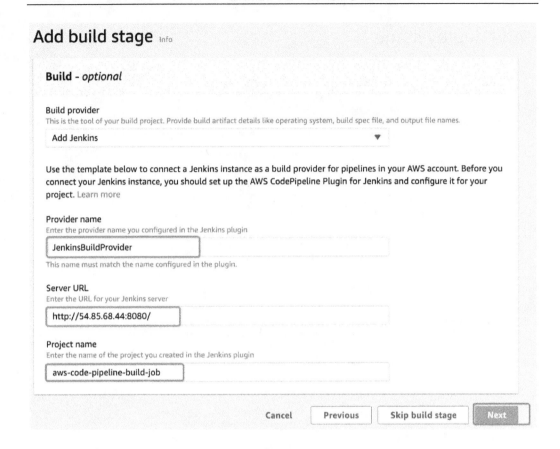

Figure 14.47 – CodePipeline Jenkins build stage configuration

7. Once you click on the **Next** button, it will ask you to provide deploy stage details. We will use AWS CodeDeploy as a deployment provider and select `chap-14-on-prem-code-deploy` as the CodeDeploy application and `chap-14-on-prem-deploy-group` as a deployment group. Then, click on the **Next** button.

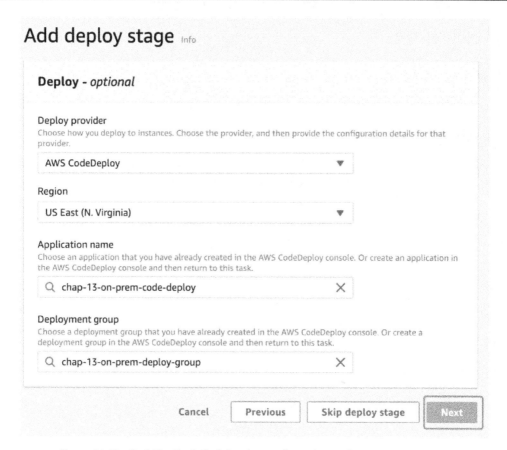

Figure 14.48 – CodePipeline's CodeDeploy configuration to the on-prem instance

8.   On the next screen, CodePipeline will ask you to review and confirm the changes. Click on the **Create Pipeline** button near the bottom of the screen. This will create the CodePipeline. We need to add a **Test** stage to this pipeline, as we want to run the Jenkins test job to perform validation when the deployment is completed. So, let's click on the **Edit** button to modify the pipeline:

Figure 14.49 – The CodePipeline edit screen

9. Click on the **Add stage** button after the deploy stage, provide a name for the test stage, and then click on the **Add stage** button. This will add the empty stage to the pipeline:

Figure 14.50 – The CodePipeline Add stage screen

10. Once the stage is added, we need to add an action to it, so click on the **Add action group** button. This will open the test stage configuration screen, similar to the following. Select the provider as Jenkins, and we will use `aws-code-pipeline-test-job` to perform a test of the deployed service. You need to ensure that the provider name matches the provider you specified during the Jenkins job definition. We will use `JenkinsTestProvider`, as we configured in previous sections. Now, click on the **Done** button to save the changes:

Action name
Choose a name for your action

Test

No more than 100 characters

Action provider

Add Jenkins   ▼

Input artifacts
Choose an input artifact for this action. Learn more ☑

BuildArtifact   ▼

Add

No more than 100 characters

Use the template below to connect a Jenkins instance as a build provider for pipelines in your AWS account. Before you connect your Jenkins instance, you should set up the AWS CodePipeline Plugin for Jenkins and configure it for your project. Learn more

Provider name
Enter the provider name you configured in the Jenkins plugin

JenkinsTestProvider

This name must match the name configured in the plugin.

Server URL
Enter the URL for your Jenkins server

http://174.129.86.205:8080/

Project name
Enter the name of the project you created in the Jenkins plugin

aws-code-pipeline-test-job

Variable namespace - *optional*
Choose a namespace for the output variables from this action. You must choose a namespace if you want to use the variables this action produces in your configuration. Learn more ☑

Output artifacts

Figure 14.51– CodePipeline's Jenkins test stage configuration

11. Now, we have added our pipeline stages, so click the **Save** button, start the pipeline execution by clicking on the **Release** button, and watch the progress of the pipeline. Once the pipeline is finished, you can see confirmation that all steps executed successfully.

12. If you go back to the Jenkins server, you can see that both of our Jenkins jobs were also executed as part of the CodePipeline stages:

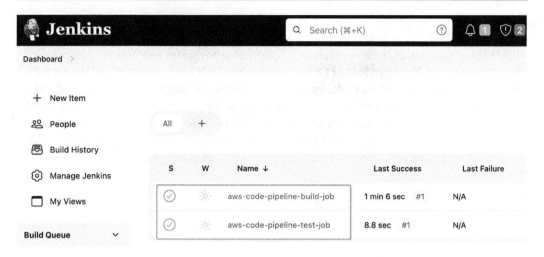

Figure 14.52 – A job execution in Jenkins

13. Now, let's go to our on-prem server's IP address and validate that the application was deployed successfully:

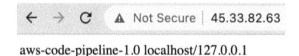

aws-code-pipeline-1.0 localhost/127.0.0.1

Figure 14.53 – The aws-code-pipeline service response from the on-prem instance

Now, we have completed the deployment of the application to an on-prem server, and any time a change is made to the source code and pushed to our Bitbucket repository, CodePipeline will automatically deploy this change to the Linode server. If you have any trouble in the deployment, you can log in to the on-prem server, check the CodeDeploy agent logs at `/opt/codedeploy-agent/deployment-root/deployment-logs`, and appropriately take action, depending on the errors you get.

Although we used the Bitbucket Source code repository here, the process to connect with GitHub is the same. You can simply select GitHub as a source provider and select the repository in your GitHub account, and the rest of the steps will work exactly the same.

Summary

In this chapter, we learned about GitHub and Bitbucket repositories and how we can connect CodePipeline to them. We expanded our CodePipeline beyond AWS infrastructure and connected a pipeline to a Jenkins server, used the CodeDeploy server to deploy our sample application to an offering, and learned details about the Lambda functions, as well as its limitations and benefits. We deployed our Lambda function manually and then created a pipeline to deploy the changes to a function. In the next chapter, we will learn more about CodePipeline and take our knowledge beyond the AWS infrastructure, integrating with other tools and deploying code to on-prem systems using CodePipeline.

Appendix

In this appendix, we will cover how to set up a user and create credentials for the user. We will also cover how to set up the AWS **Command Line Interface (CLI)**, Docker, and Git, which are all needed for the development.

In this chapter, we will cover the following setups and installations:

- Creating an **Identity and Access Management (IAM)** Console user
- Creating a user for Terraform authentication
- AWS CLI installation
- Creating a **Simple Notification Service (SNS)** topic
- Git installation
- Docker Desktop installation

To follow the instructions in this book, you must have a user account that has appropriate access to perform the tasks required. We will go through the process to create the required user account.

Creating an IAM Console user

Perform the following steps to create an AWS IAM user using your root account:

1. Log in to your AWS Console using your root account, search for `IAM Service` in the search bar, and click on **Identity and Access Management (IAM)**. We need to create a new IAM user, so click on the **Users** link in the left panel and click on the **Add users** button:

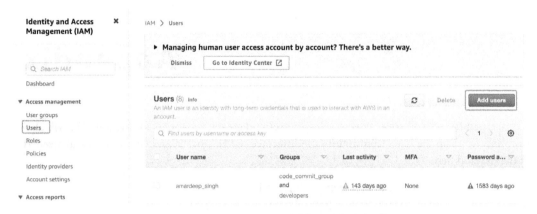

Figure 15.1 – The IAM Users screen

2. Provide a name for your new user. We are using `console_user` for this user. Check the option **Provide user access to the AWS Management Console - optional**. Under the **Are you providing console access to a person?** question, select the **I want to create and IAM user** radio button. Click **Next** to provide appropriate access to this new user:

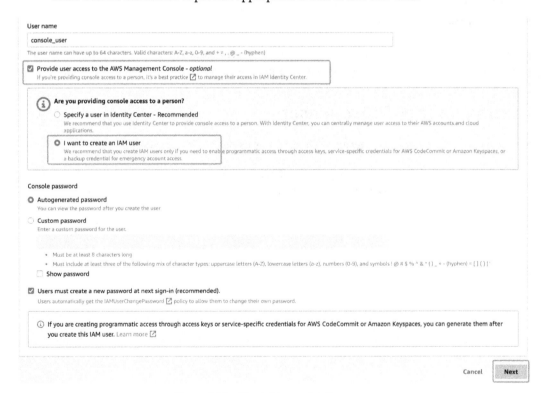

Figure 15.2 – The Add user details screen

3. On the next screen, we need to choose what kind of access this new user will have in the AWS environment. We can do that by creating a user group. Click the **Create group** button. This will open up a pop-up window:

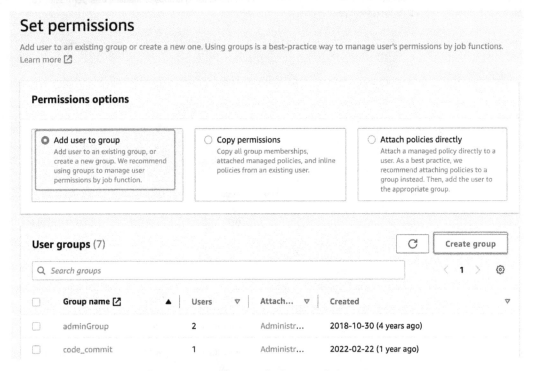

Figure 15.3 – Add user – the Set permissions screen

4. In the pop-up window, provide a name in the **Group name** field for this user group. We are using `console_user_group` here. In the **Policy name** column, select **AdministratorAccess**. These permissions should be restricted in a production environment, and we should provide only the access needed by a user, but for simplicity, we are choosing **AdministratorAccess** here:

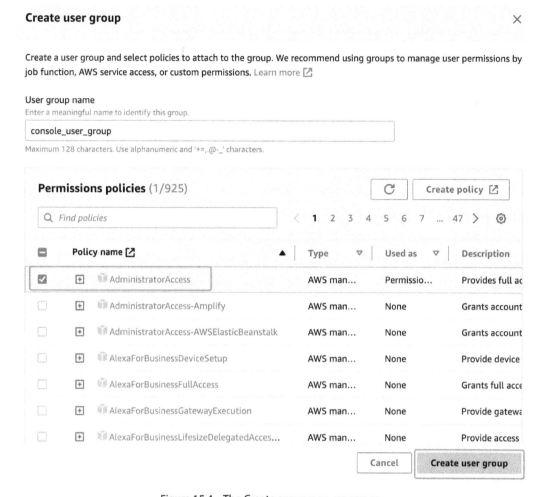

Figure 15.4 – The Create group pop-up screen

5. Click on the **Create user group** button. The pop-up window will close, and you will see a screen similar to the following screenshot, select the newly created user group and click on the **Next** button:

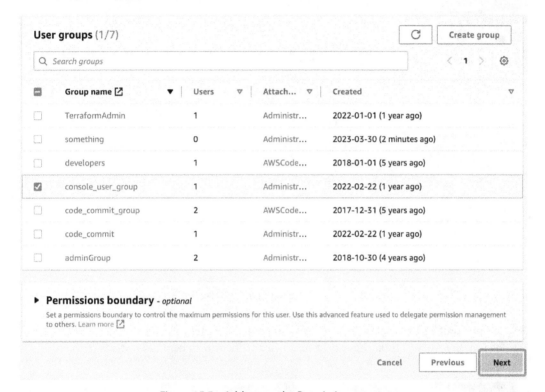

Figure 15.5 – Add user – the Permissions screen

6. In the following screen, we can review the information and provide any appropriate tag related to user. Click on the **Create user** button to create this new user:

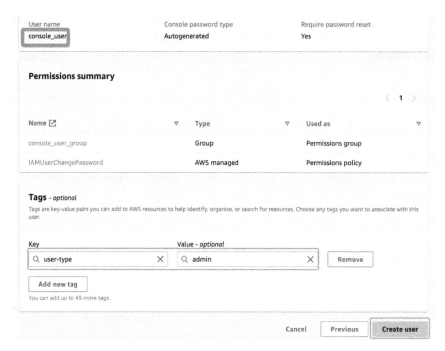

Figure 15.6 – Add user – the Add tags screen

7. At this point, our new `console_user` user is created, and you can use the credentials to log into the AWS Console. Your password will only be shown here once, so make a note of it to refer to later:

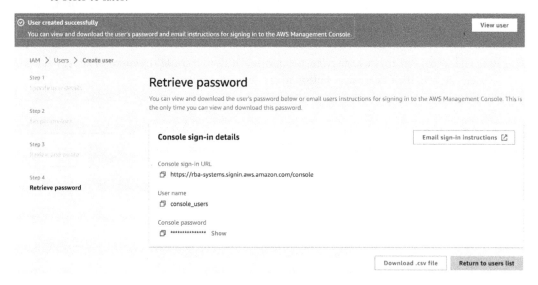

Figure 15.7 – Add user – the Confirmation screen

At this point, we are done creating our administrative console user, and we will use this account for any configuration we need to do.

Creating a user for Terraform authentication

For Terraform to provision resources in your cloud environment, you must provide access to Terraform with appropriate permissions. Terraform needs to authenticate with our AWS platform to create the resources that we will declare in our template. We specify that authentication by providing an AWS access key and secret key. If you have these keys already, then you can skip this section. In this section, we are going to create a new pair of AWS access key and secret key with administrator privileges so that Terraform doesn't run into any permission issues:

1. Log in to your AWS Console, search for `IAM Service` in the search bar, and click on **IAM**. We need to create a new IAM user to gain an access key and secret key. So, click on the **Users** link in the left panel and click on the **Add users** button. Provide a name for your new user. We can use `command_user` or `author` as a name for this user. We have not selected the **Provide user access to the AWS Management Console** checkbox here as we wouldn't be using this user to login to the AWS console, we will be creating this user to programmatically connect to AWS services. These types of credentials are designed for making API calls using AWS CLI or third-party tools like Terraform. Click **Next** to provide appropriate access to this new user:

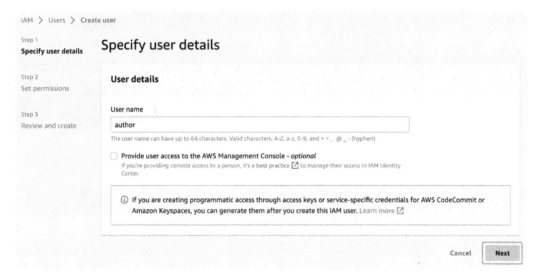

Figure 15.8 – The Add user details screen

2. On the **Permissions** screen, click on the **Create group** button; this will open up a pop-up window. In the pop-up window, provide a **Group name** value for this user group. We are using TerraformAdmin here. In the **Policy name** column, select **AdministratorAccess**, these permissions should be restricted in a production environment, and we should provide only the access needed by a user, but for simplicity, here we will choose **AdministratorAccess**:

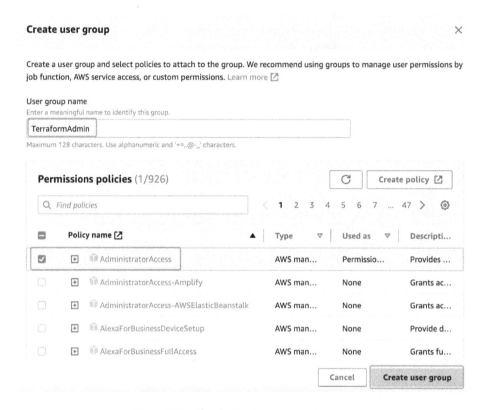

Figure 15.9 – The Create group pop-up screen

3. Click on the **Create user group** button. The pop-up window will close, and you will see a screen similar to the following screenshot. Now, select the newly created user and click on the **Next** button:

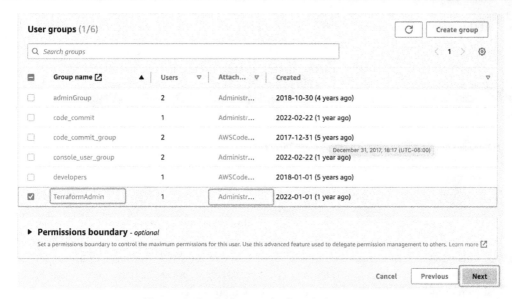

Figure 15.10 – Add user – the Permissions screen

4.   Validate the user information and you can specify any tags related to the user on the **Review and Create** screen. Click on **Create user** button to create the new user:

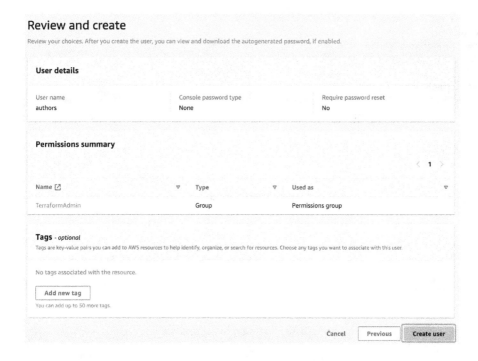

Figure 15.11 – Add user – the confirmation screen

5. At this point, a new user is created, and you can see a confirmation. Click on the **View user** button to see the details of the user:

User created successfully
You can view and download the user's password and email instructions for signing in to the AWS Management Console.
View user
IAM > Users
Managing human user access account by account? There's a better way.
Dismiss
Go to Identity Center
Users (9) Info
An IAM user is an identity with long-term credentials that is used to interact with AWS in an account.
Delete
Add users

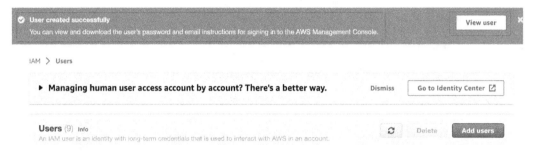

Figure 15.12 – The confirmation message after a user is created successfully

6. At this point, a user is created but doesn't have access to the AWS console or an access key. To provide access to **Access key ID** and **Secret key ID**, click on the **Security credentials** tab as shown in the following screenshot and click on **Create Access key** button.

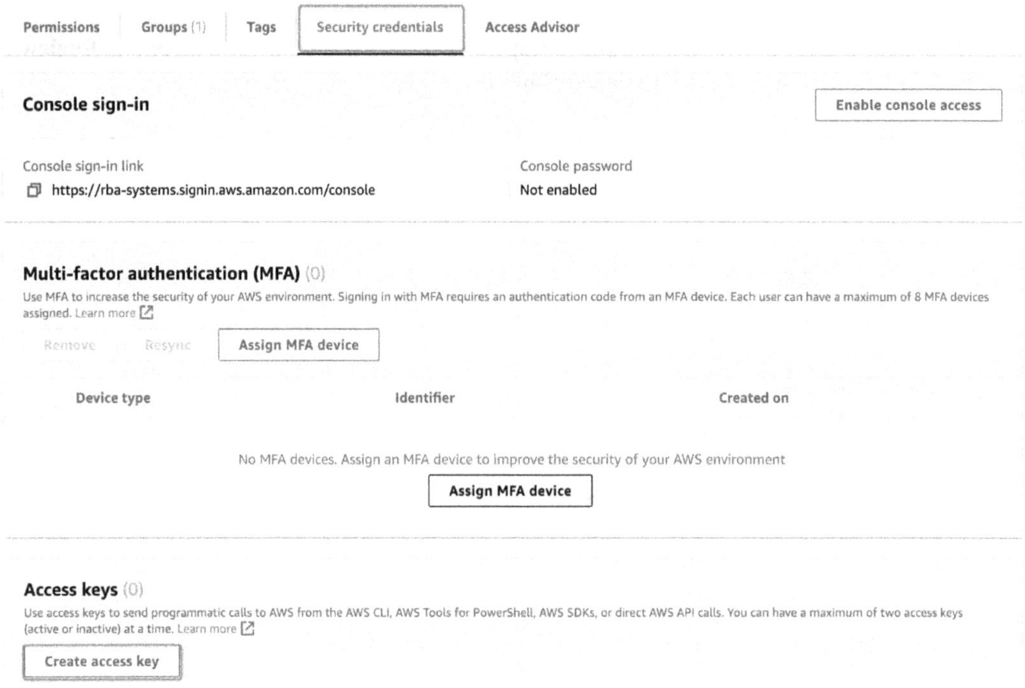

Permissions
Groups (1)
Tags
Security credentials
Access Advisor
Console sign-in
Enable console access
Console sign-in link
https://rba-systems.signin.aws.amazon.com/console
Console password
Not enabled
Multi-factor authentication (MFA) (0)
Use MFA to increase the security of your AWS environment. Signing in with MFA requires an authentication code from an MFA device. Each user can have a maximum of 8 MFA devices assigned. Learn more
Remove
Resync
Assign MFA device
Device type
Identifier
Created on
No MFA devices. Assign an MFA device to improve the security of your AWS environment
Assign MFA device
Access keys (0)
Use access keys to send programmatic calls to AWS from the AWS CLI, AWS Tools for PowerShell, AWS SDKs, or direct AWS API calls. You can have a maximum of two access keys (active or inactive) at a time. Learn more
Create access key

Figure 15.13 – Creating the access key

7. The following screen will ask for a use case for which you will be using the access key created in the preceding step. Select the **Command Line Interface (CLI)** radio button and click on the checkbox as shown in the following screenshot. Now, click on the **Next** button:

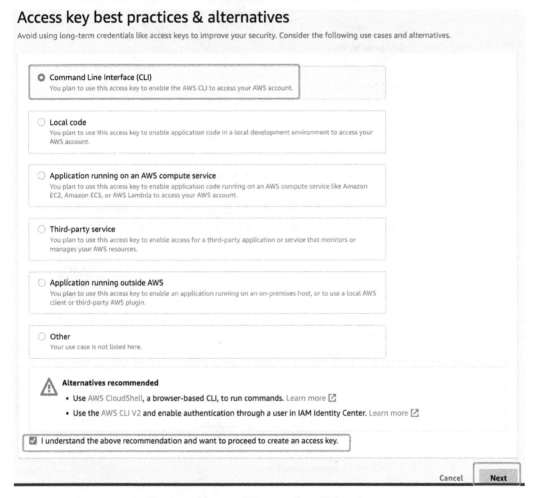

Figure 15.14 – Choosing Command Line Interface (CLI) as the necessary use case

8. In the next screen, provide a description and click on the **Create access key** button to create the **Access key ID**:

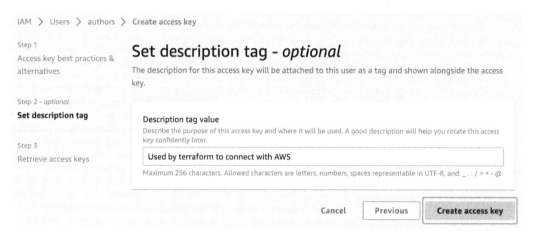

Figure 15.15 – Generating the Access key ID

9. Now **Access key ID** and **Secret key ID** are created and you can see the values on your screen. Download this information or copy it to a secure place as we will need this later. This is the only time you will be able to see the information:

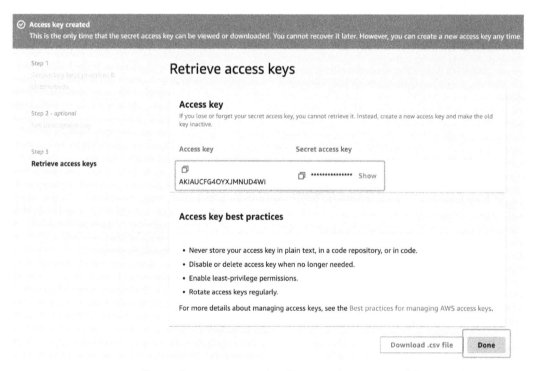

Figure 15.16 – The confirmation screen after the Access key is successfully created

At this point, we are done creating our user and will use this information to configure Terraform. Save this information and keep it secure, as these IDs are sensitive, and leaking this information can incur additional costs to your account.

AWS CLI installation

The AWS CLI is a tool provided by AWS to execute and manage your AWS resources using commands. The AWS CLI allows you to automate resource maintenance using scripts. To install the AWS CLI, perform the following steps:

1. Download the AWS CLI from https://docs.aws.amazon.com/cli/latest/ userguide/getting-started-install.html and double-click the executable to install the CLI. Click on the **Next** button:

Figure 15.17 – The AWS CLI installer License Agreement

2. You can select what feature of the AWS CLI you want to install by expanding the tree displayed in the following screenshot, but let's keep the default settings and click the **Next** button:

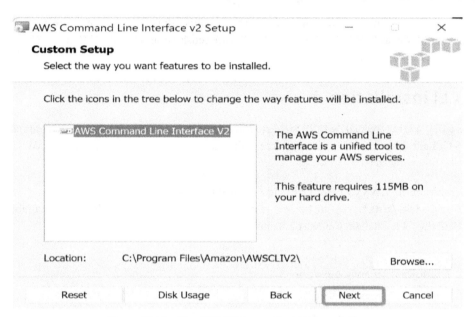

Figure 15.18 – The AWS CLI feature installation choice

3. On the next screen, confirm the installation by clicking on the **Install** button and clicking **Yes** on the Windows confirmation prompt:

Figure 15.19 – The AWS CLI installation confirmation

4. This will start the AWS CLI installation on your system, and you will see a confirmation message similar to what is shown in the following screenshot:

Figure 15.20 – The AWS CLI installation completion

Your AWS CLI is now installed. You can run the aws -version command to confirm that the CLI is installed. To configure the AWS CLI and connect to your account, use the aws configure command.

Creating an SNS topic

Amazon SNS is a notification service provided by AWS for triggering notifications to various devices and platforms.

SNS provides push-based, many-to-many messaging between distributed systems, microservices, and event-driven serverless applications. SNS also supports SMS texts, push notifications, and email. To set up SNS for your account, perform the following steps:

1. Log in to your AWS account, search for Simple Notification Services (SNS), and then click on the **Create topic** button:

Figure 15.21 – The Amazon SNS Topics home screen

2. Provide a value in the **Name** field for your SNS topic and select the **Type** option for the message. The **FIFO (first-in, first-out)** value for the topic **Type** option enforces exactly one delivery of a message and preserves the message ordering. In our use case Standard will be good enough so let's select **Standard**:

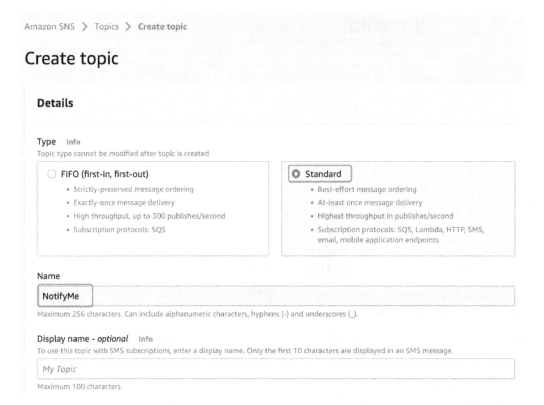

Figure 15.22 – The Create topic SNS screen

3. In the following screen sections, you can expand each section and configure different things, such as **Encryption**, **Access policy**, and **Data protection policy**. For our use case, the default options should be good enough, so let's click on the **Create topic** button:

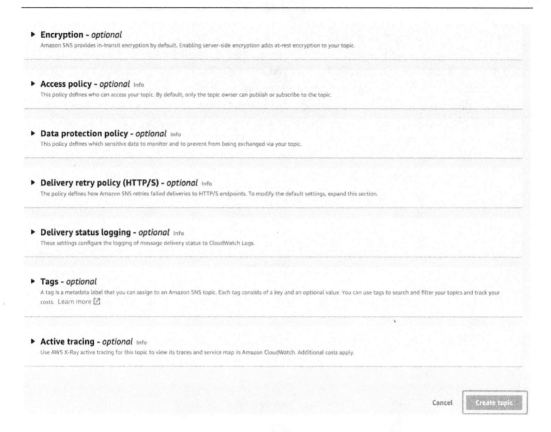

▶ **Encryption -** *optional*

Amazon SNS provides in-transit encryption by default. Enabling server-side encryption adds at-rest encryption to your topic.

▶ **Access policy -** *optional* Info

This policy defines who can access your topic. By default, only the topic owner can publish or subscribe to the topic.

▶ **Data protection policy -** *optional* Info

This policy defines which sensitive data to monitor and to prevent from being exchanged via your topic.

▶ **Delivery retry policy (HTTP/S) -** *optional* Info

The policy defines how Amazon SNS retries failed deliveries to HTTP/S endpoints. To modify the default settings, expand this section.

▶ **Delivery status logging -** *optional* Info

These settings configure the logging of message delivery status to CloudWatch Logs.

▶ **Tags -** *optional*

A tag is a metadata label that you can assign to an Amazon SNS topic. Each tag consists of a key and an optional value. You can use tags to search and filter your topics and track your costs. Learn more ☑

▶ **Active tracing -** *optional* Info

Use AWS X-Ray active tracing for this topic to view its traces and service map in Amazon CloudWatch. Additional costs apply.

Cancel   Create topic

Figure 15.23 – Create topic advanced configuration

4. Once you click on the **Create topic** button, it will create an SNS topic and present a confirmation message. As a next step, you need to subscribe to this topic to get alerts, so click on the **Subscriptions** link in the left-hand menu and click on the **Create subscription** button:

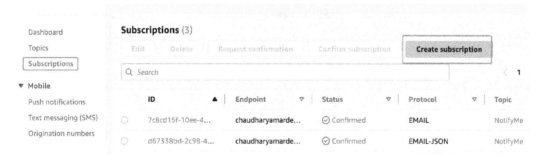

Figure 15.24 – The AWS topic Subscriptions home screen

5. In the **Create subscription** screen, select the newly created topic and choose an appropriate **Protocol** option for notification delivery. We have selected **Email-JSON**, but there are multiple options to subscribe for notifications. Once you select the **Email-JSON** option as the **Protocol** field value, you must provide your email address in the **Endpoint** field. **Subscription filter policy** allows you to filter certain messages based on message attributes or the message body, while **Redrive policy** allows you to configure a dead letter queue. Let's go ahead and click on the **Create subscription** button:

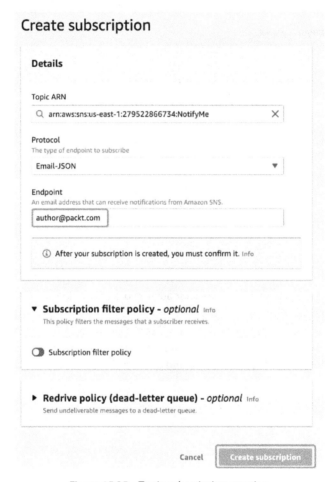

Figure 15.25 – Topic subscription creation

6. Once you click on the **Create subscription** button, it will present you with a confirmation message, but the status of your subscription is **Pending**. To complete the subscription, go to your email inbox and click on the link you received from AWS. Once you have provided confirmation through the link, your subscription status email will be confirmed and reflected in the AWS Console:

Figure 15.26 – SNS topic subscription confirmation

Git installation

To install a Git client on a Windows system, perform the following steps:

1. Go to https://git-scm.com/downloads and click on the **Download for Windows** button. This will open up a new page. Now click on the **Click here to download** link.

2. Double-click the downloaded binary file to install Git on your Windows system. This will ask you to confirm the installation of Git; click the **Yes** button:

Figure 15.27 – Windows Git installation

3. Review the license terms and click **Next** to select the installation location. Keep the default installation location for Git and click **Next**. On the next screen, select the options shown in the following screenshot:

Figure 15.28 – The Git installation component selection

4. In the next few screens, you will be asked to select the start menu folder, the default Git editor, and the default Git branch. Keep everything as the defaults and click the **Next** button:

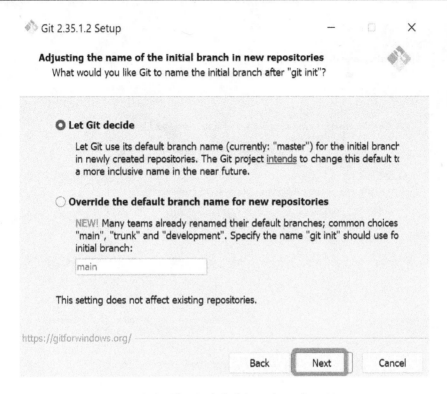

Figure 15.29 – The Git default branch configuration

5. In the next screen few screens, you will be asked to choose whether or not to add Git to your Windows PATH variable, shell program, text file line ending, terminal emulator, Git pull command default behavior, credential manager, and system-level caching. Keep the settings as their defaults on each screen and click the **Next** button to continue. The next screen asks you to select experimental support, but we are not selecting anything here. Click the **Install** button to install Git:

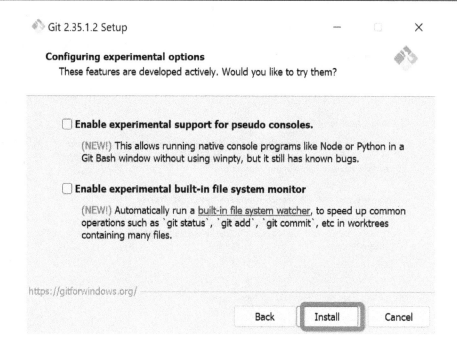

Figure 15.30 – The Git experimental feature selection

6. This will start installing the Git components on the Windows system. Once Git is successfully installed on your system, click the **Finish** button. Now that you have Git installed successfully on your system, you can configure the default user name and email address using the `git config - - user.email "EMAIL_ADDRESS"` and `git config - - user. name"USER_NAME"` commands.

Docker Desktop installation

Docker Desktop is a desktop application provided by Docker for easily managing the Docker applications for macOS, Windows, and Linux distributions. Docker Desktop includes the Docker daemon, the Docker client, Docker Compose, credential helpers, and other utilities, so you don't have to install each one separately. Follow these instructions to install Docker Desktop:

1. Go to `https://docs.docker.com/get-docker/` and download Docker Desktop for your operating system. Once the binary is downloaded, double-click on the installer. It will open up a confirmation prompt; click on **Yes**:

Figure 15.31 – Docker Desktop installation

2. The system will verify the package and ask you to use choose the virtualization type and shortcut. Keep the default settings and click the **Ok** button:

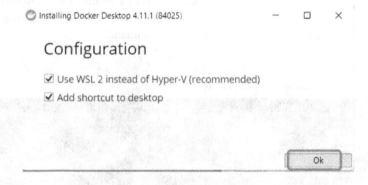

Figure 15.32 – Docker Desktop installation options

3. Once Docker is installed, it will ask you to restart the system. Click on the **Close and restart** button and save any unsaved work. Once the system is restarted, click on the Docker Desktop icon on your system and accept the terms and conditions by checking the **I accept the terms** checkbox, and then click on the **Accept** button:

Figure 15.33 – The Docker Desktop service agreement

4.   Now, Docker is ready to use, and you can see any running containers and downloaded images by clicking on the left navigation panel on the **user interface** (**UI**). To validate the installation of the Docker client, run the `docker version` command:

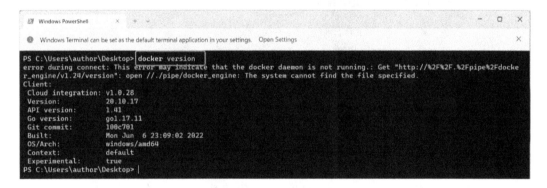

Figure 15.34 – The Docker version check

Now, your system is ready for building and running Docker images.

Index

D

Packtpub.com

Subscribe to our online digital library for full access to over 7,000 books and videos, as well as industry leading tools to help you plan your personal development and advance your career. For more information, please visit our website.

Why subscribe?

- Spend less time learning and more time coding with practical eBooks and Videos from over 4,000 industry professionals

- Improve your learning with Skill Plans built especially for you

- Get a free eBook or video every month

- Fully searchable for easy access to vital information

- Copy and paste, print, and bookmark content

Did you know that Packt offers eBook versions of every book published, with PDF and ePub files available? You can upgrade to the eBook version at packtpub.com and as a print book customer, you are entitled to a discount on the eBook copy. Get in touch with us at customercare@packtpub.com for more details.

At www.packtpub.com, you can also read a collection of free technical articles, sign up for a range of free newsletters, and receive exclusive discounts and offers on Packt books and eBooks.

Other Books You May Enjoy

If you enjoyed this book, you may be interested in these other books by Packt:

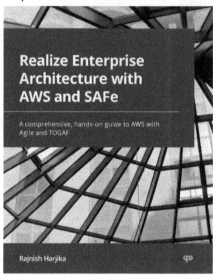

Realize Enterprise Architecture with AWS and SAFe

Rajnish Harjika

ISBN: 9781801812078

- Set up the core foundation of your enterprise architecture
- Discover how TOGAF relates to enterprise architecture
- Explore AWS's EA frameworks and find out which one is the best for you
- Use SAFe to maximize agility in your organization
- Find out how to use ArchiMate to model your architecture
- Establish proper EA practices in your organization
- Migrate to the cloud with AWS and SAFe

Industrializing Financial Services with DevOps

Spyridon Maniotis

ISBN: 9781804614341

- Understand how a firm's corporate strategy can be translated to a DevOps enterprise evolution
- Enable the pillars of a complete DevOps 360° operating model
- Adopt DevOps at scale and at relevance in a multi-speed context
- Implement proven DevOps practices that large incumbents banks follow
- Discover core DevOps capabilities that foster the enterprise evolution
- Set up DevOps CoEs, platform teams, and SRE teams

580

Packt is searching for authors like you

If you're interested in becoming an author for Packt, please visit `authors.packtpub.com` and apply today. We have worked with thousands of developers and tech professionals, just like you, to help them share their insight with the global tech community. You can make a general application, apply for a specific hot topic that we are recruiting an author for, or submit your own idea.

Share Your Thoughts

Now you've finished *Building and Delivering Microservices on AWS*, we'd love to hear your thoughts! Scan the QR code below to go straight to the Amazon review page for this book and share your feedback or leave a review on the site that you purchased it from.

https://packt.link/r/1803238208

Your review is important to us and the tech community and will help us make sure we're delivering excellent quality content.

Download a free PDF copy of this book

Thanks for purchasing this book!

Do you like to read on the go but are unable to carry your print books everywhere?

Is your eBook purchase not compatible with the device of your choice?

Don't worry, now with every Packt book you get a DRM-free PDF version of that book at no cost.

Read anywhere, any place, on any device. Search, copy, and paste code from your favorite technical books directly into your application.

The perks don't stop there, you can get exclusive access to discounts, newsletters, and great free content in your inbox daily

Follow these simple steps to get the benefits:

1. Scan the QR code or visit the link below

https://packt.link/free-ebook/9781803238203

2. Submit your proof of purchase
3. That's it! We'll send your free PDF and other benefits to your email directly

www.ingramcontent.com/pod-product-compliance
Lightning Source LLC
Chambersburg PA
CBHW081449050326
40690CB00015B/2733